Railroads in the Old South

Railroads in the Old South

Pursuing Progress in a Slave Society

AARON W. MARRS

The Johns Hopkins University Press
Baltimore

© 2009 The Johns Hopkins University Press
All rights reserved. Published 2009
Printed in the United States of America on acid-free paper
2   4   6   8   9   7   5   3   1

The Johns Hopkins University Press
2715 North Charles Street
Baltimore, Maryland 21218-4363
www.press.jhu.edu

Library of Congress Cataloging-in-Publication Data

Marrs, Aaron W., 1976–
Railroads in the Old South : pursuing progress in a slave society /
Aaron W. Marrs.
      p.   cm.
Includes bibliographical references and index.
ISBN-13: 978-0-8018-9130-4 (hbk. : alk. paper)
ISBN-10: 0-8018-9130-2 (hbk. : alk. paper)
1. Railroads—Southern States—History—19th century.
2. Southern States—History—1775–1865.   I. Title.
HE2771.A13M37 2009
385.0975′09034—dc22   2008022572

A catalog record for this book is available from the British Library.

Special discounts are available for bulk purchases of this book. For more information,
please contact Special Sales at 410-516-6936 or specialsales@press.jhu.edu.

The Johns Hopkins University Press uses environmentally friendly book
materials, including recycled text paper that is composed of at least 30 percent
post-consumer waste, whenever possible. All of our book papers are acid-free, and
our jackets and covers are printed on paper with recycled content.

For Melissa Jane

CONTENTS

Acknowledgments   *ix*

List of Abbreviations   *xiii*

Maps   *xiv*

Introduction   1

ONE   Dreams   11

TWO   Knowledge   31

THREE   Sweat   55

FOUR   Structure   84

FIVE   Motion   106

SIX   Passages   135

SEVEN   Communities   162

Epilogue: Memory   192

Notes   *199*

Essay on Sources   *255*

Index   *261*

ACKNOWLEDGMENTS

||||||||||||||||||||||||||||||||||||||||||||||||||||||||||||||||||||||||||||||||||||||||||||||||||||||||||||||||||||||||||||||||||||||||||||||||||||||||||||||||||||||||||||||||

Like most first academic books, this work began as a doctoral dissertation. Before that it was a seminar paper, and before *that* it was a conversation with Mark M. Smith, my dissertation director at the University of South Carolina. I owe much of my intellectual and professional development to his constant wise counsel. He has been tireless in reading and rereading drafts, offering advice as I ventured out onto the job market, and helping me keep a healthy perspective on my professional career. I am fortunate to have earned my doctorate under his tutelage and look forward to continuing our conversations. I offer hearty thanks to the other members of my dissertation committee: Lawrence Glickman, Paul Johnson, Peter Coclanis, and Ann Johnson. Their advice and tough questions helped me write a better dissertation and transform the dissertation into this book.

Many, many people contributed to the wonderful time that I had at Carolina. Connie Schulz recruited me to USC and has remained a good friend ever since. It was a pleasure and a privilege to work with Walter Edgar on the *South Carolina Encyclopedia* and numerous other projects, and I am grateful for his enthusiasm for my work. Among my fellow graduate students, Kathy Hilliard, Mike Reynolds, and Rebecca Shrum have proved their endurance by listening to innumerable stories about railroads. I count myself fortunate to have been at USC at the same time they were, because they provided a model of scholarly cooperation and friendliness that is exactly how the life of the mind should operate. USC's history department and public history program allowed me to be a historian in a variety of settings: researcher, classroom teacher, editor, and archivist. I can't imagine a better way to learn about the breadth of our exciting profession. Thanks to all the other friends and teachers who have helped me laugh and learn at various stages of life: Michael Hittle, Kevin Allen, Paul and Laura Keefer, Susan Asbury-Newsome and Brian Newsome, Tom Downey, and the remarkably supportive and steadfast friends I have kept since my days at Lawrence University. You will always have a place to crash in D.C.

Having been trained as an archivist, I am well aware of how obnoxious I must have been while doing research, but I am grateful to the following archives and their staffs: South Caroliniana Library; Historical Collections Department, Baker Library, Harvard Business School; Virginia Historical Society; Wisconsin Historical Society; Special Collections department, Newman Library, Virginia Tech; Hargrett Library, University of Georgia; and the Middle Georgia Archives. I would like to thank the same institutions for permission to quote from their materials or use images, in addition to the Southern Historical Collection, Wilson Library, University of North Carolina at Chapel Hill; R. G. Dun; the CSX Corporation; the United States Trade and Patent Office; the Smithsonian; and the South Carolina Historical Society. The Norfolk Southern Archives now houses material that was at Virginia Tech when I did my research, and I am grateful to Kyle Davis of Norfolk Southern for facilitating permission to quote from these materials. At the University of South Carolina, I owe thanks to Marna Hostetler and the superb interlibrary loan staff for tracking down numerous reels of microfilm and printed diaries that helped me round out my research. Some portions of this book have also appeared in the journal *Enterprise and Society,* and I thank that journal for permission to reprint them here.

Travel funding to archives came in the form of an Alfred Chandler Traveling Fellowship from the Harvard Business School and a Mellon Fellowship from the Virginia Historical Society. I had a wonderful time researching at each of these institutions, which would not have been possible without the funding provided. My travel to archives was also made possible by old and new friends who let me stay with them. So thanks to Miriam Ledford-Lyle and Scott Lyle, Scott Trigg, Randy and Irene Leech, and Eric Plaag. I also thank the USC College of Liberal Arts (later the College of Arts and Sciences) for awarding me the Newton Fellowship during my last three years at the university.

I have benefited immensely from scholars who have provided formal and informal commentary at conferences: Michael Bailey, John Brown, David Carlton, Albert Churella, Colin Divall, Colleen Dunlavy, John Lauritz Larson, William Rorabaugh, and Kathryn Steen. Travel to these conferences was made possible by several grants from the USC Department of History, the USC Graduate School, and the Society for the History of Technology. Bob Brugger and the staff at the Johns Hopkins University Press have been of invaluable assistance as I worked to turn the dissertation into a book. I am deeply appreciative to the two anonymous reviewers for their insightful comments. Many thanks to Bill Nelson for designing the wonderful maps, and to David Prior for tracking down some last-minute citations. Obviously, all errors in the text are my own.

I am proud to be a historian in the Department of State's Office of the Historian. I thank the office's scholars and staff for welcoming a nondiplomatic historian into their midst to participate in the important and vital work the office does. The views expressed in this book are solely my own and are not necessarily the official views of the Office of the Historian, the U.S. Department of State, or the U.S. government.

I can't repay the love—and patience!—shown by my family as I ventured through graduate school and beyond. My parents, Margie and Bob, and sister, Kristin, have always been supportive as I've wandered farther and farther from the heartland. But as for my biggest debt: I'm sure Melissa Jane Taylor is wondering how she ended up married to someone who doesn't drink coffee, doesn't like the beach, and prefers cats to dogs. I know that I am fortunate for her indulgence in these matters, and there are not words I can write or actions I can undertake that will adequately repay her love and kindness. Telling people that we married as two academics usually evokes looks of pity, but I wouldn't have it any other way. We slogged through the dissertation process together, survived the job hunt together, and it is with joy and anticipation that I look forward to our future together. She is the perfect intellectual sparring partner as well as the love of my life, which makes me the luckiest guy I know.

BRRR	Blue Ridge Railroad (of South Carolina)
CRRG	Central Railroad of Georgia
CSCRR	Charlotte and South Carolina Railroad
ETVRR	East Tennessee and Virginia Railroad
LCCRR	Louisville, Cincinnati and Charleston Railroad
NCCRR	North Carolina Central Railroad
NCRR	North Carolina Railroad
RDRR	Richmond and Danville Railroad
RFPRR	Richmond, Fredericksburg and Potomac Railroad
SARR	Savannah and Albany Railroad
SCRR	South Carolina Railroad
SURR	Spartanburg and Union Railroad
VTRR	Virginia and Tennessee Railroad
WMRR	Wilmington and Manchester Railroad

Map Key

1. Baltimore and Ohio Railroad
2. Winchester and Potomac Railroad
3. Orange and Alexandria Railroad
4. Richmond, Fredericksburg and Potomac Railroad
5. Virginia Central Railroad
6. City Point Railroad
7. Petersburg Railroad
8. Richmond and Danville Railroad
9. Seaboard and Roanoke Railroad
10. Virginia and Tennessee Railroad
11. Norfolk and Petersburg Railroad
12. Southside Railroad
13. Raleigh and Gaston Railroad
14. Wilmington and Raleigh/ Wilmington and Weldon Railroad
15. North Carolina Railroad
16. Western North Carolina Railroad
17. South Carolina Railroad
18. Charlotte and South Carolina Railroad
19. Wilmington and Manchester Railroad
20. Greenville and Columbia Railroad
21. Blue Ridge Railroad
22. Spartanburg and Union Railroad
23. Lexington and Ohio Railroad
24. Louisville and Nashville Railroad
25. East Tennessee and Virginia Railroad
26. East Tennessee and Georgia Railroad
27. Georgia Railroad
28. Central Railroad of Georgia
29. Macon and Western Railroad
30. Western and Atlantic Railroad
31. Southwestern Railroad
32. Tallahassee Railroad
33. Saint Joseph and Iola Railroad
34. Montgomery and West Point Railroad
35. Tuscumbia, Courtland and Decatur Railroad
36. Memphis and Charleston Railroad
37. Southern Railroad
38. Raymond Railroad
39. Mobile and Ohio Railroad
40. Mississippi Central Railroad
41. Grand Gulf and Port Gibson Railroad
42. Natchez and Jackson Railroad
43. West Feliciana Railroad
44. Clinton and Port Hudson Railroad
45. Pontchartrain Railroad
46. Mexican Gulf Railroad
47. Baton Rogue, Grosse Tete and Opelousas Railroad
48. New Orleans, Opelousas and Great Western Railroad

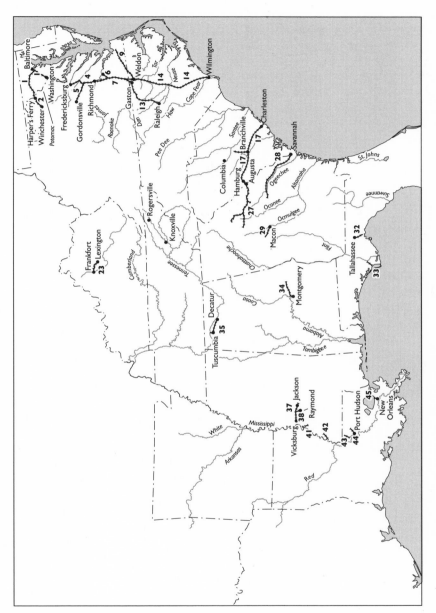

Map 1. Early developments: Railroads in the South, 1840.

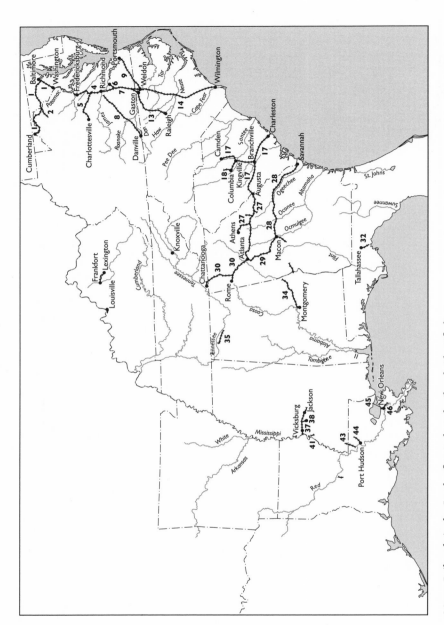

Map 2. Railroads in the South, 1850 (selected railroads identified).

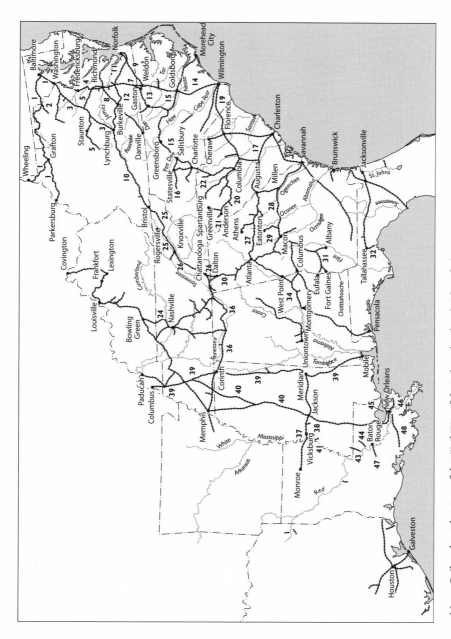

Map 3. Railroads at the start of the Civil War, 1861 (selected railroads identified).

Railroads in the Old South

Introduction

While writing his memoirs in the 1870s, Kentucky papermaker Ebenezer Hiram Stedman recalled the rumors spread about railroads in the mid-1820s: "For More than two years we heard most Remarkable Storyes about Rail Roads. Some People Said that They had Seen Cariges drawn on a Rail Road by Steam. He was put down as a Munchawson."[1] Some of the stories were so fantastical that locals were probably right to doubt them: "Another Said he had Road on a Coach that went so fast that he had to Breath Through a Brass tube made on purpose So that the Speed woold not take their Breath away. & Some told Such Storyes that people woold not Believe anny Thing they woold Say." To Stedman's neighbors in the 1820s, such rumors were beyond belief.

But in 1828 the doubters saw proof they could not deny when a model railroad arrived in Kentucky. Stedman reminisced about its effect on the community:

> If i Rember Right This Summer, Cox, of Louisville of the Firm of Cox & Bridgford, Made a Small locomotive & portable Rail Road to Exhibit through Ky. He Come to Frankfort then to Georgetown. He laid his Rail Road in the old Masonick lodg Room & had it on Exabition for Sevral days. "Admitance one dollar to Se the Great wonder." The Rails ware laid So that one Person Could Ride Round the Room. Evry one Must Ride By Steam & Such talk & Excitement at this time about Rail Roads. From hear he went to Lexington. The Excitement Got up By this little Moddle of a Rail Road In Lexington did not Stop till a Company was formed & a Charter obtained for the Lexington & Louisville Rail Road. A Flying Maching in this Day woold not cause one half the Excitement that [the] Rail Road [did.]

While Stedman would not live long enough to see a "Flying Maching," he captured the wonder and excitement that accompanied the railroad. Rumors of an incredible machine had become a reality.

Just over thirty years after the "Moddle" caused a sensation in Kentucky, Samuel Edward Burges rode the railroad network of South Carolina to oversee his agricultural holdings, to work as a collector for several newspapers, and to attend horse races. During the first nine months of 1860, he was carried all over the state by seven different railroads: the Cheraw and Darlington, Wilmington and Manchester, South Carolina, Charleston and Savannah, Greenville and Columbia, Northeastern, and Blue Ridge. Yet Burges's journal makes it clear that his encounters with railroads were very different from Stedman's. Instead of describing his travels in breathless language or being mesmerized by a "Great wonder," Burges perfunctorily recorded the distances traveled and time required. On February 9, he wrote: "We reached Gourdin's T.O. a little before 2 A. M. . . . took the down train about 4:30 A. M. reached Charleston at 8 A. M. Stage 42 miles; N E R R 78 miles." Eleven days later he traveled on two railroads: "W & M R R 22 miles; C & D R R 40 miles." On June 14: "Left on N. E. R. R. at 2:30 A. M. reached Charleston at 8 A. M. N E R R 102 miles."[2] Other details were not worthy of his attention—the railroad simply did its job of carrying Burges around the state, at speeds and across distances unimaginable by Stedman just a few decades previously. In less than forty years, the railroad went from a curiosity that excited the imaginations of Kentuckians to an unquestioned, almost banal, part of the southern landscape.

This transition from novel to normal took place in a region that has not always been noted for its technological transformations. For many historians, the Old South's foundation—slavery—prevented the region from reaping the benefits of modernization in the early nineteenth century. *Modernization* and *progress* are terms found throughout the historical literature, although they can be problematic because of the potential value judgments attached to them. These judgments stem in part from the course of American history: as Eugene Genovese has noted, the North's victory in the Civil War "sealed the triumph of the association of freedom and progress over an alternate reading," with slavery characterized as neither "progressive" nor "modern." Thus, *modernization,* a term with a positive, forward-looking connotation, is more easily applied to the winners than the losers in the Civil War. And historians often use the general rubric of modernization to describe the transformations in politics, communications, social attitudes, technology, and transportation that were under way in the antebellum era. Broadly speaking, the South has not fared well compared to the North under such points of comparison. The North's Civil War victory, combined with its innovations in these areas, helped lock in the meaning of modernization in the way that Genovese described. With its comparative lack of industry and an economy based on staple agriculture, the antebellum South seems underdeveloped. Perhaps most important, because the South's wealth was in labor, successful planters plowed their money back into purchasing more slaves instead of investing capital in infrastructure or other industries that would create a diversified economy. Such lack of investment meant that railroads—although acknowledged to be present—are generally portrayed as less advanced and less important in the antebellum South than they were elsewhere in the United States.[3]

Some historians have argued that the comparative economic "failure" of the Old South extended into the realm of southern imagination as well. In his classic work on the power of technology in the American imagination, *The Machine in the Garden,* Leo Marx argued that the locomotive constituted "the leading symbol of the new industrial power" in antebellum America. But Marx did not believe that the railroad's development had any particular meaning for the southern states. Rather, the South remained enamored of a preindustrial, premodern "pastoral ideal," which southerners used as a "weapon against industrialism." Putatively poor economic performance and supposed ideological conservatism have combined to exclude the South from the story of the economic and technological transformations taking place in antebellum America. Southerners, so the argument goes, rejected "innovation, enterprise, and reform," whereas their northern counterparts pursued these goals enthusiastically. One early historian of modernity in American life, Richard Brown, wrote that the Civil War provided "unexpected liberation" for southern whites as they were set free to join the North on the road to modernity. Thus, slavery in the South not only damaged the millions of African Americans who toiled under the slave regime but also prevented white southerners from enjoying the full fruits of technological development that were blossoming in the North.[4]

Such an analysis appears borne out when comparing northern and southern railroads by several statistical measures: southern railroads are the laggards. Tables 1 and 2 demonstrate how nonslaveholding states outpaced their southern colleagues. The states of the future Confederacy contained railroad companies that built fewer miles, hauled fewer goods, and earned less money. The South certainly played an important early role in the country's railroad development with the advent of the South Carolina Canal and Railroad Company, but economic depression in the early 1840s slowed growth substantially. As a result, construction in the 1840s took place mostly in the North while that in the South stagnated. Contemporaries noticed the comparatively slow construction in the South during the decade: Ralph Waldo Emerson criticized slavery in 1844 by declaring that slavery was "no improver; it does not love the whistle of the railroad." The result was that southern states fell behind. By 1850 Georgia had the most mileage of any southern state, with 643 miles of track. This placed Georgia fourth nationally, behind New York, Pennsylvania, and Massachusetts. But it was a distant fourth, far behind Massachusetts's 1,035 miles. Tiny Rhode Island had nearly as many miles (68) as all of Mississippi (75). Taking a cue from the comparative performance of northern and southern railroads, most accounts of southern railroading have concentrated

TABLE I
Comparative Statistics on Railroad Output by Region, 1839–1859

Region	1839	1849	1855–1856	1859
		Total receipts ($000)		
New England	1,224.9	8,832.2	17,722.9	16,938.1
Middle Atlantic	4,278.2	14,471.2	42,050.6	43,292.8
West	175.0	1,796.8	28,558.1	34,508.1
Total	5,678.1	25,100.2	88,331.6	94,739.0
South	1,474.0	3,618.7	10,573.8	14,412.7
Southwest	216.5	532.0	3,317.1	9,519.5
Total	1,690.5	4,150.7	13,890.9	23,932.2
California				174.5
National total	7,368.6	29,250.9	102,222.5	118,845.7
		Net earnings ($000)		
New England	562.5	4,310.5	6,137.9	7,219.8
Middle Atlantic	1,837.1	7,276.8	19,495.5	19,159.0
West	90.5	885.8	14,005.9	13,067.2
Total	2,490.1	12,473.1	39,639.3	39,446.0
South	466.1	1,661.5	5,395.4	7,110.5
Southwest	77.8	243.0	1,538.1	4,868.0
Total	543.9	1,904.5	6,933.5	11,978.5
California				92.8
National total	3,034.0	14,377.6	46,572.8	51,517.3
		Freight receipts ($000)		
New England	377.8	3,727.9	8,106.4	8,493.0
Middle Atlantic	1,234.3	7,041.9	25,352.4	26,441.8
West	83.0	872.1	13,888.7	18,201.5
Total	1,695.1	11,641.9	47,347.5	53,136.3
South	697.6	2,246.2	6,228.8	8,398.5
Southwest	69.5	169.0	1,668.0	4,847.9
Total	767.1	2,415.2	7,896.8	13,246.4
California				94.5
National total	2,462.2	14,057.1	55,244.3	66,477.2
		Ton-miles ($000)[a]		
New England		82,842.2	227,498.1	253,991.7
Middle Atlantic		190,041.7	1,022,489.3	1,159,541.6
West		21,802.5	431,616.9	791,369.5
Total		294,686.4	1,681,604.3	2,204,902.8
South		49,915.6	174,175.2	248,476.3
Southwest		2,414.3	48,172.5	123,043.1
Total		52,329.9	222,347.7	371,519.4
California				1,260.0
National total		347,016.3	1,903,952.0	2,577,682.2

Source: Albert Fishlow, *American Railroads and the Transformation of the Ante-Bellum Economy* (Cambridge, Mass.: Harvard University Press, 1965), 322, 326, 328, 337. Fishlow groups states into regions as follows: New England (Maine, New Hampshire, Vermont, Massachusetts, Connecticut, Rhode Island), Middle Atlantic (New York, New Jersey, Pennsylvania, Delaware, Maryland), South (Virginia, North Carolina, South Carolina, Georgia, Florida), Southwest (Kentucky, Tennessee, Alabama, Mississippi, Louisiana, Texas), and West (Ohio, Indiana, Illinois, Michigan, Wisconsin; Missouri added in 1855–56, Iowa added in 1859).
[a]I used Fishlow's figures for "1849" (not the "upper bound") and his figures for "1859" (not the "lower bound"). For an explanation, see Fishlow, *American Railroads*, 325, 339–40.

TABLE 2
Railroad Mileage by State

State	1840	1850	1860
Alabama	46	183	743
Arkansas	0	0	38
California	0	0	23
Connecticut	102	402	601
Delaware	39	39	127
Florida	(22)	21	402
Georgia	185	643	1,420
Illinois	(24)	111	2,790
Indiana	(20)	228	2,163
Iowa	0	0	655
Kentucky	28	78	534
Louisiana	40	80	335
Maine	11	245	472
Maryland	213	259	386
Massachusetts	301	1,035	1,264
Michigan	59	342	779
Mississippi	(25)	75	862
Missouri	0	0	817
New Hampshire	53	467	661
New Jersey	186	206	560
New York	374	1,361	2,682
North Carolina	53	283	937
Ohio	30	575	2,946
Pennsylvania	754	1,240	2,598
Rhode Island	50	68	108
South Carolina	137	289	973
Tennessee	0	(9)	1,253
Texas	0	0	307
Vermont	0	290	554
Virginia (includes West Virginia)	147	481	1,731
Wisconsin	0	20	905
Total	2,899	9,030	30,626

Source: John F. Stover, *Iron Road to the West: American Railroads in the 1850s* (New York: Columbia University Press, 1978), 7, 11, 26, 61, 116.
Note: Figures in parentheses represent Stover's estimates for mileage drawn from evidence suggesting railroad construction in those states, despite it not being so listed in *Poor's Manuals* or the U.S. Census for 1860. See Stover, *Iron Road to the West*, 11, 13. The totals include the parenthetical figures.

on the poor performance of southern railroads. Ulrich Bonnell Phillips's account of railroads as haulers of cotton and little else has remained the dominant interpretation for decades and is now conventional wisdom.[5]

This image of beleaguered southern railroads seems odd when juxtaposed with the excitement that Stedman and his neighbors felt over the railroad's arrival or the routine travel that Burges undertook on the eve of the Civil War. While in a broad comparative sense the South's economy was underdeveloped compared to that of the North, historians have long wrestled with contrary evidence in terms of both economic performance and planter ideology. Some of this evidence comes from railroads, which underwent remarkable transformations in the 1850s. From 1850 to 1860 southerners began committing more money to railroads, and an explosion of mileage resulted. Here, the southern experience was closer to that of the Old Northwest, which also saw a dramatic increase in mileage. Ohio and Illinois displaced New York and Pennsylvania as the states with the most mileage in 1860; Virginia and Georgia leapfrogged Massachusetts to claim sixth and seventh place and Tennessee's 1,253 miles trailed Massachusetts by only 11 miles. If the mileage gains were impressive, so were the percentage increases. Florida's percentage increase was the most spectacular among southern states, from 21 miles in 1850 to 402 in 1860, a 1,814 percent increase. Like every state in the Old Northwest, every southern state more than doubled its mileage in the 1850s, a feat accomplished only by New Jersey, Pennsylvania, and Delaware outside of those two regions. State aid continued to flow throughout the decade, a successful one for southern railroads.[6]

Southern railroads' extraordinary expansion during the 1850s demonstrates that the South's ideological landscape contained more than a simple commitment to a "pastoral ideal." Southerners themselves were embracing, demanding, and funding this development. Historians seeking to reorient our understanding of the southern economy have recently explored this different perception of southern attitudes. Enough acquisitiveness and industry were present in the South to spark a lengthy historiographic debate over how capitalist or noncapitalist the Old South was. Historians are now beginning to reframe the question: Walter Johnson, for example, has argued that historians need to stop treating capitalism and slavery as wholly distinct entities and to recognize their "dynamic simultaneity" in the eighteenth- and nineteenth-century Atlantic economy. Slave labor in America produced cotton, which was transformed into cloth by wage laborers and then purchased with those wages. Slavery and wage labor were not antithetical but were part and parcel of a larger system that called on the strengths of both when required.[7]

Framed in this way, the dichotomy of a "modernizing North" versus a "premodern South" holds less currency. Southerners certainly recognized that the times were changing and that they would need to alter their behavior to fit the times. "If we are content to remain stationary, while all others are on the advance," railroad boosters warned Charleston residents in 1835, "we must of course be left far behind." Historians have demonstrated that such attitudes stretched back for decades. Planters in the late eighteenth century, for example, formed societies to acquire and disseminate knowledge about agriculture. After the turn of the century, planters experimented with steam mills for processing the rice that their slaves grew. Planters refined their management methods, embracing the discipline that time management encouraged on southern plantations and northern factories alike. Merchants and other "men of capital" flourished in unlikely places, successfully "sculpting the agrarian landscape" to meet their own needs. Southern reformers eagerly adopted such traits as "system, uniformity, technology, organization, and bureaucratic control" in their quest for economic excellence, and progress extended into the social arena as southerners pursued moral reform movements such as temperance. Although moral reform obviously never extended to the abolition of slavery, planters in the early nineteenth century did begin reorienting their relationships with slaves from one of pure force to a more "organic" vision that led them to consider slaves as members of their extended family. In sum, southern planters were seeking their "own vision of a healthy modernity" throughout the antebellum era. Slaveholders adapted to keep pace with the changing economy in which they operated but held free wage labor at bay. We, of course, do not characterize slavery itself as modern or progressive, but white southerners did not see it in the same way. White southerners were interested in modern developments, either on a large scale such as railroads or a small scale such as management techniques on individual plantations. Yet they took these steps on their own terms, accommodating both slavery and their agricultural economy.[8]

Such an understanding of southern planters—one that allows for a willingness to reform coupled with an unwillingness to sacrifice slavery—alters our view of southern railroads and, in turn, our understanding of the Old South. Rather than simply functioning as markers of the South's relative economic success or failure, railroads constitute the ideal prism through which to view how white antebellum southerners married conservative social ideals with forward-looking technological advancement. Although most historians have seen the South's reliance on agriculture as preventing the region from pursuing innovation, the very success of cotton production drove planters and businessmen to push for the development of railroads, the most modern form of transportation available. To be sure, contemporaries were well aware of problems with southern railroads: civil engineer John McRae once complained that the South Carolina Railroad's "Depots and workshops would be a disgrace to any company & ought to be burnt & would be if the present vigilance were not used."[9] But McRae's lament—to the exclusion of other narratives—has prevailed in the limited historiography and left us with a warped and incomplete view of railroads in the antebellum South. Railroads had a far deeper meaning to southerners and southern society than as simple haulers of goods and people. As Stedman's account demonstrates, railroads also excited southern imaginations. They could create or break communities, transform the ability of southerners to travel, bring white families together, shatter slave families, or carry sons off to war. Moreover, the presence of railroads in the antebellum South, as well as the ease with which slavery was integrated into railroad development, reminds us that modernity and progress cannot automatically be associated with freedom. Statistical analysis may allow us to understand some aspects of the South's railroad history, but it cannot capture the full impact that railroads had on southern society. To fully appreciate the braiding of premodern and modern that the railroads represented in the antebellum South, we need to move beyond the traditional framework of business and economic history.[10] Whatever financial problems may have plagued southern railroads, the expansion of the 1850s makes clear that before the Civil War southerners were on a trajectory that embraced railroads. In overemphasizing the economic troubles that railroads faced, historians have slighted their very real impact on the society that they served.

A fuller investigation of southern railroads can help illustrate the tension between the modern goals of antebellum slaveholders and their determination to achieve those goals while retaining and even bolstering their conservative social order. In order to understand southern railroads on their own terms, I explore four major themes. The first theme is that, to a large degree, southern and northern railroad experiences paralleled each other. Although one region based its labor system on wage labor and the other on slavery, when it came to the experimental technology of railroads both the North and the South found themselves in similar situations. Both regions of the country experimented with a new technology that they jointly imported from England. Both regions had a shortage of labor, skilled and unskilled, to build these massive works. Both regions faced engineering challenges and opposition to railroad development. Civil engineers were in short supply in the early republic, and qualified men moved around the country in search of employment without regard to the region in which they worked. Although there were unique challenges to construction in each region (on matters such as labor and topography), both were linked to a national framework of internal improvements, one that consciously excluded politics at an early date in order to preserve unity.

Although northern and southern experiences paralleled each other in important ways, one critical difference was the presence of slavery in the South, and this labor option forms the second theme. Upon close examination, railroads fit well with what Joyce Chaplin has termed the South's "yes-and-no response to the modern age": southerners pursued railroads vigorously but integrated this modern development into their slave society. While the broad range of reactions to the railroad means that it makes little sense to speak of an explicitly "southern" way of understanding the railroad, it does make sense to emphasize that the white South desired to modernize on its own terms. Slaves were integral to every aspect of railroad operation in the South, excluded from only a small minority of skilled positions, such as engineer or stationmaster. Slaves also rode the railroads and, in so doing, threw into relief southern norms over race. Despite the South's commitment to racial slavery, railroads demonstrated that racial barriers were not entirely stable. Certain slaves, such as female slaves accompanying a mistress, were allowed to travel in coaches with white passengers, whereas others, such as gangs of agricultural laborers, were forced into seatless boxcars.[11]

Third, southerners had a variety of responses to the railroad's introduction. There is no single "southern" response to railroads because multiple groups of southerners encountered them in multiple ways. Religious leaders condemned trains as violators of the Sabbath. Whereas planters saw railroads as a boon to business and land values, horse-cart drivers viewed freight trains as competition for carrying goods between cities. To pedestrians, train tracks provided a well-marked path (albeit a potentially dangerous one). And to travelers, trains offered convenient routes to their destinations. Railroads were an integral part of southern communities, but it is important to remember that members of these communities interacted with the railroad in different ways. Southerners may have worried about the railroad's implications or its potential for changes in social relations, but railroads themselves remained popular in the antebellum South, mimicking the reaction of the nation at large to machines. At bottom, the attempt to discern a "southern"—or, for that matter, "northern"—way of understanding the railroad obscures the diversity of reaction found in antebellum America.[12]

One of the crucial ways of understanding these reactions is by examining time, and the importance of time forms the fourth theme of this study. Railroads have a firm place in the historiography of time as the exemplars of scheduling and timetables.[13] Yet the railroad's time was far more complex. Railroads promised regularity but were unable to fully overcome the constraints imposed by nature. Railroad companies, never in full control of their own time, argued about the value of time with outside groups such as the U.S. Post Office Department (which demanded that trains run on Sunday) and Sabbatarians (who demanded that they did not). Workers in this ostensibly clock-driven enterprise found that their work was managed by the task at hand as well as the clock. Time was not simply a way to manage the safety of trains but also figured prominently in the elaboration of power relationships within the corporation as well as between the corporation and the communities it served.

By exploring railroads through these four themes—the parallelism of northern and southern experience, slavery, community relationships, and time—we reach a better understanding of southern railroads and of the complex society that they served. While these four themes inform the entire study, the seven chapters are arranged in a roughly chronological fashion. The first chapter, "Dreams," recounts how southerners encouraged the development of railroads and also places southern railroad development in the context of the national internal improvements movement. "Knowledge" describes the engineering efforts that went into railroad development and investigates the work done by civil engineers and contractors to begin the construction process. "Sweat" examines labor arrangements during construction and charts the widespread use of slave labor on southern railroading projects. "Structure" details the corporate hierarchies of southern railroads. "Motion" looks at the challenges that railroads met in operation, in particular their handling of accidents. "Passages" investigates the travel experiences of southerners, white and black. The final chapter, "Communities," looks at the different constituencies that railroads created, influenced, argued with, and even destroyed. The epilogue, "Memory," assesses the place of railroads as a marker of the antebellum era in the memories of southerners after the Civil War.

A comprehensive look at railroads, one that encompasses both Stedman's excitement and Burges's indifference, will move us substantially beyond our current superficial understanding of this critical technological development. More important, railroads in the South illustrate the dual nature of the Old South's society: striving for technological advancement while wholly committed to slavery. Indeed, the Old South was neither fully premodern nor modern but interwove aspects of both conservatism and modernity into its social fabric.

CHAPTER ONE

Dreams

Business leaders in 1820s Charleston, South Carolina, faced a problem. Just a few decades removed from the city's colonial position of economic preeminence, they worried that land values were falling, industry was stagnant, houses sat unoccupied, and grass grew "uninterrupted in some of her chief business streets." Although rival cities such as Savannah were drawing away precious trade, these business leaders expressed hope that their own city still possessed outstanding "commercial advantages" that could restore Charleston to its former prominence. After forming an organizing committee, they urged their fellow residents to join them in promoting a railroad to the town of Hamburg, 136 miles distant. It was, in retrospect, a brave decision. Although Charleston's economic decline was "relative rather than absolute" before the Civil War, businessmen were worried enough to take a gamble on an untested and unproven technology. The railroad would not ultimately save Charleston's fortunes, but the scale and success of the Charleston-to-Hamburg railroad helped spark a powerful phase of the antebellum transportation revolution.[1]

While the size of the Charleston project had few parallels in the South, enthusiasm for railroads throughout the region was undeniable. Before the 1830s, most railroads in the United States were small projects built by private companies such as quarries. In 1830 James Holmes reported to the *Southern Agriculturalist* his opinion of one of these small railroads in Massachusetts, believing it to be "valuable to all who are compelled to remove earth to a distance." That same year, two men built an experimental railroad in Nashville, in which a single rail was elevated on posts. A person in a carriage suspended from the rail turned a crank to drive two wheels on the rail and thereby propel the carriage. The inventors claimed that speeds of fifteen or sixteen miles per hour could be achieved. State legislatures set their sights on larger projects, hoping to achieve more than simply moving earth. In 1825 Georgia passed "An Act to lay out a Central Canal or Railway through this State." Five years later, the West Feliciana Railroad was organized in Mississippi. An early railroad broke ground in Muscle Shoals, Alabama, in 1831. General boosterism found a home in southern states as well. Railroad boosters in Rogersville, Tennessee, published their own newspaper, the *Rail-Road Advocate,* one year before the creation of the national *American Railroad Journal.* The *American Railroad Journal* itself reached every state of the future Confederacy within the first six years of its existence. In 1833, just five years after Charlestonians worried about "uninterrupted" grass, the railroad to Hamburg stood completed—at 136 miles, the longest railroad in the world at the time. Railroads were a fledging technology in the 1830s, and some southerners were eager early adopters.[2]

"The rail road is the topic of the day"

Although these early efforts testify to southerners' willingness to pursue this new technology, not everyone was instantly swayed. Boosters still had to convince their fellow southerners that railroads were an appropriate investment for their personal—or governmental—funds. Southern boosters used three primary arguments to push for railroads in the early national period, ranging from utilitarian to ideological: promoters argued that railroads were superior to alternative forms of transportation, would improve the commerce of the South, and would bind the country together to preserve both the union and slavery.

Railroad boosters believed that trains possessed decided advantages over other types of travel. Water travel was a frequent target of criticism, because the variable and uncontrollable level of rivers could make travel difficult. When water levels were low, cotton shippers had to transfer their cargo to additional boats, spreading their load to make travel possible. Cotton was more likely to be damaged because of the increased handling, and long delays could ensue. If too little water was problematic, too much water created problems as well. "A superabundance of water in the Savannah river," noted one South Carolina committee, "always puts a stop to loading or unloading of boats for several days." Dependence on rivers also made planning difficult because boats were beholden to the tides. Even years after the introduction of railroads, steamboats continued to make newspaper announcements that declared a "CHANGE OF HOUR" because of shifting tides. For boosters, railroads held an advantage in that they were masters of their own time. Planters anxious to see their goods shipped would not have to wait for nature's cooperation but could immediately ship by rail.[3]

Water travel on canals did not present as great a challenge to railroad boosters in the South because canals were not as well developed there as elsewhere. Of southern states in 1830, only South Carolina had more than fifty miles of canals (New York had more than ten times the mileage), and Virginia, Tennessee, North Carolina, Alabama, Mississippi, Louisiana, and Florida had none. Although southern states added canal mileage during the antebellum era, canals were rapidly overtaken by railroads. By 1840 only one southern state had more canal mileage than railroad mileage, and the difference was negligible: Alabama had fifty-two miles of canals and fifty-one of railroads. Nevertheless, even in that state boosters argued that land transportation with railroads had clear advantages over water. After citing examples of railroad development in northern and southern states, boosters in Alabama declared in 1831 that "we consider it well established that railways for traveling and transportation are greatly preferable to any other kind of artificial improvement, or of steam navigation upon our best rivers." Presented with the option of railroad travel, even some dedicated boaters chose railroad transit: when the SCRR first began operating, the company reported that raftsmen on the Edisto River boated down to the coast and made the return trip by railroad with their boats shipped as freight.[4]

Having established the railroad's superiority as a mode of transportation, boosters made a second principal argument that revolved around the railroad's economic benefits, which took several forms. Railroads brought the advantage of division of labor: planters would no longer have to worry about transporting their own goods but could depend on the railroad to do it for them. Economic benefits would manifest themselves in terms of time savings: "To the Merchant 'time is money,'" declared railroad supporters in Charleston. SCRR president Elias Horry argued that time-conscious planters could attend to their own business in Charleston and "return . . . to their homes, all in a short space of time." Thus, railroads would remove the worry of managing transportation and allow planters to conduct their business more efficiently.[5]

Another commercial benefit boosters predicted was that railroads would improve the amount of trade conducted by the South. Some of this anticipated traffic was between the coast and the interior of the South; the availability and speed of transportation would make it easier for merchants to do business with remote areas. Horry explicitly developed this point in 1833, stating that "every agricultural, commercial, or saleable production" would be taken from the interior to Charleston, and goods desired in the interior "could be forwarded with great dispatch and economy, thereby forming a perfect system of mercantile exchanges, effected in the shortest possible time, and giving life to a most advantageous commerce." The Montgomery Railroad Company claimed in 1836 that its "great source of revenue . . . will be from the transportation of up freight, such as groceries, merchandise, etc. for the supply of an extensive interior," and it cited the SCRR's interior trade as a positive example. Similar arguments were made at a Knoxville convention that same year to support a proposed (and never completed) railroad from Charleston to Cincinnati; boosters hoped that railroads extending into the interior would lead to "consequent extension of the *production* as well as the *consumption* of the people along the whole line."[6]

But railroad boosters had a grander vision than simply the southern interior. Supporters also argued that southern railroads held the key for the South's connection to the burgeoning West. While historically southern railroads have been characterized as purely "limited, local, and conservative enterprises," southerners were clearly capable of dreaming big. It is also misleading to suggest—at least in the southern case—that early railroad advocates "did not foresee the future value of long-distance railway transport," which has been cited as a reason that standard gauge was not adopted. To the contrary, boosters were quite interested in long-distance transport but may have seen no reason to standardize with competing lines, because each group of boosters hoped to control this transport independently. Exhortations to look to the West began early in railroad development. In 1831 Horatio Allen and Henry N. Cruger told a railroad meeting convened at Estillville, Virginia, that obtaining the trade of West and South would allow them to "realize the emoluments of being both sellers and buyers." That same year the *Charleston Courier* opined that a railroad should be extended to the Tennessee River, which would allow Charlestonians to get their "Corn at a much cheaper rate from the Western States, than we now do from North-Carolina." Boosters thought that railroads would allow the South and West to pursue their particular economic advantages. As a citizens' meeting in Charleston in 1835 declared, "The great productions of the South, are COTTON and RICE, articles that can only be produced by *slave labour.* The West may be appropriately designated, as a PROVISION COUNTRY, producing mostly by free labour, grain and meat, in the greatest abundance, and on the cheapest possible terms." By allowing each region to produce its own specialized goods—and using the railroad to facilitate exchange—the South could protect its slave economy.[7]

Southern railroad backers warned that competition for the western trade would be fierce. Southern subscribers of the *American Railroad Journal* learned about competition in the North and saw that southern states were held up as viable competitors. One editorialist, pushing the New York and Erie Railroad in 1835, argued that "not only are Pennsylvania and Maryland moving shoulder to shoulder in the cause of internal improvements, but Virginia is also aroused from her long sleep." The author begged New Yorkers to get behind the New York and Erie effort before Virginia and other states completed their works. When talk of the New York and Erie line revived in 1837, its proponents characterized the competition among Pennsylvania, New York, Maryland, and Virginia as a "struggle." Such struggles continued throughout the antebellum era. The Pennsylvania Railroad, for example, told its stockholders in 1848 that "competition for the trade of the West is vigorous, and the stake is immense." While northern railroads ultimately won the battle for western trade, southerners clearly recognized its importance.[8]

The practical arguments for the railroad's superiority as a form of transit and its financial benefits were joined by a third argument: that railroads would bind the country together. D. K. Minor, the editor of the *American Railroad Journal*, was a notable proponent of these views on the national stage. With a national system of railroads, Minor declared, "we should have little apprehension of a dissolution of the *Union*." The possibility of disunion weighed heavily on Minor's mind in 1832, and he believed that railroads would "have a great influence in removing the prejudices now cherished by one section of the country against another. They will enable us to visit different sections, to compare our *own* faults with theirs, and to find that there is not, after all, so much difference as we apprehended." Two years later Minor still saw the "natural result" of railroad linkages as "better understanding, and knowledge of the character of each other. It will then be ascertained that we of the north are not all 'pedlars of wooden nutmegs and horn gunflints,' and they of the south not all 'negro-drivers, nor hard masters.'" To accomplish that end, Minor regularly reported on southern railroads in his newspaper. He praised the SCRR, noting in May 1833 that "we have of late heard many inquiries relative to the condition and prospects of this road." The SCRR was even held up as an example for northern roads to emulate. Reporting on the SCRR's dramatic increase in profits from 1834 to 1836, Minor declared: "Thus it is on the Charleston Road, and thus it *will* be on the [proposed] New-York and Erie Road."[9]

Minor was not simply projecting his own goals onto the southern populace. Writers in southern states saw the *Journal* as an appropriate outlet for news about their railroads, thus making themselves part of a national movement. Southerners wrote with pride about their own accomplishments and explicitly placed these in the context of national, not regional, success. From Alabama, one railroad man recognized that the New York–based paper was the appropriate "medium . . . to give publicity to the designs and wishes of so many people of this state who are anxious for a railroad between Tuskaloosa to Tuscumbia." Under the pseudonym "North Carolina," another citizen wrote that a planned railroad connecting Wilmington with the West "will be found entitled to rank with the greatest, now carrying on in the United States." Another North Carolinan declared impending work to connect Raleigh with Virginia would "form one of the grandest thoroughfares in the United States." Even if we discount booster hyperbole, it is clear that southerners were proud of their accomplishments and were anxious to express that pride in a national forum.[10]

Southern acceptance of the newspaper is all the more remarkable when one considers that the *American Railroad Journal* began publication on the heels of the nullification crisis. We cannot know if Minor lost southern subscribers because he disapproved of disunion. We do know that Minor decided at the end of 1833 to stop printing material on political affairs. Opening the third volume, Minor declared that the *Journal* would henceforth be "avoiding every thing like *partizan* politics."[11]

Southerners agreed with Minor that railroads held political and social advantages. Advocates of a rail link between Cincinnati and Charleston argued that people "who, but a short time since, were strangers to each other" would be "brought into neighbourhood" by the railroad. Trade would take place on the basis that they were "FRIENDS; the citizens of one common country, brethren of the same political family." These boosters considered it a "painful reflection" that southerners knew their northwestern neighbors as poorly as they did. However, they were also positive that a railroad could not fail to help the residents achieve a "greater intimacy." Evoking the usual imagery of binding the country together, the promoters hoped that when southerners and nonsoutherners were able to visit each other, "those cords of sympathy, by which men's hearts are united," would increase as railroads clearly demonstrated "the influence of social intercourse in smoothing disparities, removing prejudices, and binding us together, by those *social ties,* which are among the strongest bonds of society." In states like Ohio, Indiana, and Illinois, "where the fanatical anti-slavery spirit is as yet almost unknown," the railroad would serve to create bonds to the South that might keep these states out of the rabid antislavery camp. In 1836 one orator argued that economic self-interest would prevent northwestern farmers from supporting abolitionism, because either disunion or emancipation "will give the death blow to their prosperity." The speaker had no doubt that northwestern farmers would see the folly of sacrificing their interest, but " 'to make assurance doubly sure,' our road is wanted. Make it, and you . . . may rest assured, that your institutions are secure, your property safe, and that your repose will not be disturbed." He had no need to be more explicit about the "institutions" and "property" to which he was referring. Southerners agreed with northerners like Minor that railroads could smooth the political tensions of the early republic.[12]

Thus, railroad advocates employed a three-pronged attack when arguing for railroads in the South: railroads were technologically superior to other forms of transport, they would benefit the region commercially, and they would bind the union together. Southerners were receptive to this message. "The rail road is the topic of the day," wrote a young Alexander Stephens, future vice president of the Confederacy, in 1834. "Some think it will be a profitable investment of capital. Others fear to run the risk with their own pockets, while all seem very anxious it may be effected by some means [or] other." Having summed up the range of opinion, he concluded that the "greatest obstacle is the greatness of the enterprise." Stephens was not among those dissuaded by the greatness. "Speed to the work," he concluded. Other southerners demonstrated their curiosity by demanding that their friends keep them abreast of developments. A letter writer in Florida requested Hugh McLean of Columbia to "let me know . . . how far the Rail Road is finished and all the news about town." When the SCRR was first being built, one Charlestonian wrote that "to vary the monotony we go and look at the rail road and anticipate the pleasure of riding on it one of these days, but I am very much afraid we shall all be dead before we realize it." Completion would happen much sooner than that writer realized. Clearly there was genuine interest in railroads as construction began.[13]

One of the strongest signs of popular approval of the railroad was the ease with which railroads gained their land. The process is even more remarkable when one considers that railroads, whose ability to claim land was backed by the state, were exercising powers on a scale hitherto unknown. In South Carolina, for example, canals had previously been able to appeal to court-appointed commissioners for a fair valuation of land when property owners proved recalcitrant. But few canals were successful, so the ability of the SCRR to claim land represented the boldest private use of the state's eminent domain privilege. Yet courts in South Carolina upheld the SCRR's right to take land, recognizing that the corporation, although private, represented a public improvement. Landowners across the South agreed, as they turned over land without protracted court battles. Southern railroads reported that they encountered little difficulty in securing the route that they desired. When the SCRR opened in 1833, its president, Elias Horry, thanked landowners for giving up their land without charge to the company and also for allowing the company to make use of the timber found along the route. He characterized these acts as "generous as to the Company, and patriotic as regards the State." The CRRG also reported little trouble in the early years. Although the process of obtaining right of way was referred to as "troublesome and vexatious" in 1842, it was also reported that there were "very few cases remaining unsettled" and there would not be any "serious difficulty in arranging them." The Greenville and Columbia Railroad noted in 1849 that it had secured more than fifty-five miles of right of way without having to pay compensation. The president particularly thanked one individual in Richland District for giving up six miles, which suggests that most of the remainder was pieced together in smaller sections from multiple landowners.[14]

Such generosity continued into the 1850s, when railroads were no longer a novelty. Indeed, communities continued to show their support for commerce and technological progress as they allowed railroads to traverse the landscape. The North Carolina Railroad encountered little opposition when it began to secure right of way in 1851, and the Thomaston and Barnesville Railroad in Georgia was equally successful. In 1852 the president of the Spartanburg and Union Railroad reported that he passed over the section to be surveyed from Spartanburg Courthouse to the Tyger River and "met with only two persons who refused to sign the Right of Way to the company free of cost." He also noted that only four landowners below the Tyger had objected to the road. The Blue Ridge Railroad in South Carolina secured thirty-seven miles from Pendleton, South Carolina, to the North Carolina border at the cost of only $518. Citizens in other states were also generous to the company. The seventeen miles in Georgia were secured at a cost of $3,640. Of the fifty-seven North Carolina landowners, forty-three released their land by November 1855, with very few claiming compensation. In May 1857 the citizens of Walhalla, South Carolina, gave the same company twenty acres of land for a depot. Whether this generosity was inspired by self-interest or altruism, communities clearly wanted to take advantage of the railroad. Perhaps some could afford to sniff at the railroad's benefits and the horrors of modernization, but numerous landowners across the South disagreed.[15]

The Problems of Funding, Land, and Rivalries

Although the widespread enthusiasm for railroads cannot be discounted, railroad advocates were not given a free pass in the antebellum South. Railroads constituted an entirely new creation on the southern landscape. Not only was the railroad as a form of transportation subject to debate, but a host of other debates on topics such as funding sprang up around these new entities. As railroad advocates soon found out, not everyone shared their excitement, and opposition could be formidable. Railroads faced three significant challenges: the battle to secure adequate funding, stubborn individual landowners who could stymie progress, and the rivalries that inevitably sprung up when railroads were constructed. These problems required the special attention of railroad advocates.

Adequately funding these expensive works would prove to be one of the most significant challenges that southern railroads faced. Corporations attempted to find local subscribers who would lend their financial support to the new enterprise. Some large planters eagerly invested in railroads and their potential for exporting cotton. Planters such as Benjamin Sherrod of Alabama, James Everett of Georgia, and Paul Cameron of North Carolina all lent their fortunes to railroad development. Others, however, were wary of gambling on such a new technology. J. Newton Dexter reported in 1830 that his nephew found "but little spirit amongst the stockholders" of the SCRR. Editor William Woods Holden of North Carolina had a similar complaint in 1849: "Men who ought to take thousands put down Hundreds; but we shall hope on, and therefore work on." Holden understood why potential investors might be scared off: "The losses herefore incurred by our citizens by the Gaston Road, is the great drawback." Bitten once, the citizens were reluctant to engage in another project.[16]

As a result of the difficulty of finding investors, state spending was critical to railroad construction in the South. Yet it also launched a host of opponents.[17] However much boosters claimed a railroad fit into a national system of improvements, the battle for funding required thoroughgoing involvement in state politics. While state funding would ultimately form a critical component of railroad finance in the South, it required constant advocacy on the part of boosters.

State governments were leery of the large expenses that railroad projects required, particularly when the benefits were not demonstrable statewide. The Blue Ridge Railroad in South Carolina, for example, was particularly dependent on state aid because of the massive expense associated with tunneling through Stumphouse Mountain in modern-day Oconee County. In 1859 the credit reporter from R. G. Dun wrote, "A strong effort will be made at this session of the Legislature to obtain an appropriation, but it will in all probability fail, as the upper sections of the state (except in the immediate vicinity of the road) are opposed to it."[18] If railroads could not argue for their relevance beyond the immediate area of construction, they could be met with staunch opposition in state legislatures.

Jealousies already present in southern states were magnified in any competition for state funds. Such problems, of course, predated the railroad era. In South Carolina, for example, distrust between the small farmers of the interior and the wealthier lowcountry planters stretched back for decades. Interior farmers had long chafed under a political system that left them severely underrepresented in state government despite their growing wealth and population. In Virginia, tidewater planters were loath to spend money on projects that would benefit western Virginians. Eastern planters in North Carolina were also unwilling to loosen the state's purse strings in the 1830s. Railroads, then, not only had to make a compelling case for an unproven form of transit but also had to navigate preexisting rivalries and resentments. As the railroad era progressed, railroads made substantial headway in securing state funding for their works despite old rivalries.[19]

Companies worked hard to make their case to local governments and individuals. Horatio Allen evidently built a model railroad in the South Carolina capital, Columbia, when the legislature was considering a bill funding the SCRR. He recounted that he "constructed at Columbia about 400 feet of rail, upon which one horse moved effectively in either direction (it was level) 12 to 13 tons at 4 to $5^1/2$ miles per hour. We effected a very great change in the opinion of the House of Representatives, but the Senate was obstinate and ignorant." Despite Allen's efforts, the company received only $100,000 of the hoped-for $250,000. The RDRR seems to have inspired particular hostility in certain counties; one annual report described the efforts of three men, including an "aged and experienced minister of the gospel," who took a "regular and constant canvass against the county subscription."[20]

Opponents to aid worried that if states were to extend their favor upon certain roads, they would find themselves besieged by endless projects of dubious value. Others feared that governments would sink into debt. Writing in 1847 under the name "Anti-Debt," James Henry Hammond of South Carolina detailed several objections along these lines. Hammond was no Luddite: he helped organize the celebration when the SCRR reached Columbia in 1842. And he recognized the extraordinary power of railroads. "Experience has not yet taught us how far they are destined to revolutionize the world," he wrote. "They have already, in their infancy, wrought great changes, and will undoubtedly effect others yet far more wonderful." He recognized the futility of opposing railroads: "I might as well be opposed to all locomotion—to commerce—to steam—to agriculture itself." But he wanted railroad expansion to proceed in a reasonable manner and saw "no reason why the world should go mad about them, and mankind bankrupt themselves to force them into premature existence." Hammond believed that the state would be entrapped in debt if it were required to support internal improvements. Hammond did not oppose all state spending; he lauded education as worthy of the state's attention. "But to bankrupt herself by Canals and Railroads," he concluded, "enterprises designed solely to facilitate trade, and in which money mongers and speculators alone usually invest, for the mere sake of gain, would be not only the sheerest folly, but disgraceful and disgusting." Individual capitalists would be the gainers from railroads, and therefore they should be the ones to foot the bill.[21]

Some capitalists agreed. Renowned industrialist William Gregg protested South Carolina's consideration of giving money to the Blue Ridge Railroad, arguing that individual capitalists were better suited than the government to run such enterprises. He cautioned against state investment: "We want rest and repose from this eternal ding-dong on the State for aid, and what we, and the whole people of the United States want, above all other things, is, that the idea that Railroads are not in future to look to *private capital* for their erection should be dispelled." Some railroads even opposed state aid. The president of the Central Railroad of Georgia felt that giving other railroads state aid would be unfair to "that portion of her people . . . who by their enterprise and their private pecuniary means first brought to the State her glory, and honor, and power." The "portion" to which the president referred comprised those who had the foresight to invest in early projects. Railroads already in existence were not anxious to see the government fund their competition.[22]

But state spending had powerful advocates as well. Some worried that private individuals would be too stingy to support railroads. "Railroads ought to be built, but I think they ought to be built by the public," wrote one North Carolinian. He complained about those who lived along the line of the NCRR and refused to contribute to the effort, yet received all the advantages of the railroad's presence. Men like *"Judge Ruffin,* who is worth $200,000 has a mill and farm and who uses the road once or twice a week himself to go to Raleigh and elsewhere and who's family or some member of it is on the road every few days. He who has these great advantages never paid one cent, and persuades his neighbors not to take any stock, saying to them that they could have plenty of stock if they wanted it at fifty cents on the dollar after the road is built."[23] Concern about such freeloading led some to support state aid in order to ensure that work would continue.

In the end, a dearth of private investors and the sheer size of projects led state governments to play a crucial role in funding railroad projects. Of course, most railroads in the country received some sort of public assistance, in the form of bonds, land, materials, or the privileges provided by a charter. Reliance on state-level aid meant that some aid fluctuated with political winds. In North Carolina, for example, the Whig Party "won the governorship in 1836 largely because of its advocacy of internal improvements," and it continued to support improvements when it controlled that state's government. North Carolina Whigs consistently supported railroad projects more than their Democratic rivals. Other states followed different paths to the same result. South Carolina lacked a robust Whig Party, but the state still established a revolving fund in 1847 to assist new railroad corporations. In Georgia, Democrats and Whigs worked together to support the Western and Atlantic Railroad, and the state contributed substantially to internal improvements. By the eve of the Civil War, the critical role of public financing was clear in the South: more than half of the amount invested in capital stock in the region came from public treasuries. By comparison, public funding nationally had contributed approximately 25 percent of the $1 billion in capital stock. States proved to be a critical solution to the problem of railroad financing in the South.[24]

After addressing the problem of financing, some railroads had to wrestle with the difficulty of securing the route that they wanted. Despite the support that most railroads received, any civil engineer could tell stories of headstrong landowners. The president of the CSCRR informed stockholders that many of the citizens of Winnsboro, South Carolina, objected to the road passing through the town, and so the railroad was forced to go another route. The East Tennessee and Virginia Railroad complained that many landowners "refused to settle on reasonable terms" or "had imperfect titles." Some landowners feared the railroads would do more damage than good. One landowner along the route of the SCRR opposed it because he feared trains would kill his slaves and that the noise would be a "nuisance." Somewhat more dramatically, William Adkins, executor on Ballard's land required by the same company, informed civil engineer John McRae that he "intends to appeal from the assessment of the commissioners & that he will *shoot* any one who attempts to . . . go on this place under orders from the Rail Road Company until he is paid." McRae confessed that he was unaware that he had to consult Adkins, because "Jno. Ballard spoke to me on the subject of right of way through that place as if it were entirely his own. But I think it better [to] arrest operations until I hear from you." McRae knew enough to take certain threats seriously.[25]

Landowners convinced that the railroad would bring economic benefits wanted to squeeze every bit of that benefit out of the companies. The president of the Hiwassee Railroad (later the East Tennessee and Georgia Railroad) noted in 1838 that most farmers believed that the railroad would increase the value of their land, and thus they would treat the company liberally. Instead, he reported, "landholders seem not only determined to exact heavy damages but require immediate payment." As did other railroads, the Hiwassee Railroad appointed a committee to assess the land damages and make the company's case. Sometimes the gulf between the two was large. For example, Charles Rice asked for $500 if he was to build and maintain his own fences around the area where the railroad passed his farm; the committee decided that he should be given $300. Although all railroads had to address landowners who opposed the railroads' claiming their land, railroads also successfully overcame this obstacle, as frustrating as it was. According to a report of the Tuscumbia, Courtland and Decatur Railroad, "Opposition to the location of the Road is fast giving way & that there will not be as much difficulty as was at first anticipated." Railroads were generally successful in obtaining the land they wanted.[26]

Rivalries constituted the third major difficulty that railroads encountered. This was a difficulty laced with irony. As southern railroads grew, they became victims of their own success. Antebellum railroad development soon demonstrated that railroads did *not* necessarily bind the country together but could initiate rivalries between cities and states or exacerbate existing tensions. Excitement for one road did not necessarily mean that people supported all roads. Explaining why more people did not subscribe to the *American Railroad Journal,* John McRae wrote to D. K. Minor, "I was in some hopes that before now I could have sent you some additions to your subscription list but cotton is low, times are hard, & though our Road penetrates a country entirely new to the Railroad, those who are most deeply interested in its success are so from personal feeling & not from any general concern for the Rail Road cause."[27] Although some railroad boosters would push their own project, they were not attached to the general cause.

Minor recognized there could be value in competition and occasionally played states against each other to spur on greater development. When a railroad project was announced in Florida, Minor asked, "Is it not time for Georgians to strain every nerve, if they do not wish to see one of the Old Thirteen outstripped in enterprise by the citizens of a Territory but lately acknowledged as a part of our domain." When Alabama began to construct railroads, Minor hoped that its works would "stimulate those of *older* states to action." After praising Virginia for committing funds to internal improvements, Minor asked if his own New York would "stand with folded arms and see other states going so far ahead of her?" And in one issue of the *American Railroad Journal* that was particularly flush with news from the South, Minor wrote that "it will be seen that the Southern States are even taking the lead of the North." Minor hoped to tap into a competitive spirit and urged states into greater efforts.[28]

Despite Minor's enthusiasm, not all the effects of rivalry were positive. Railroads did not guarantee long-term benefits to a city. Citizens of a town might rejoice at being selected as a terminus for a road, but if the company desired to extend beyond their city, the town could find itself transformed into a mere way station. Internal improvement advocates recognized this problem. Simeon Colton, writing to the North Carolina Board of Internal Improvement in 1840, noted that the new availability of goods in the interior of the state would increase the number of stores but also decrease the power of former market centers. "Intermediate towns" on the route of the railroad would suffer if they could be easily bypassed.[29]

Such a dangerous transition could affect even railroad pioneers, such as Charleston. Merchant Thomas Napier foresaw the danger of allowing a railroad to be constructed from Wilmington, North Carolina, directly to the interior of South Carolina: "Should Charleston fail in constructing the lower road from Wilmington to Charleston they may lose much which it may never be in their power to regain." And that is precisely what happened: in 1854 the Wilmington and Manchester Railroad was completed, linking Wilmington to Manchester, South Carolina, located in southwestern Sumter District. A connection was soon effected with the SCRR, and now goods arriving by steamer in Wilmington could be sent along by railroad to the West and bypass Charleston completely. McRae felt that allowing the WMRR to gain the upper hand was a tremendous error on Charleston's part and proposed as remedy another railroad, arguing that "the only hope of restoring, at least a portion of the lost ground, depends at present upon the speedy completion of the North-Eastern and Charleston and Savannah Rail Roads." Such a route would draw traffic away from the WMRR back toward Charleston and present a speedier route to Savannah.[30]

In the competitive realm of transportation, it seemed that a city could never stand still for fear of being left behind. When the RFPRR was completed, Fredericksburg found itself less significant to north-south trade than it previously had been. Some forward-looking citizens of Fredericksburg began to push for the town to establish a western railroad that would join the RFPRR at Fredericksburg, transforming the town into a junction and restoring its importance. But soon there was a rival for the trade of the West: residents of Louisa County also agitated for a western connection. The Louisa County group was more successful, and its railroad joined the RFPRR south of Fredericksburg. Fredericksburg's own project never got off the ground, and the city was unable to capture the western trade.[31]

Railroad companies tried to put the best face on rivalries. The Louisville, Cincinnati and Charleston Railroad noted in its annual report that, although there were indeed rival projects, stockholders should not worry that their funds had been misplaced. Rather, it was better to take an enlightened view: all projects were part of one vast system, like veins in a human, and the health of the system depended on distribution through all of the veins. The completion of these projects would result in "a stream of South-Western trade and travel," which would not be "a subject of strife, between sister cities and Roads, but of most abundant participation for all." In a similar fashion, the SCRR attempted to assure its stockholders that the development of railroads in Georgia did not mean a danger to the interests of their road. Rather, the multiplication of roads in the country could benefit only Charleston because Charleston was a link to the sea: each railroad west of Charleston "in turn may become the parent of others, indefinitely reproducing others and spreading their arms in every direction, gathering and bearing on their tracks the products of countless farms and manufactories, to be exchanged at our Seaport for the merchandize of other lands, which in turn are to be re-conveyed by the same channels to the places of their consumption."[32]

In reality, however, such poetic hopes were overly sanguine. Rivalries could be damaging to companies and the areas they served. Rivalries could even prevent rail connections across states. John McRae feared in 1846 that the "projects of connecting the SoCa and NCa Railroads are likely to be sacrificed to the local interests of the State in the other projects or to fall along with them." The SCRR did not achieve its goal of an uninterrupted link with Georgia until 1853, when the town of Augusta finally allowed the company to extend its tracks across the Savannah River. Southerners had to balance the local benefits of railroads with the ever-present danger that those benefits could easily be whisked away.[33]

Finally, people feared that railroads would create new rivalries by drawing business away from existing forms of transportation. Businesses also had to consider what railroads would do to their existing business practices. In Columbus, Georgia, business leaders feared that railroads would destroy their businesses done on the Chattahoochee River, yet ultimately they decided that being bypassed by the railroad altogether was a far worse fate. James Henry Hammond argued that the wagon trade "brings with every wagon four or five horses to be fed, two or three purchasers, black and white, with countless orders from their neighbors, and fancies of their own to be charmed by the retailers." By contrast, the railroad simply encouraged people to pass through without spending money: "Formerly every passenger who traversed South-Carolina, deposited by the way from fifteen to twenty dollars—now he rushes through for six or eight. Here is a clear loss." While Hammond had little evidence to back up his assertions, his fears of rivalry are telling. When launching into railroad projects, southerners had to worry about rivalries on a variety of fronts: fears that railroads would eventually bypass a city, or criticism that railroads would only supplant existing trade, not add to it.[34]

Southern Exceptionalism?

While railroad development in the South engendered opposition, this opposition was not uniquely southern. Northern railroad advocates faced similar opposition in state legislatures, in the press, and from the populace. Commenting on a potential railroad in Connecticut, Myron Webb wrote in 1841, "A survey of different lines was made last winter and it was expected to be put under contract this spring but has not for the present want of Subscribers to the Stock. . . . There is a good deal of opposition by the farmers whose lands it will cross in some places. But I think it will be built in time." Railroad conductor Charles George characterized his native state of Massachusetts as "one of the most conservative in adopting the railroad."[35]

Opposition to railroads could be found throughout the antebellum North. Citizens of Illinois protested the creation of the Galena and Chicago Union Railroad, threatening to "let their cattle run loose on the tracks" if construction took place. In Pennsylvania, farmers in the 1830s opposed the use of locomotives on the Philadelphia and Washington Railroad and urged the legislature to make it a public highway. Opposition was also extensive in New Hampshire, where radical Jacksonians launched a campaign against railroads. This "Railroad War" was set in motion in part because of displeasure over the power of private corporations had with the state's eminent domain authority. The lashing out at corporate power culminated in the Railroad Act of 1840, "which effectively halted all railroad construction in the state for five years."[36]

Other states threw up roadblocks in different ways. In the early 1850s, Ohio and Pennsylvania railroads used seven different gauges. Naturally, the constant gauge changes forced trains to stop and ensured that communities would not be bypassed. The leadership in this battle came from Pennsylvania, hoping to protect the interests of the town of Erie. A law enacted in 1851 "directed that tracks west of Erie had to match the gauge prevalent in Ohio, while that east of Erie should conform to New York's usual gauge." The Ohio-Pennsylvania rivalry eventually found its way into the courts. When the Franklin Canal Company attempted to build a railroad that would connect with a railroad from Cleveland, Pennsylvania took the company to court and won. The Supreme Court of Pennsylvania "spoke darkly of perverting privileges to aid rival states" in its 1853 decision. That same year, however, the Pennsylvania state legislature allowed companies to determine gauge. The citizens of Erie then sprang into action. The city council outlawed a gauge change, and residents—including the mayor—destroyed "newly laid track" in December 1853. Only after a few more years of legal wrangling did railroads get their desired connection. There was also opposition to railroad construction in antebellum Philadelphia. The City Railroad was prohibited from using steam locomotives between Vine and South Streets until after the Civil War. From 1840 to 1842, residents of that city objected to the Philadelphia and Trenton Railroad's laying track in the city. Local citizens tore up track and threatened workmen four separate times to express their opposition. Opposition continued in the late antebellum era as citizens protested horse-drawn railroads when charters for such companies began to be considered in 1857.[37]

Northern companies also faced difficulties in securing the right of way. The Boston and Worcester Railroad reported that between Brighton and Newton, Massachusetts, "the proprietors in general are unwilling to make any abatement on account of any advantage to result to them from the construction of the road," and as a result the amount spent on land damages would "considerably exceed the original estimate." An agent of the Western Railroad in Massachusetts also lamented the extensive problems he encountered in attempting to secure the right of way. In one town the landowners had "combined to prevent work, till their damages are paid," and one owner in particular "is so violent that he swears that the road shall not cross his land, & will not permit the level to be taken, pulls up every stake & level pin, as fast as driven, even before our men are out of sight." The president of the Western responded with caution, urging the agent to have the contractors only do work on "peaceable ground."[38]

Like their southern counterparts, northerners worried mightily about competing lines cutting into their business. A New York, New Haven and Hartford Railroad report urged the corporation to strike first when it became clear that a rival company would construct a rail link that would cut off a portion of its business: "It is no longer a question whether this intervening piece of Road is to be built, but by whom shall it be controlled." The board was told not to worry about expense: "What apology could we offer to our Stockholders if by a narrow and short sighted policy of saving a trifling expenditure we were to entail upon them an evil of such magnitude as the establishment of an independent line of Road between Hartford & New Haven of which they had failed to secure the control entirely through our inaction and neglect?"[39] Rivalries were a considerable worry, regardless of geographic location.

Given the range of opinions considered here, it hardly seems appropriate to characterize southerners as uniquely opposed to railroad development in general. While individual projects may have struggled to get off the ground (as happened in Fredericksburg) or landowners may have proved troublesome, these were debates that played out in other states across the country. Some were uneasy with the financial commitments necessary to construct railroads, yet railroads themselves remained popular. Southerners were early converts to the promise of railroads and prosecuted them with vigor when labor and finance allowed.

"A perfect fever on the subject of Railroads"

Whatever opposition Hammond, Gregg, and others may have had to the zealous prosecution of railroads, their warnings were clearly not heeded by southerners in the 1850s. That decade was one of tremendous expansion and growth: every southern state that had some mileage in 1850 more than doubled it by 1860. Contemporaries recognized that enthusiasm was high beginning in the late 1840s. McRae reported as much in South Carolina: "The people in the upcountry & Columbia are in a perfect fever on the subject of Railroads," he wrote in June 1847. "The Charlotte Road is the all absorbing topic," he continued, charting the fever in North and South Carolina towns. "Camden has fairly succeeded in rousing the jealousy of Columbia & the battle is to be commenced I understand in Charleston on next Thursday." McRae informed D. K. Minor the next month, "We have railroad mania here now. Railroads are projected every where but whether the fever will last long enough to build one remains to be seen." Despite his own cautiousness, McRae wrote that he had "not the slightest doubt" that North and South Carolina would soon be linked by railroads. Others noted the enthusiasm as well. Lawrence Branch wrote of North Carolina in 1855 that the state legislature was "running wild about Rail Roads. . . . all parties are rising with each other in the Internal Improvement race."[40]

Railroads were so exciting because they tapped into a deep strain of southern optimism about the spirit of the age. Newspapers, for example, encouraged railroad development. "Safety Valve" urged residents of Edgefield District, South Carolina, to support railroads in 1847: "We live in a wonderful era of the world. *The go ahead age,* as it has been called by some, is not a very inappropriate style of the present condition of the world." "Steam power" had "given such a momentum to the affairs of men that it calls for a correspondent action in devising and executing plans," Thomas Napier wrote that same year. The short-lived Charleston journal *Self-Instructor* proclaimed itself in 1853 to be "devoted to southern education and to the diffusion of a knowledge of the resources and power of the South, as represented by the negro, the rail and the press." Railroads clearly represented progress. Jennie Speer wrote that although Greensboro, North Carolina, "does not boast of anything strange or marvelous" in 1851, she hoped that shortly residents could "hear the [sounds] of the great horse with his lungs of iron and breath of fire," as a sign of the town's improvement.[41]

Some individuals transformed their support into more practical action. Mississippi doctor Elijah Walker wrote that he was "decidedly in favor of the railroad Scheme," with such benefits as "enhancing the value of lands 100 percent and afford[ing] safe and speedy [passage] to all things grown by a farmer to a safe and redy market for every variety." Walker demonstrated his belief by organizing a barbecue to promote the Mississippi Central. On June 8, 1852, he recorded in his diary that he "rode all day trying to make a collection of vegetables and other eatables to make a Rail Ro[a]d barbecue at this place on the 19th Inst— I succeeded very well and there no doubt will be a fine company of people assembled on that day."[42] He detailed the success of the barbecue on the 19th:

> At an early hour this morning the citizens of this county began to assemble, both male and femal[e] and a host of children— At 11 O'clock Col [Harvey W.] Walter addressed the audience consisting of about five hundred persons, in a speech of an hour and a half's length, his speech was eloquent and to the purpose fitting in eve[r]y particular the peculiar views of the people of this vicinity— After the speech the books were opened for the subscription of stock. I [*unreadable*] and soon succeeded in getting up about six thousand dollars worth of stock. After that we all repaired to the table where was well prepared an ample sufficiency.[43]

Walker's hard work paid off. At the election on June 26, Walker's precinct voted 50–1 to allow a tax for the benefit of the Mississippi Central.[44]

The interests of consumers continued to be important to transportation boosters in the late antebellum period. In 1852 a railroad booster in Tuscaloosa, Alabama, claimed that, thanks to the railroad, a woman could order a hat from New York on Monday and "get the hat in time to wear it to church the next Sunday." In the late 1850s, boosters continued to use familiar arguments of economic and social benefits. However, because railroads had been operating for years, advocates could now point to real, enduring advantages. Speed and better communication remained powerful arguments. Advocating South Carolina's Blue Ridge Railroad in 1860, a pamphleteer wrote, "The States of the Mississippi Valley are not more remote from Charleston now, than our own back country was thirty years ago; you can reach Memphis in a shorter time, and at less expense, than you could the village of Greenville at that period." The writer also noted that the railroad had not damaged slavery; rather, the number of slaves in the South Carolina upcountry had increased since railroads were first built in the area.[45]

Finally, by the close of the antebellum era, railroads had developed their own expansionist logic. Boosters urged southerners not to abandon what they had already done so much to support. Urging completion of the Blue Ridge Railroad, Benjamin Perry appealed to the honor of his audience as well as their financial sensibility. The state and investors had already expended a considerable amount of money on the railroad; to finish the job at the Blue Ridge would require more. Did it make sense to essentially squander all work done to this point? "Shall this unfinished railway, with all its grading, embankments, bridges, aqueducts and the long tunnel into the mountain on which we stand, be pointed out to our children and their descendants for centuries to come, as an eternal monument of the State's folly, and the childish fickleness of their ancestors?" Perry asked. The answer was clear: "Never! never! never!" Perry concluded that life was indeed better for South Carolinians in the upcountry than it had been at the beginning of the century, and the credit for these changes went in part to the state's "intercourse with the world." The railroads—what Perry called "the great civilizers of the world"—had brought not just higher prices for Greenville's corn and flour but increased opportunity for men to travel and see the world. Moreover, the material improvement that trains brought was evident: "We see the comforts, and luxuries, and elegancies of life amongst you. You are no longer shut out from the world—the busy, industrious, commercial, manufacturing world." Railroads had fulfilled their promise, and Perry hoped that his audience's positive experiences would sway them to support yet another project.[46]

Railroads never lacked for advocates in the antebellum South. Southerners quickly grasped the utility of this new mode of transportation. Boosters employed a three-pronged attack when urging construction, pointing out the railroad's technological advantages over other forms of transportation, the economic benefits the South would receive, and the fact that railroads would bind the country together, preserving both the union and slavery. To be sure, there was some opposition. The expansion of railroads helped trigger debates about the propriety of government expenditure on private enterprise, some landowners were reluctant to part with their property, and southerners also worried that commercial rivalries could leave their towns bypassed. But southerners were hardly unique in having these debates. Railroad projects in the North also faced political and economic challenges and sparked sharp dissent. It is not possible to speak of "southern" opposition to technological advancement. Rather, southerners embraced railroads, demonstrated by the rapidly accelerating construction during the 1850s.

Accomplishing such great works required the substantial dedication of time, money, and personnel. Southern corporations also had to take risks, given the uncharted territory that railroads represented in the 1820s and 1830s. In order to chart this new territory, southern corporations turned to civil engineers. These men created the knowledge base that drove the country's tremendous antebellum expansion of railroads. The dreams of railroad promoters like D. K. Minor, Benjamin Perry, and Elijah Walker would become concrete reality through the labors of these engineers.

CHAPTER TWO

Knowledge

When Charleston attempted to improve its fortunes with a railroad, Georgians did not sit idly by. In July 1833 a public meeting in Augusta agitated for a railroad, and citizens successfully incorporated the Georgia Railroad Company that December. The company was given the authority to construct a railroad leading out of the town, linking the interior to the Savannah River. While the road was to be constructed of rails, the corporation's seal suggested that the road would not necessarily be driven by steam power: at the center was a horse's head (figure 1).[1]

The prominence of the horse on this young company's seal reflects the uncertainty that surrounded early railroads. Augustans were hardly shy or tentative in their pursuit of this transportation link, but the use of steam was not yet a settled question: early trains could just have easily as been pulled by horses. Georgians were not alone in weighing their options. Comparing the early days of the SCRR to the early days of the Baltimore and Ohio Railroad, civil engineer John McRae reflected that "both commenced before it was determined whether Horse or Steam power was the most advantageous & both have had to wade through a series of most expensive experiments which other companies of more recent date have benefited by without the cost." It was a time when "engineers vigorously debated all aspects of railroad design" and had to do so in the context of young corporations attempting to turn a profit. Balancing these several and sometimes competing concerns would be a task that fell to civil engineers.[2]

"A Locomotive sort of character"

As a profession, civil engineering was in its infancy in the United States during the early nineteenth century. Opportunities for formal training were slim, but the growth of internal improvement projects created substantial demand. In the late eighteenth and early nineteenth centuries, canal builders faced tremendous challenges as every aspect of the process—how to retain water in the canal, how to build locks—was unfamiliar to them. As the nation turned from canal building to

Figure 1. Seal of the Georgia Railroad. Reproduced with the permission of the CSX Corporation.

railroad building, a new round of technological questions sprang up. The shortage of engineering talent led companies to turn to the federal government for assistance: engineers from West Point surveyed more than twenty railroads across the nation from 1832 to 1836. The national demand for engineers was reflected in their salaries, which rose dramatically from 1835 to 1849.[3] Companies hoped to find engineers with technical qualifications and skills. When the SCRR was considering hiring Horatio Allen as chief engineer, board member E. L. Miller wrote to another civil engineer, John Jervis, inquiring after Allen's qualifications. The first item requested by Miller was an assessment of Allen's "scientific qualifications as a Civil Engineer." The appeal of proceeding from a "scientific" base was clear enough to the board—it wanted its expensive project to be handled with care and expertise. But with few formal training opportunities available, solutions to problems were developed from on-the-job experience, not simply copied out of books. Civil engineers became skilled in a wide range of tasks. "You would marvel to me what a wonderful machinist I am becoming," Henry Bird, an engineer working in Virginia, wrote to his fiancée in Pennsylvania in 1832. "I can make all sorts of cars & I could make a tolerable steam engine by this time. Warehouses, bridges & weighing houses, cast iron wheels, wooden wheels, spinning jennies &c. &c." Civil engineers remembered with pride what they had accomplished through ingenuity. McRae recalled in 1849 that engineering was "very laborious business & to meet with success requires either a long continued application or an unusual share of talent. The book knowledge as to any scientific profession is very important but it is not all that is necessary." Even if comprehensive formal training would have been available, any knowledge had to be leavened with appropriate understanding of local conditions.[4]

Civil engineers responded to increased demand for their services and readily moved around the country. "A Rail Road Engineer is a Locomotive sort of character," McRae wrote in 1849.[5] Engineers were not always concerned where they worked, if they were able to find good employment and exciting engineering challenges. Northern projects attracted men who had worked on southern works, and vice versa. As historian Raymond Merritt has observed, "Nineteenth-century engineers . . . were so rootless in their employment, so flexible in their work, and so variable in their associations that the geographical and social stereotypes commonly used to identify outstanding men often do not apply to them."[6] Examples of engineers who were unrestricted by region abound through the antebellum era. In its first annual report, the Pennsylvania Railroad announced with pleasure that it had hired John Edgar Thomson, "a gentleman of enlarged professional experience and sound judgment, who had obtained a well-earned reputation upon the Georgia Road, and in whom the Board place great confidence."[7] Horatio Allen, the SCRR's chief engineer during construction, also did work on the Delaware and Hudson Railroad. William Gibbes McNeill and E. S. Chesbrough worked on the Louisville, Cincinnati and Charleston Railroad in 1839. They also worked on the Boston and Providence Railroad; Chesbrough went on to become the city engineer for Chicago. Engineers in Georgia hailed from the North, such as William Wadley (from New Hampshire) and L. P. Grant (from Maine). Alfred Sears was working in Massachusetts when he attempted to secure a position on the Charleston and Savannah Railroad. Alexander Cassatt worked on railroads in Georgia and Pennsylvania. John Childe worked on the Mobile and Ohio Railroad, Tennessee and Alabama Railroad, and the New Orleans and Ohio Railroad, among others. Simply put, engineers followed work wherever it went.[8]

There were some calls for more permanent training facilities for engineers in the South. A journal called the *Self-Instructor* published an appeal in 1853 for engineering training: "Let us support our military schools as near as possible on the model of West Point, and there will be no want of mechanical talent and theoretical knowledge among our youth, to supply any demand we may make." The advantages to the region were clear: "When the South supplies her own working-men, in the shop, the school-house, the pulpit, the field and the bar, *as the North does,* then, and not before, may we hope to command the wealth and use the strength which is ours, by the gift of God." Georgia civil engineer William Mitchell pushed for a school for engineers at the University of Georgia. Despite his efforts, one did not open until 1866. Efforts to open engineering programs at the University of Alabama and University of Virginia also faltered in the antebellum era.[9]

The lack of formalized training in the South reflected the state of the profession. Significant, formalized training and professional associations did not emerge in the United States until the decade preceding the Civil War and did not solidify until the years after it. Schools began to add engineering to their curricula in earnest in the 1850s—while southern schools did not join this movement as eagerly as the

Self-Instructor would have liked, this disadvantage did not prevent southern states from pursing railroads aggressively in the same decade. To make up for the dearth of formal training opportunities, engineers established intellectual networks with their counterparts working on projects across the country. Sometimes this interchange came from direct observation. A Mr. Dod asked the SCRR's board of directors in 1836 for permission to travel north for the "purpose of procuring men, and examining rail roads." He was granted a leave of absence for five or six weeks. Consulting also extended to internal management practices. In 1851 Herman Haupt noted that in order to set up the accounts of the Pennsylvania Railroad he visited several railroads throughout New England to learn about "everything connected with their business operations." He concluded that "no mode of keeping accounts exceeded that of the Georgia Railroad in its simplicity," demonstrating that southern practice could influence northern practice. Given England's importance to railroad development, trips overseas were also warranted. England was, as Allen told Jervis in 1828, "the land of Rail Roads," and Allen's own trip to England exemplifies the lengths to which Americans went to gain information about this new technology. Americans were reliant on England through published literature, letters, and personal accounts of visits such as Allen's.[10]

In addition to traveling, engineers wrote frequently to each other to track the newest developments in a rapidly changing field. As they traveled and worked on different projects, engineers built networks that they maintained for information and support. The topics ranged widely. McRae shared his information on contracts and rates with L. D. Fleming of the Wilmington and Manchester Railroad. The next year, McRae was on the other side of the table; he wrote to several railroads requesting "any printed regulations in use on your road & copies of such blank forms for recording the operation &c as it may be convenient for you to send." These letters were sent to engineers on the Boston and Providence, the Baltimore and Ohio, the Central of Georgia, and the Baltimore and Philadelphia railroads. McRae wrote to Benjamin Henry Latrobe to ask about inclined planes and railroad track. Latrobe responded with information from his own travels, reading, and discussion with other engineers. Latrobe's responses to McRae's inquiries demonstrate the importance of experience and experimentation and how experimental knowledge was transferred. Referring to a method of preserving wood, Latrobe wrote that "the application of this process has not been invariably successful as I have had opportunities of observing in some experiment upon it I have made upon my own work." Later, Latrobe wrote that his "experience confirms this view of the superiority" of a certain type of rail. While Latrobe cited publications, McRae was clearly interested in Latrobe's opinion, which was informed by his personal observations and experimentation. Not all engineers were forthcoming. E. S. Chesbrough wrote to McRae that he had given up trying to get some Massachusetts reports because previous attempts had been "useless." Because word about their unhelpfulness spread through the network, uncooperative engineers snubbed colleagues at their peril; to justify his response, Chesbrough noted that "other engineers tell me that like efforts on their part have met with like success."[11]

Engineers also kept abreast of the literature in their field. General works on civil engineering were advertised regularly in Charleston, South Carolina's bookshops. We know about McRae's reading habits in part because of notes he sent to publishers. He once complained to a book supplier that books and journals that he had ordered months ago had yet to be sent. From another publishing house, McRae ordered an article on bridges from the *Encyclopedia Britannica* and a report "'Embankment of the River Adige in Tyrol' drawn up by 'Court Counsellor Dassette' (not Gazette) which is probably printed in Italy and of course in Italian." The *American Railroad Journal* also served as an important conduit of information. McRae wrote to the *Journal*'s editor, D. K. Minor, that "I sent you a short time ago a list of five new subscribers and $20 to pay for the subscriptions of four of them. I hope they arrived safe. I hope in the course of a month or two to send you some more. We are just beginning to waken up on the subject of R.R.s." He also asked Minor to send some extra copies of the *Journal* so that he could take them to a meeting of the stockholders of the Greenville and Columbia Railroad. Engineers made up for a lack of formalized training and professional credentials by maintaining a network of fellow engineers, by remaining conversant with engineering literature, and with practical knowledge and hard-won experience. Because of networking, literature, and the mobility of the engineers, we cannot characterize southern projects as purely of "limited" or "local" interest. However small, these projects were tied into the developing national engineering culture through their engineers.[12]

Engineers had to be willing to improvise and experiment. In 1829 the SCRR's board of directors reminded the stockholders that constructing railroads was still a "novelty." Railroad construction in the South followed a relatively simple plan, as outlined by N. J. Bell: "A piece of hewed timber was laid on each side of the roadbed, lengthwise, and the crossties laid on the sills—called mudsills. A stringer was let into the ties; the stringer was a square piece of long sawed timber; the rail was a flat iron rail with spike holes in the center of the rail, and was spiked on top of the stringer. The ends of the rails came together with a little neck and groove that made the joint. At the joint of the stringers was a wooden wedge driven in to keep them in their places." This simple, strap-iron construction was followed by many railroads in the country, particularly at the beginning of railroad development.[13]

In his memoir, Allen described the unique process used to build the SCRR, revealing how cost considerations came into play: "Confidence and capital had not yet reached the growth to make an iron track of the most modest weight per yard a possibility, and steel rails were as unthought of as the telegraph. On timber rails, six-inch by twelve-inch section, iron bars two and a half inches by half an inch were spiked. The wood was the Southern pine, the hard, resinous surface of which was as suitable for the iron bars as wood could be. I desired to use iron of the same width and thickness, but with a flange on one edge," but that proved too expensive. In this case, Allen demonstrated that he had not only mastered the technical concepts associated with railroad building but was also profoundly aware of the financial constraints.[14] Perhaps the most notable experiment on the SCRR was constructing the railroad on wooden piles instead of embanking the route. By elevating the entire road, Allen hoped to accomplish three goals: "permanent solidity of foundation, uniformity of surface and accuracy of direction." Moreover, the method of construction was cheaper than embanking, and Allen believed that the method used by the road would prove popular across the country. The piles could reach a rather dizzying height. "In some places the road is raised upon wooden stilts 25 feet high," commented traveler James Davidson in 1836. The unfamiliarity of the work created problems, though. Workers attempting to drive piles found themselves hampered by quicksand in some places and hard clay in others. The construction of the inclined plane (a source of much future frustration and controversy) near Aiken was undertaken by men faced with the "novelty of construction . . . who only begin to be expert when the work is done."[15]

The harsh reality of such experiments is that they could fail. The SCRR almost immediately recognized that the piling system was inadequate. A period of dry weather in 1833 followed by harsh rains "has presented a combination of circumstances tending thoroughly to disclose all places where the supporting structure has been wanting either in solidity of foundation, or substantial workmanship." As a result the railroad announced that "the Embankments will be gradually introduced, before the natural decay of the material, will render them or a re-construction indispensable." By the end of 1835, the company could report that seventy-seven miles of the road had been embanked.[16] By 1839, according to a traveler, embanking "had just been completed" but the road was still high enough that "looking from those elevated structures down into deep chasms" caused a "shudder." Failure, of course, presented opportunities for others to learn. Soon after the SCRR realized the error of its ways, a committee in Columbia, South Carolina, argued that a proposed road to its city should follow a "more substantial and perfect plan." Yet the committee acknowledged that the road from Charleston to Hamburg was constructed "at that early period in the history of Rail Roads," and given the "limited resources of the company," it was "the *only one which could have succeeded at the time*." The report was made in 1834, yet the SCRR's initial construction of a few years previous was already labeled an "early period" in railroad history. Experimentation led to error but also learning opportunities.[17]

Other experiments consumed Allen's time. His efforts to secure satisfactory wheels from J. and J. Townsend of Albany, New York, demonstrate the problems he faced. He complained that, in one shipment of wheels and axles, "the journals have had nothing around them to protect their surfaces from rust. I must beg your attention to these particulars." He also made it clear that he expected the company to adjust the wheels to the SCRR's specifications as they investigated the proper dimensions. "We find it necessary to make some slight alterations in the cast iron box in which the brass bearing is secured, and will send a pattern by mr David Brown," he wrote the Townsends much later. "The alteration is required to suit our arrangements for springs." Experimentation and a willingness to fail and learn from that experience were necessary qualities for nineteenth-century engineers.[18]

If experimentation was a major characteristic of engineering life, variety was another. Engineers had to address a range of nontechnical aspects of their work. Antebellum engineers dealt not only with the mathematical and scientific challenges of constructing a railroad but also with the political world of directors, recalcitrant landowners, and the like. "Making railroads forms but a small item in my business," Henry Bird reported to his fiancée in 1832. "I have not only to assist in making the road, but to arrange wagons & engines; to commence the transportation of passengers & goods, and to keep dozens of people daily from breaking their necks." Thus, in addition to their technical work, engineers battled weather and disease, worked with contractors as the road was built, handled corporate bureaucracy, and dealt with landowners. This wide array of tasks did not appeal to all engineers, and some demanded that their jobs be better defined.[19]

Battling Nature

Despite the promises of boosters that the railroad would conquer nature, in the construction process engineers found themselves at the mercy of the weather. Construction on the SCRR did not begin in earnest until February 1831, but even then the coldness of the weather "render[ed] the labour of blacks totally inefficient," according to an early report. Given such problems, civil engineers wanted to take advantage of good conditions whenever possible. For example, Andrew Talcott, chief engineer of the RDRR, informed his bridge contractors, "I wish it were practicable for you to be here now with a strong force as the river is unusually low and will probably continue so for some weeks & I should be glad to see the foundations done of a portion of the piers at least." Such conditions made it imperative to hire workers quickly.[20]

High water presented its own set of problems. On the SCRR, McRae reported that one group of workers in 1847 "crossed the river just before the present flood & cant get back now they are consequently idle." The necessary trestle and bridge-work over South Carolina's Congaree Swamp could be acquired by the LCCRR only at a high price. Building the railroad to Columbia was admittedly difficult: in addition to the work over the swamp, it required, in places, deep cuts and embankments of up to fifty feet. Torrential rains in 1840 and 1841 destroyed much of the progress that had been made: to complete their work, the contractors had to wait not only for the rain to stop but also for the water to drain away so that foundations could be laid.[21]

Wet weather plagued McRae in 1853. His lengthy letter of complaint illustrates just how badly work could be set back:

> Since I last wrote you but little work has been done. The ground has been so thoroughly saturated with water that for nearly two months the work done by carts would not pay for the wear and tear. The ground is now beginning to dry up & most of the contractors have got to work again. The effects of the weather on the cutting have been very bad. The slopes are every where caving in & without much regard to the degree of slope given to the cuts. One cut on Carter's section near Jamestown where the slope had been changed has slipped more than 10 ft I should think outside the slopes. Harris' cut on section 10–4th Division had just been finished & ditched when several hundred cart loads slipped. The cut is long and wet & it will be nearly impossible to get this earth out before spring until it dries somewhat, which will not be before spring.[22]

Of course, northern railroads also faced weather problems. While building the Boston and Worcester, contractors struggled in the face of "the severity of the winter, and the long continued cold and stormy weather." Samuel Nott reported that the "great proposition of mild, wet weather, having very much increased the difficulties of the Contractors, and the consequence is that the estimate is not so large by far as is desirable." Wary of such rising costs, railroads tried to beat out bad weather when they could. George Bliss urged the president of the Western Railroad to appropriate funds for the immediate laying of foundations for bridges and culverts, because masonry was better done "before the Wet & cold weather."[23]

In addition to the challenges posed by weather, engineers faced disease. Sickness could have a powerful effect on communities of the early nineteenth century. Reporting from Knoxville, contractor A. L. Maxwell wrote that shortly after he arrived in the city he found it nearly deserted, four thousand people having fled disease "within three days." Everything was closed, and people were "badly frightened." When attracting laborers from outside the South, southern railroads had to address the region's reputation as an unhealthy place. An Alabama railroad advertised for workers in 1836 with the promise that work would be in an area "of the most pleasant and healthful character."[24]

The experience of the northern construction firm Stone and Harris, building a bridge in Richmond, demonstrates concern about disease. Initially in June 1849, J. R. Anderson informed the company, "There have been a few cases of cholera here, the greater number whilst you were here. Yesterday there were reported 2 cases, no death except of a patient who had previously been taken. Our city has never been more healthy. . . . We dont feel any apprehension of danger here." Anderson gave another positive report a few weeks later, noting that six workers freshly arrived in Virginia "say they are not at all afraid of the Cholera & should like to be employed as early as practicable." But the luck of Stone and Harris's workers eventually took a turn for the worse. In October, cholera in Richmond gave "cause of alarm to northern laborers" and delayed the work on the James River bridge for a while until a "large and efficient" force could be fully secured.[25]

Indeed, the timing of sickness could affect the completion of a railroad, which made engineers nervous. McRae wrote to a northern colleague that he was "getting very anxious" about completing his road, adding "if the sickly season next summer should catch us it will be very bad especially if it should prove as sickly as this season." Of course, engineers themselves were not exempt from sickness. When he discovered that he was to have a new assistant, McRae's only regret was that "he comes here at a bad season & may get sick."[26]

Engineers and Contractors

Once a route was surveyed, the engineers began the process of supervising construction. They set the terms for the contracts, took bids, and ensured that the work was done to the contract's specification. Contractors submitted bids for work, stating what rates they considered reasonable for the work to be done. The best bid was then selected by the company. Of course, factors could come into play besides the financial appropriateness of a proposal. When A. W. Craven proposed to work on the Camden branch of the SCRR, his colleague and road employee John McRae discovered after talking with one of the road's directors that "your bid was rejected with it being stated that it was invited by the President & that this might have altered the decision some what." Contractors were required to provide security to demonstrate that they had the financial means to finish the work and sustain themselves through economic hardship. Such a demonstration could prove critical. The LCCRR extended a contract to a man to build their depository in Orangeburg, South Carolina. The winning bidder, "though not considered the most responsible man," received the contract "after giving security to the amount of $2500."[27]

Contracts laid out specifically which parties would bear which responsibilities. William S. Mills submitted the following bid to the Spartanburg and Union Railroad in 1854:

> We will furnish hands to lay Track & to do any grading that may be necessary, to clear the way for the same as follows. Board &c &c and be at all expenses of Hands for one dollars per day overseer and Cook to be counted hands. Mule and cart to be counted equal to a hand. Cart-Boys half-price, wet days to count half time. We will loose going down and returning and all sick time, we will find a sett of Blacksmith tools & all other tools necessary for grading except shovels, the Company to find shovels and all other tools necessary for laying Track. There will be a Blacksmith and Two pretty good carpenters in our lot of hands. Waggon-Team and Driver five Dollars per day. We will leave Alston the 1st day of June and return there the final week in October.[28]

Rates for different workers (and ages of workers), timing of the work, allowance for sickness, provision of equipment—all these were accounted for in the contract. Other contracts were executed in a similar fashion. In a contract for embanking, the Raleigh and Gaston Railroad agreed to furnish "a Locomotive and train of gravel cars" as well as "the rails spikes sills, frogs, levers and switches for any side tracks that the Engineer may adjudge necessary for the performance of the work." The contract detailed how the account would be paid and the time by which the work must start. The engineer was also acknowledged to have "final and conclusive" authority over "any dispute which may arise between the parties to this agreement relative to or touching the same."[29]

Contractors were both local southerners and northern businessmen. Southern railroad companies actively solicited northern contractors, although it could be difficult for contractors to compete with planters. While working on the SCRR, McRae invited Craven to place a bid on the Wateree Swamp trestlework. McRae warned Craven, however, that it was not worth his while to make any sort of offer for the grading, because "you could make nothing on it." The fact that planters along the road intended to offer for the grading meant that "there is little or no chance" that anyone else would get the contract. This demonstrated the crucial advantage that planters had in their ability to provide slave labor. McRae wrote to another potential contractor, James Herron of Philadelphia, and warned him that "if you trust to getting hands here it is not unlikely that you might be disappointed & you might find differently in prevailing upon hands to come from the North at that season of the year." Herron assured McRae that he was not scared by the prospect, but "engagements here will preclude the possibility of my compassing yours." Later, McRae told Herron that he made the right decision by not taking the job: "The work you would have to do would come in the most sickly season of the year. If you had trusted to getting hands here you would probably have been disappointed, had you brought hands with you they would have got sick or frightened."[30]

Such problems notwithstanding, some northern companies did substantial business in the South. As we have seen, the RDRR contracted a Massachusetts construction firm, Stone and Harris, to erect a bridge in 1849. Stone and Harris's agent, J. R. Anderson, indicated that he was also in negotiations with Virginia's Louisa Railroad regarding bridgework. Although Stone and Harris sent its men throughout the South, their experiences were not always positive. Working on a railroad in Knoxville, A. L. Maxwell wrote to the home office that the chief engineer was the "poorest pay master we have anywhere, and the most troublesome Eng[inee]r withall. . . . I would not do any more work under him at any price." Arriving in Richmond, William Birnie complained that so little had been done that he feared they would not have the road graded and track laid for "*six months.*" Moreover, an inexperienced "young Irishman" was now responsible for the engineering drawings, and Birnie "felt very much" like informing the lead engineer "that I would go home and wait a few months untill he got his road in better shape."[31]

Because of their proximity to the work and their ability to provide slave labor, southern planters played a critical role in constructing southern railroads. Although they initially lacked experience in railroad building, this hardly made them different from anyone else at the beginning of the railroad era. Planters brought with them significant advantages. In 1839 the CRRG reported that it preferred to use planters as contractors, because it would "enable us for the future to keep up a more uniform scale of operations during the whole year, and also to render the work more popular, by diffusing the benefits attending its construction, more generally among our own citizens, than if the labor were performed by strangers." Planters would not stand in the way of the modernizing South; the CRRG argued that planters could actually facilitate this transformation. Moreover, the company realized that having planters involved would make the work more "popular," and give the community around the road a greater stake in the road's success.[32]

The LCCRR's construction of the railroad from Branchville to Columbia, South Carolina, also demonstrated the practicality of using planters as contractors. Although the LCCRR admitted that the planters did not always posses the most experience, it was easily gained, and any lack of experience was more than compensated by the fact that the planters possessed "labor, provisions and quarters of their own, in the immediate neighborhood of the work." Indeed, planters did better than their nonplanter counterparts. Those contractors who lacked their own capital and were dependent upon the company's monthly payments ran into difficulty when the weather and sickness brought delays, but "those who had their planta-tions to resort to, were scarcely, if at all embarrassed." The company noted with sat-isfaction that the contractors, "as a body, were very efficient and gave full satisfac-tion, as might have been expected from the high standing of most of them. The interest which some of them took in the workmanlike completion of their sections is not only highly commendable to themselves, but ornamental to the Road."[33]

Other railroads agreed. As it prepared to begin construction, the Spartanburg and Union Railroad approved a new bidding process: "Land holders shall be pre-ferred as contractors for Grading the Road through their own Land provided their Bid be equal" or lower than that of any other bidder. The Blue Ridge Railroad of South Carolina switched to having planters work as contractors after an unfavor-able experience with all-white labor. The company president reported in 1854 that the "mixed population of our Northern seaports . . . was not found to answer" and so the road had been "sub-let to parties chiefly on the line of the road." As might be expected, there was some delay in getting the work started as the switch was made near the harvest season and planters were unable to put their slaves on the line until the crops had been taken in. For planters, agriculture remained the top priority, but the success with which planters were employed as contractors also shows their willingness to diversify and improve the South's transportation infrastructure.[34]

Managing the work of contractors could be a difficult task. Civil engineers and contractors were driven by very different desires. Engineers obviously wanted to have the highest-quality construction; contractors hoped to keep costs to a mini-mum if there was to be any chance of profit. McRae's constant struggles with con-tractor Thomas Stark on the SCRR demonstrate the problems that civil engineers faced. McRae informed Stark in July 1846 that it was "many weeks since I in-formed you that the contract for grading the first section of the Camden Branch was awarded to you & I have not yet seen any steps taken towards a commence-ment." With only three months remaining on the contract, McRae warned Stark that "to do the grading in this time will require a force of 70 hands even if there were no more work than shown by the estimate. If you do not expect or rather are not prepared to put this force on in a week or ten days, I think you had better give up the work."[35]

In August the story was the same; McRae again had to write to Stark demand-ing that the work be taken seriously. Although Stark had promised to secure as many hands as possible with a minimum of forty, McRae found that only seven-teen were working on the road. In September, McRae wrote that only half of what should have been done had been completed. McRae reminded Stark that the president of the company had extended a contract to him "out of kindly feelings." The difficulties with Stark led McRae to take his own initiative in the search for workers to complete the grading of the road. Even when McRae was able to locate these workers, the job did not necessarily go more smoothly. "I wrote to you on the 18th ult. offering to aid you in procuring Mr. Cordes hands & I have heard nothing from you since," McRae wrote to Stark on October 3. "I am somewhat surprised at this as time is now so precious." Unable to get hands from an expected source, McRae forwarded Stark hands from the "reserve" force, although the price of hiring such hands would be higher for Stark.[36]

Stark was also unable to keep his workers under control. McRae informed him of a rather serious offense: "I am informed that your overseer Mr Moye with some others on Sunday last took one of the R R Company's cars which was on the Camden Branch & ran it on the Columbia Road, and not satisfied with this trespass whereby they laid themselves liable to prosecution they cursed and abused one of the officers of the company who was coming up the road on an Engine. If this is so I am left no alternative but to order that Mr Moye be discharged from employment on the road. I shall regret very much if this will subject you to inconvenience as I do not of course consider you in any way to blame." When the cost and trouble of reletting the work was greater than the cost of retaining contractors already engaged in the work, engineers were forced to deal with such incompetence.[37]

Contractors bore fiscal responsibility for the work that they were to complete, but they were also required to complete the work to the satisfaction of the railroad. Understandably, this could lead to disputes. Companies demanded adherence to the contract price, which could lead to ruin for contractors who were unable to keep their own costs down. Engineers regularly inspected railroad work in order to judge progress as stipulated by contracts. "Had OBrians hands (2) cleaning dirt off of & raising retaining wall that he had covered up," noted Orange and Alexandria Railroad engineer Thomas Shaw in 1857. This error would cost the contractor: "Shall not allow him anything for them as he should not have had it covered up." Engineers also held firm to the estimates. "You will observe that the contract fixes the price you are to be paid," McRae informed a contractor, "under that contract you cannot claim more; in a legal point of view your objections are of no avail." In settling a dispute, engineers were quite specific about enforcing the formal and detailed requirements of their contracts. McRae once referred a negligent contractor to "page 5, specification 7," and "on page 7 in two separate paragraphs" in order to make a point. In response, contractors sought to keep their expenses down and could be tenacious in holding to their claims; contractor Anson Bangs once told another contractor that "he would not pay 5 cents if it cost him $50,000." The opposing demands of contractors and engineers could make for hard battles.[38]

If contractors were not happy with the estimates for the work given by the engineer, they could express such dissatisfaction by removing their laborers from the work. Others protested the low rates that were allotted for their work directly to the engineers. McRae wrote that one contractor was particularly out of sorts. "This morning I mentioned to Mr Shaver (it having escaped me till now) the price you fixed on the masonry of the street Bridge Salisbury ($5.50 per cy). It put him so much out of temper that I thought it better to make no remarks or explanations. He says the price will scarcely pay his masons & swears he will sooner tear the work down than take it."[39]

Managing the costs of railroad work could be perilous, and some contractors fell on hard times. McRae once considered a few of his contractors to be in dire straits: "I do not know any contractor more in need of money than Mr Murdoch. I have heard several contractors, H. C. Jones, Saml. C. Harris & others, complain of want of money but cannot say who are more needy." Legal consequences were the result for those who fell deeper into financial misery. McRae notified the president of the SCRR that contractor William Bowen was "in jail for debt," and McRae had to take over Bowen's work himself.[40]

Such difficulties reveal the risks involved in railroad construction and were not peculiar to the South. Northern contractors were also capable of mismanagement and falling into bankruptcy. Surveying the state of ten numbered sections on the Schenectady and Troy Railroad, an observer noted that on section one the grading was "indifferently managed" and that the contractor for the depot "made very little." Although the contractor for section two made a profit of $1,500, it could have been "double" with "proper management." The contractor for section three made out "about even." Section four was "managed indifferently" with a loss of $3,000. Sections six and seven had "medium" management; section eight had "bad" management.[41] On another railroad, Samuel Nott complained that one subcontractor was in so much debt that he would not "be able to secure himself, if he work[ed] there forever." One contractor on a Vermont railroad worried about a $10,000 debt that was "frightful to think of it but we must make the best of it." And northern contractors protested when their finances were inadequate to their needs. "The Contractors complain loudly of their estimates," Wilson Fairfax reported on the Philadelphia and Reading Railroad in 1836. "Wages are very high $1.00 to $1.12 for hands is enough to destroy all their profits entirely. Many advances were consequently made to them by Mr. [Moncure] Robinson who is very kind & feels much for their situation."[42]

Financial risk and cost overruns were not unique to southern projects. Indeed, in our own time, large transportation infrastructure projects have "widespread" cost overruns. As with nineteenth-century railroads, new projects that involve techni- cal innovations are susceptible to expenditures that quickly outpace estimates. Such problems "haunt" major projects, almost exclusive of time and place. Ante- bellum engineers faced similar problems that modern engineers do—unfamiliar technology, difficulty in procuring materials, fluctuating labor force, the vagaries of nature—and estimates can easily be made irrelevant by one or a combination of such factors. It may simply be in the nature of such projects—and not a pecu- liar "southern" incompetence—that led to inadequate projections and financial difficulties. According to planning professor Bent Flyvbjerg and his colleagues, writing in 2003, "Cost overrun today is in the same order of magnitude as it was ten, thirty or seventy years ago." Moreover, they discovered that publicly and pri- vately undertaken projects exhibited similar "patterns" of overruns, even if causes were different. Rather than seeing widespread mismanagement in southern im- provement projects, we may simply be seeing the real costs of working at the cut- ting edge of technological change and wrestling with construction projects of enor- mous scope.[43]

Railroads had to design strategies for working with contractors who fell on hard times. Some companies tried positive incentives. In order to get the Southwestern Railroad completed before the cotton crop was ready for transport, the company promised "additional prices . . . to several of the contractors, with the condition that they complete their contracts by a specified time." When contractors fell delin- quent it was often difficult to reassign the work; McRae noted that it was usually "a most expensive and troublesome business to relet work." When contractors did not complete their work in time, the railroad occasionally had to take steps itself to complete the job. Such was the case on the RDRR in 1850. The chief engineer, An- drew Talcott, reported that contractors for one portion of the road had completely neglected it, "though repeatedly directed to prosecute it with greater vigor." Talcott asked the permission of the board of directors to put his own force on the road should the contractors continue to delay.[44]

In other cases, the railroad found itself stuck with the contractors. Talcott rec- ommended allowing Robert Harvey and Company additional time to complete its contract instead of disposing with the firm and reletting the work:

> Shall we re-let it at as favorable prices? It is not reasonable to conclude so, as labour
> is in great demand and all prices are ranging higher. In such a state of things econ-
> omy & the true interests of this company in my opinion require that we should go on

with the present contract by which we make it only a question of time. Is it then better for this company to wait the completion of this contract twelve months, or by declaring it abandoned arrest the work and run all the hazards of increased expense,
with the additional hazard of not getting it sooner. I think the policy is plain, I think
a delay of a few months is not to be compared with the hazard, rigors and casualties
we should encounter by an opposite policy.[45]

Harvey illustrated the bind that railroad companies found themselves in: the contractor was not performing well, but starting over from scratch would mean an
enormous loss of time and would not necessarily be less costly.

Managing a New Bureaucracy

Engineers found themselves at the center of a rapidly growing corporate bureaucracy. Engineers were responsible for a vast amount of paperwork associated with
operating the railroad. Contractors needed detailed information about the work expected of them for their sections. The engineer collected information from contractors in order to judge the progress of the road and provided assessments to the
executives of the company. Engineers also oversaw financial matters, with a dizzying array of items required to run the company properly. Expenses for surveying a
railroad in 1836 included such items as advertising, horse hire, food, slave hire,
firewood, pencils and drawing paper, clothes for slaves ("Shirts for George," "Great
Coat for Daniel"), wages, and "cooking furniture" for the camp. Standardized
forms helped routinize information management but could also make work mul-
tiply: a contract for slave hire on the SCRR in 1846 was to be "*Executed in Tripli-
cate.*" Given the need to track all of this paper, combined with the fact that offices
could move as the work went on, it is little wonder that occasionally engineers
would take time out to catch up with paperwork. On October 25, 1839, the senior
resident engineer of the LCCRR began to organize his office; from the 28th to the
30th he and his underlings were "arranging and labelling the papers, drawings
and books, and overhauling and storing away the instruments, camp equipage &c
of the company."[46]

The workload could lead to long nights and weekends. Closing a letter to a
friend, McRae noted that it was "Sunday evening & I have yet several letters to write
for tomorrow mornings mail." Some engineers did not appreciate night work. Edward St. George Cooke complained about his boss for that very reason: "I will cite
an instance of his annoying me. He told me that he desired me to work regularly
in the office at night. I told him that when any work was pressing, I would willingly

do so, but when there was such a lack of employment, that I was idle half the day, I must decline any such arrangement."[47]

In the process of overseeing construction, engineers had to make sure that the appropriate materials were ready when needed. Timing was critical when doing construction, because not working meant that workers were idled, could get restless, and leave (if not already held in bondage). Examining one week of construction on the LCCRR easily demonstrates the logistical management required. Contractors began laying mudsills near Branchville on December 23, 1839. After working for about a half mile, however, it was discovered that many of the sills were "very defective" and "frequently very crooked." The rail layers were already on the construction site but could not begin their work until the sills were down. Without work, they would not be paid and would likely leave. On the 25th, the contractors were instructed to cut down trees near where the road was being built and to fashion their own sills to replace those that were defective. That same day, the engineer learned that the rails, spikes, and plates were all waiting in Charleston and had yet to be sent up the road. The next day the resident engineer wrote to the contractor responsible for supplying the sills and informed him that "he must immediately send horses and hands to supply the deficiencies." On the 27th, the engineer instructed the idle rail layers to take over the process of cutting down trees and fashioning sills for the company's use. Engineers had to juggle labor forces and schedules when equipment deliveries were not timely.[48]

There were similar problems on other roads. In 1849 a contractor in Richmond wrote hurriedly, "We have been looking for some time rather impatiently for a reply to some one of our last four or five letters. we wrote you twice within ten days to know what had become of the cars that Wasson made for the Richmond & Danville RR co. not hearing any thing from you we Telegraphed you on Friday. Every thing will be ready in a day or two to commence hauling stone from the quarry but no cars *heard from*." One week later, G. H. Burt complained, "The car man has arrived but not the cars & *we are ready for them*." One small delay could have a damaging effect on other parts of the work.[49]

In addition to overseeing the office, managing contractors, and overseeing procurement, engineers also had to work around the demands of their superiors. John Smedberg, working on a surveying crew in 1837, wrote that he "was hard at work till 10 every night . . . doing my share of the estimates to be ready for the meeting of directors." When the engineering drawings of the LCCRR were "hung up" for inspection by the stockholders at their annual meeting, the engineers could do only "very little in the way of drawing." Boards could also demand work to be redone when they received complaints from citizens. Regarding competing claims from residents of Kershaw and Sumter, South Carolina, McRae noted that he would "be delayed a little in consequence of the consternation which the proposed change in the location has caused among the people of Sumter. . . . I have in consequence been directed to make more extended surveys as the matter will be brought before the Board again." Engineers learned how to communicate needs in language the board would appreciate, couching their own arguments in the language of the business needs of the road. "An immediate movement and the utmost despatch in the laying of the rails will now be necessary, to open the road in season for the fall business of the West," reported the engineer of the Northwestern Virginia Railroad Company in 1856. The engineer doubtless knew that his superiors would be thinking along the same lines.[50]

Engineers also had to be attuned to the political machinations of the boards of directors. McRae wrote to a friend in 1846, "Ker Boyce & his friend have been turned out of the Board of Directors, a happy riddance in the opinion of all friends of the road." Engineers could also try to influence votes themselves. "I understand that an effort is to be made next meeting of the stockholders to turn [James] Gadsden out" of the presidency of the SCRR, McRae wrote at the beginning of 1847. He encouraged his correspondent: "If you know any who hold the stock use your influence." Shortly thereafter, McRae would discover that the politics would end up influencing his own career. When Gadsden was thrown out, McRae wrote that he had but one "close friend on the Board [and] it is not improbable I will have to follow." Indeed, he soon informed a friend that he was attempting to find employment elsewhere. Years later McRae still realized that his reputation was tied up with that of Gadsden, and because Gadsden was "unpopular with the moneyed interests in Charleston . . . his unpopularity is visited upon my shoulders." Engineers could not separate themselves from the politics and nonengineering aspects of railroad construction.[51]

Engineers and the Public

Finally, civil engineers had to address the needs of individual citizens along the route of the railroad. Most often, these encounters came when the railroad was attempting to secure land. Sometimes, these encounters were made more complex by the legal standing—or lack thereof—of the landholder. Mr. Sheppard of the RFPRR paid $200 to the "friend and agent" of "Mary Harris a lunatic" who was to receive land damages for construction of the railroad. A few days later Sheppard asked to have payment made to a representative of a landowner who was under twenty-one years of age. Land damages could not always be settled between the engineer and the owner. After talking to landowner Thomas Seay,

McRae reported that "we both came to the conclusion that the law must take its course."[52]

Once railroads received land, they tenaciously defended their claims. "We met the assembled wisdom of the Dutch & Dutch Reformed Churches here yesterday," McRae wrote in 1851. "They had been quietly laughing in their sleeves at the Company with the belief that their property was sacred, but have been forced to admit that they could not prevent the Company from going through & have concluded to be satisfied by getting all out of the Company they can." The railroad would not give up its land easily, and those who stood in the way only garnered McRae's sarcastic contempt.[53]

As construction progressed, engineers fielded complaints from landowners. In January 1840 the engineer for the LCCRR was still working with one landowner to settle claims for damages done to his crops in 1838. McRae received a complaint from a landowner in 1846 that his crops had been damaged. He expressed surprise that the landowner intended on suing the contractors, because "your overseer informed me that no damage was done to your crop." McRae reminded the landowner that the contractors had delayed the work in part to allow him to do work necessary on his crops. Once again, engineers found themselves to be the public face of the corporation.[54]

The wide variety of work and its frustrating nature led some engineers to cry out for a better definition of their duties. Engineers argued that their work was professional enough that they should not have to trifle with petty concerns. Likewise, they felt that important decisions should not be left to laymen. After wrestling with two claims of landholders and facing the recent resignation of an assistant engineer, the resident engineer of the LCCRR wrote a letter to the company president "relative to the necessity *of defining the duties and* responsibilities of the Eng[inee]rs." After speaking with a member of the board of directors, the resident engineer agreed to draw up a set of regulations that might alleviate this problem. McRae had similar questions more than a decade later and wrote to a colleague to gauge what his duties should be: "How should you occupy your time on the survey and location? Would you consider the procuring of the right of way a part of your duty professionally?" Evidently McRae did not believe that his feelings were adequately addressed, because he resigned his position on the Charleston and Savannah Railroad after a disagreement with the president as to his duties. McRae considered it "reasonable" that an engineer of his position "should be subject *only* to the *general* instructions of the Board through their President, and that as to all matters of detail such as the point at which his services are needed whether in the field or in the office, the disposition of the parties and the duties of his Assistants he should be himself the judge." By 1855 engineers had professionalized to the point where McRae felt he could claim such privilege.[55]

Stumphouse Mountain

All of the challenges that engineers faced in the antebellum era could be encapsulated in the Blue Ridge Railroad's effort to build a tunnel through Stumphouse Mountain in northwestern South Carolina. Ultimately unsuccessful, the tunnel would have been 5,863 feet long when completed and would have trimmed seven miles of off the BRRR's route. The portion of the tunnel that still stands is a testament to the multiple frustrations engineers faced.

From the perspective of civil engineering, it was a massive undertaking. Numerous people visited Stumphouse Mountain and commented on the extraordinary magnitude of the work. Dr. J. T. Craig visited in 1854 and observed that there were "about 200 cabbins put up, two stores, & two very good Hotels. . . . The supposition was that there would be about 1500 work hands before the summer ended." John Hamilton Cornish visited a few months later. "We . . . drove 5 miles to the top of Stump Mountain, through which the Blue Ridge R. Road Company are cutting a tunnel—1 and $1/4$ mile long—through solid granite or Ness Rock. The East side of the mountain is faced down and the head of the Tunnel cut in about 60 feet, and the whole cleaned out some 15 or 20." Cornish then described the process by which the tunnel was being constructed. "There are to be four shafts sunk a thousand feet apart. Shaft No. 1 is now about 60 feet deep. They are working in it, as in the Tunnel—night and day. Shaft No. 2—the water has stopt further progress till they get a steam pump, the depth of this shaft will be about 200 feet. There is a hotel and quite a village of Cabins already on the mountain." These four shafts dropped down to grade level were to speed the work. Workers could then go down the shafts and dig outward in each direction. Thus, the workers could work on ten surfaces at once—inside four shafts and at the two openings of the tunnel. The work required a large commitment of machinery and men and led to the deaths of nine imported Irish laborers during the course of the work.[56]

Soon after Cornish visited the work, tunneling was delayed because of problems with contractors. Anson Bangs and Eli Bangs (who, with other partners, formed Anson Bangs and Company) accepted a contract for constructing the BRRR from Anderson, South Carolina, to Knoxville and initially advertised for three thousand workers.[57] In November 1854 the Bangs brothers relinquished their contract and the remaining partners reformed the company under the name of A. Birdsall and Company. The railroad was displeased with the fact that Anson and Eli Bangs had left the firm, because they "were represented to be experienced

contractors who had realised a large capital in the business of rail road building."
The new company claimed the legal right to fulfill the contract. Although the Birdsall firm had no experience in building railroads, the company later said that
A. Birdsall and Company "were presented and believed to have brought into the
concern capital and credit to assist in the performance of the contract." Sadly for
the railroad, it was not to be. Indeed, as the company later claimed, it "never would
have made the Contract with Birdsall, Mather, and Bixby, or either of them" had
those three individuals alone been part of the original contract.[58]

The contractors abandoned their work around April 1856, and all the grading
that they had done to that point was "chiefly in earth, and in patches where they
found the work easy."[59] Later in 1856 Dun credit reporters reported that a $434
claim had been put out for Birdsall and Company, but "none of the firm are at present in this state, and doubt whether any or either of them return to this jurisdiction, a foreign attachment has been taken out and levied on sundry goods and chat-
tles belonging to said firm." The next year, they were further exposed by the credit
reporters: "All the members of this firm are from the state of N.Y. . . . Since dismissed by the R.R. Co. have now a suit vs said Co. in the U.S. Court for Georgia. . . .
Some 18 mos or 2 yrs since forced to leave this state on acc't of suits by sub-
contractors vs them. a large no of attachment cases, now pending in our courts."[60]

As the Dun reporters alluded, the result of these delays was a legal battle. Contractors complained that the BRRR had "refused, repeatedly, to furnish the Contractors with the location and survey of the line," and had, "on more than one occasion,
actually ordered the suspension of portions of the work then in progress." The
BRRR countered that it had hired the Bangs brothers because of the special skills
they possessed in regard to railroads and that the partners who took over were
"wholly destitute" of those same skills. The BRRR further charged that much of the
work done by the replacements was ineffective. Piles at Darricott's Bottom were "so
insufficient, and so slightly driven, that they were condemned and cut down." A
"shapeless excavation" was at the eastern end of the Stumphouse Mountain Tunnel.
Other tunnels were likewise in poor condition. Some work had been done on Middle
Tunnel, but it had since collapsed. Little had been done at Saddle Tunnel, and only
ninety men were employed at all three tunnels when the contract was dissolved. Similar complaints could be made about trestlework and masonry by several creeks. Finally, the company believed that it had been lied to regarding a potential contract for
iron in England. The railroad employed agents in England to learn about the supposed contract with a man named Parry and discovered "that Parry's last employment was that of a bookseller, and that he was declared a bankrupt in 1853."[61]

The work was relet to George Collyer in May 1856. One shaft was sunk to grade
in late February 1857, a second in September 1857. The delayed delivery of a steam

engine to sink the shafts in the spring of 1857 arrested the work, but when a report was made in November 1857, three steam engines were at work sinking the shafts and digging out the tunnel. Two smaller steam engines were driving fans for ventilation in the tunnel. Two hundred workers were employed "by relays, night and day" to dig out the mountain. Although the force applied by the contractor was large, the company still warned that it would take from three to four years to complete the work. Moreover, Collyer himself had left the work, by November 1857, complaining that he was losing too much money to attend to the work "vigorously."[62]

The BRRR was clearly losing patience by 1857. A tunneling contract issued by the BRRR that year required that work proceed at a constant pace as demonstrated by three particular clauses: "6th. The excavation shall proceed at the same time from both ends of the Tunnel—and also in opposite directions from each shaft. 7th. The Contractor shall have in the Tunnel and shafts as many hands as can be employed to advantage. 8th. The work is to progress night and day, without interruption, in the Tunnel and shafts, and to be performed by not less than two shifts, and if required by the Engineer, by three shifts."[63] Collyer was replaced by the contractors Humbird and Hunter. The BRRR's chief engineer, Walter Gwynn, noted that Humbird and Hunter had successfully worked on six other railroads, including the Baltimore and Ohio and the Virginia Central. Gwynn further noted that the "suspension of the public works at the North" should make it easier for the firm to obtain the hands it needed to complete the tunnel. In any event, the firm was fully prepared to employ African Americans to prevent any "deficiency of force."[64]

Humbird and Hunter immediately set about increasing the force. In November 1858 they had seven steam engines working at the mountain, and the third shaft had been sunk to grade. Work seemed to be progressing well; the credit agent for R. G. Dun described the contractors as "men of experience." Labor remained a problem: as the chief engineer noted, the "only impediment to the regular and uniform progress of the work has been caused by the inability of the contractors to keep at all times a full force." Yet an enormous work force was present; at the time the chief engineer made his report the contractors had brought down 832 men from the North, and the total "population" of the mountain was 1,232. The engineer believed that the difficulties northern public works were experiencing meant that the company would have no trouble attracting workers to the South. Irish laborers presented their own difficulties, however. Although the engineer believed that the work could be completed in just under two years, the fact that the Irish had a tendency to "roam among the various public works in progress, and . . . constantly arriv[e] and depar[t] from the different lines as interest or caprice dictates a change" led him to increase his estimate for completion to twenty-six or twenty-seven months. This proved too optimistic; the sheer cost of the enterprise and the inability of the company to secure governmental funding led the work to be essentially suspended in 1859. By June 1859 only 340 workers were at work on the tunnel. Soon, the Civil War would permanently end the dream of tunneling through Stumphouse Mountain.[65]

Engineering work was not for everyone. The difficulty of the work could lead even accomplished engineers to experience self-doubt. "The prospects of the Hamburgh Rail Road have brightened very much of late," Horatio Allen wrote in 1830. "I however am seriously thinking of abandoning the profession, and have already made some arrangements to that effect. I of course keep my views to myself." Allen stuck with his work, but not everyone remained satisfied with the life that engineering offered. Edward St. George Cooke left the field in 1856: "I cannot go into details but I have almost concluded to abandon Engineering and study Medicine as the quickest way of getting along. . . . I am dead broke, and would be glad if you could *lend* me a little money." But the impact of those who were willing to stay was undeniable. Engineers played a critical role in turning the dreams of promoters into reality, and a profession that was only in embryo at the dawn of the railroad era had constructed a remarkable series of railroads by the eve of the Civil War. Lacking a large variety of sources for formal training, engineers adapted well to the myriad local conditions they found. Engineers created the intellectual community they needed by traveling, keeping in contact with fellow engineers, and reading and contributing to an expanding literature. A willingness to experiment, fail, and share the results drove the country's growth in engineering knowledge in the era before institutionalized training programs.[66]

Once in the field, engineers discovered that "book learning" went only so far. Engineers had to master a wide array of nontechnical tasks to complete their works. They had to manage materials and men, chase after incompetent contractors, contend with politics and boards of directors, and address the claims of aggrieved landowners. The best-laid plans could be destroyed by a week of bad weather or the threat of disease. Through all of this, engineers had to maintain a steady hand, master the intricacies of a new technology, and preferably come in under budget.

In examining the practice of engineers who worked in the South, it is easy to see the parallels with northern developments. Although engineers in the South certainly had to adjust to southern landscapes, there were not substantial differences between North and South when it came to engineering practice. Engineers traveled the country searching for work and gladly took it where they could find it. The professional networks they built were not limited by region. Engineers North and South had to contend with problems of weather and contractors who could not meet their obligations. Just as railroad boosters could be found in all parts of the country, the experience of engineers was not generally defined by region. There was, of course, one substantial area of difference: the institution of slavery. The option of hiring or purchasing slaves was available to civil engineers in the South and the railroads that employed them.

CHAPTER THREE

Sweat

If the work of engineers and contractors illustrates the parallelism of northern and southern development, examining the labor under their control yields an important difference between North and South. Southern railroads made abundant use of slave labor. There is no better illustration of the South's attempt to integrate its preferred social order with the demands of modern technology than the degree to which slaves figured in railroad construction and operation. To be sure, the use of slaves sparked some debate, and whites remained an integral part of every railroad's work force. But slaves worked in nearly every capacity for railroads and, in so doing, demonstrated to white southerners the reliability of slave labor in non-agricultural pursuits.

Given the scale of railroad projects contemplated across the nation, procuring labor was a monumental task everywhere. Because railroads were invariably a large undertaking, demanding a tremendous number of workers, labor shortages were almost unavoidable. Throughout the antebellum era, thousands of workers were required. The initial labor force on the SCRR was approximately six hundred men. In 1832 there were two thousand workers on the road. The Louisa Railroad employed more than four hundred slaves during construction in 1836–37. The CRRG reported in 1839 that the "force at present on the line—consisting principally of blacks, with a large number of carts and horses, is equivalent to about 500 men." A contractor on the RDRR reported in 1849 that he had secured four hundred hands to work on his contract and hoped to have another six hundred within a month of his report. The following year, the Southwestern Railroad reported that 498 hands were at work on all parts of the road. In 1856 the Virginia and Tennessee Railroad had a force of 643, of which 435 were hired slaves. The Baltimore and Ohio Railroad employed "foreign labor numbering over 2,000" for building its extensions to Wheeling and Parkersburg, Virginia. As noted in the previous chapter, the BRRR created an entire community near the Stumphouse Mountain Tunnel.[1]

Such substantial work forces meant that railroads employed more workers than most plantations. The Virginia and Tennessee Railroad's force of 435 hired slaves in 1856 outpaced every Virginia plantation in 1850 save one and would have ranked sixth in 1860. Compared to Tennessee plantations, the company's slave force would have ranked second in both 1850 and 1860. The Southwestern Railroad's 498 hands in 1850 made it larger than all but two Georgia slaveholdings that same year. While railroad workers were not necessarily concentrated in a single area as they would be on plantations, railroads involved some of the largest mobilizations of manpower in southern states before the Civil War. Projects of this scale were unprecedented.[2]

The difficulty of acquiring laborers plagued the SCRR from the beginning. In 1829 Horatio Allen complained that it was difficult to find the men he needed, even when he offered thirty dollars a month and housing. When construction began in 1830, the interruptions of the work due to labor shortage were described as "frequent" in the annual report. Although the use of slaves was advantageous, it was also problematic because planters would pull them off the railroad when harvesting cotton was more profitable, thus depriving the railroad of the steady labor it desperately needed. "I fear the high price of cotton will make it difficult for him to get hands," engineer John McRae wrote in 1847 of one contractor on the SCRR.[3]

McRae echoed a common refrain. Complaints about the scarcity of labor are found throughout the annual reports of southern railroad companies. Indeed, historian Walter Licht has written that whereas most evidence for railroad labor "points to a general pattern of labor adequacy and even surplus," southern railroads "faced severe difficulties meeting both their unskilled and skilled manpower needs." He attributes this difficulty to the "strong competition for slave labor from the agricultural sector."[4]

Yet a glance at nonsouthern railroads shows that labor problems were hardly a southern phenomenon. An engineer on the Philadelphia and Reading Railroad complained in 1836 that "good mechanics are not to be had for reasonable or indeed *for any wages.*" The Pennsylvania Railroad reported in 1848 that it had difficulty securing adequate labor that year until August, when a depression in the coal industry drove more workers to railroad construction. The following year, sickness prevented the work from going forward for two entire months thanks to "the impossibility of procuring workmen." When the company attempted to build its permanent route across the Allegheny Mountains, it was again faced with a lack of workers. The workers demanded higher wages, forcing the company to bring laborers in from the east at its own expense. Herman Haupt, the chief engineer, noted that in addition to the problem of securing workers, the men who actually did show up had to be policed. Competition from agriculture was also present in the North. In the summer of 1854, a Connecticut contractor worried that his workers would leave, because "*farmers* are offering $1.50 per day & board. help is *very scarce.*" The Illinois Central faced substantial shortages in its need for thousands of workers to build the road, leading some contractors to send agents to Ireland to secure the workers. Railroads in both North and South, then, found themselves attempting to address shortages and competing for the labor they needed. Moreover, whatever particular shortages southern railroads might have faced, it clearly did not prevent them from constructing thousands of miles in the 1850s.[5]

Naturally, railroads did not *always* suffer from shortages. Working on a bridge for the RDRR, contractor William Birnie reported in 1849 that "I never in all my experience saw masons & Laborers so plenty. Laborers are arriving dayly in crowds and masons in dozens. We were going on beautyfully before the storm, working about 100 men all told and any number standing ready go to work." Regardless of the battles that railroads may have faced to secure labor, the idea that southern railroads were uniquely held back seems to stretch credulity. Companies and planters may not have always seen eye to eye, but railroads were able to locate enough workers to amass large crews and fuel the tremendous growth of the 1850s.[6]

Slaves and Railroads

Despite the importance of slave labor to railroads, it has received only scattered historiographic attention. As historian Theodore Kornweibel has ably demonstrated, most southern railroads either purchased their own slaves, hired them, or were built by contractors who hired slaves. By his accounting, 76 percent of the 118 southern railroads in operation at the start of the Civil War used slaves. Robert Starobin estimated that southern railroads employed "more than 20,000 slaves." Understanding how slaves were fully integrated into railroads demonstrates how southerners made modern technology fit into their labor hierarchy. The specific contours of the relationships of master, slave, and railroad are thus worthy of more exploration.[7]

Railroad companies certainly felt that slave labor was to their benefit. The CRRG declared it an "established fact" in 1839 that "negro labor is perfectly adapted to the construction of works of internal improvement." Indeed, some companies were vehemently opposed to the use of white labor. The surveyor of the SARR declared that relying on white labor "would be disastrous, in a high degree, and postpone the completion of your road to an indefinite period." The surveyor thought that in "all contracts made, especially for grading, a stipulation or understanding to this effect should not be overlooked." After labor disruption in Georgia, contractors on the CRRG even fired white workers in favor of blacks.[8]

Although most slaves who worked on railroads were men, work was also carried out by women and children. A slaveholder hired out 11 women and 16 men to the Mississippi and Pearl River Railroad in 1836. In 1838 one contractor on the SCRR had "140 men women & children on the road," and a commentator noted that "they carry the dirt on their heads in little trays, and of course do not make much or very profitable progress." Nelson Tift took out a contract to do work on the Ocmulgee and Flint Rail Road in 1841 and hired "2 negro men & one woman at $33 pr. mo." In 1850 the Montgomery and West Point Railroad reported that it owned "53 men, 7 women, and 11 children." Former slave Hanna Fambro recalled that when the CRRG was built, she "worked on de gradin' 'long wid de other people of de plantation. Yes, ma'am, it was hard work." Another former slave remarked that "my parents worked very hard and women did same jobs that we would think them crazy for trying now; why my mother helped build a railroad before she was married to my father." Just as slave women were not exempt from hard labor on the plantation, so too did they perform hard labor on railroad projects.[9]

While slaves performed nearly every conceivable function on railroads, there are only a few glimmers of evidence that slaves may have acted as engineers. Ezra Michener, a northerner who traveled to Virginia in 1846, rode the City Point Railroad and commented that the "road was rickety, the engine was rickety, the car was rickety and the *engineer* was rickety—an old man who had no doubt been a slave for seventy years and who served as engineer, conductor and sole manager."[10] Aside from Michener's comment, any other hints at African American engineers are only that—hints. President John Ravenel of the SCRR urged the board to consider the "expediency of running freight trains with black Engineers under the management and control of a white conductor" in 1836. At that meeting, the board approved a resolution that such a course be adopted "as soon as practicable." However, there is no additional evidence that this was ever carried out. Ravenel may have been willing to put forth this proposal regarding freight trains because the lives of white passengers would not have been in the engineers' hands; slave engineers might have been less visible to the public on the freight trains as well. In any event, the plan did not come to fruition.[11]

Although slave engineers were a rarity, slaves held a wide range of other positions in railroad work. Slaves and free blacks worked regularly as firemen on trains. A challenging position, firemen kept the fire constantly supplied with wood and were exposed to the elements as the engine charged down the track. An African American worked as a fireman on the early SCRR engine *Best Friend*. On the CSCRR there were seven "Colored firemen," each paid $180 per year. In 1857–58, twenty-one of the NCRR's thirty firemen were slaves, and free blacks constituted another six. Slaves worked in a variety of other capacities. Some served on engineering corps. When the SURR began surveying its line, the group included four slaves and two "free boys." Some slaves were hired for their specific skills. One RFPRR bond was drawn up in 1836 for "Philip a sawyer." Likewise, the company hired William, a "hewer & common carpenter." Slaves were hired as cooks. Once a railroad was complete, slaves tended the wood and water stops along the railroad's route. Traveler Joseph Wharton noted in 1853 that "at intervals I awoke when the train stopped and caught a view of a wood shed and a group of grinning negroes lit up by a blazing pine knot and then I would whirl along to see their faces no more in this world." One slave of the Mississippi Central, James Hill, even "managed to learn the alphabet and the use of figures" while working in the machine shops. Most slaves, however, were employed in the backbreaking construction work recalled by Hanna Fambro. Slaves cleared land where the rails were to run. They graded the land, sometimes carving deep cuts in the landscape to ensure a smooth path, hauling away rubble. They laid rail to the specification of the engineer. Cutting, clearing, hauling, grading, trenching—all difficult tasks needed to create the South's transportation system.[12]

The use of slaves was widespread enough that there was relatively little debate over whether slaves should be used to build railroads. Instead, discussion revolved around whether slaves were more properly owned or hired. The records of the SCRR and its predecessor companies reveal the nature of this discussion. At the SCRR's spring 1832 meeting, it was reported that because "of the extreme difficulty and expensiveness of occasional hiring, the Board judged it proper to purchase a certain number of labourers, and the Company now own sixteen." The number of slaves directly owned by the railroad constituted but a small portion of the entire force. That same year, the number of laborers working on constructing the road was estimated at thirteen hundred hands, with approximately one hundred additional laborers working on the engines and in the shops.[13]

While contractors on the SCRR certainly hired slaves, the company itself was slow to embrace purchasing them outright. The board of directors moved in 1836 to look into the "expediency" of purchasing a large number of slaves for the company, "not exceeding three hundred in number." Nothing appears to have come of this, however, for the number of slaves owned by the company remained low. In 1840 the topic was raised again when the agent of transportation recommended purchasing more slaves. In response, the president gave a comprehensive argument against purchasing slaves. He felt that the company lacked the capital to make such a purchase, but even if it were possible, he felt the move would be unwise. In the first place, hired slaves could easily be returned to their owners if a slave proved unable or unwilling to work, but this would be impossible if the company owned the slave. Second, slaves would be without the benefit of a caring master who was directly responsible for their welfare. Whereas a slave on a plantation had easy recourse for redress if he or she was unfairly treated by an overseer, the slave owned by the company would not have the same advantages. He would be likely to be mistreated, and the paternalist impulse would not be there to protect him. Finally, the president argued that the types of training and specialized work that took place on trains and in shops should be reserved for white men. This would give the railroad an advantage that reached beyond the company itself: by training and employing "able bodied white men," the company would give "strength to our militia, patrols, and police generally." The president's argument against purchasing carried the day.[14]

The SCRR addressed the issue again in 1847. The superintendent noted in the annual report that the corporation was beginning to rely upon the skills that hired slaves who had worked on the road for a number of years had acquired. Thus, the railroad was forced either to pay higher rates for hiring the experienced laborers or to train new slaves. Although it was submitted for the consideration to the stockholders that ten women and seventy-five men be purchased by the company, there was obviously still some resistance within the company. A committee report simply stated—without any additional argument or justification—that while hiring slaves may have its "inconveniences . . . the disadvantages of an opposite course are perhaps as great." Having seen no clear advantages of purchasing over hiring, particularly given the capital outlay required, the committee found it "inexpedient" to purchase. The result was a compromise: a resolution allowed for the purchase of "twenty negro men." The motion passed, along with an amendment that "if a sufficient number of women can be employed with the men who may be purchased, in order to prevent the demoralizing effects of separating them, it would be advisable to do so."[15]

The company slowly began to purchase slaves. At least one slave was purchased the following year. And in 1850, the railroad spent $10,000 for the purchase of "12 Negroes," and listed a total of nineteen slaves as the property the company, valued at $15,036. Another slave, William, was purchased for $800 in 1851.[16] Whatever problems SCRR stockholders may have had with owning slaves before 1852, the objections evaporated after that point. In 1852, the railroad purchased "fifty young negro fellows, practiced and experienced in Rail Road work," and set them about replacing rails on the original road from Charleston to Hamburg. The slaves were purchased for $7,439.60, bringing the railroad's total holdings at the end of the year to seventy-two slaves valued at $59,485.34. Engineer George Lythgoe appreciated the presence of the company-owned workers that same year: "Such have been our difficulties that had it not been we had recourse to the Company's own hands, my opinion is it would have scarcely been possible for us to keep the Roads in working order."[17]

The SCRR purchased more slaves as the 1850s progressed. In 1856 the slaves owned by the SCRR were valued at $71,727.89, including one, Prince Jordan, worth $968.95. The following year the company reported that it owned eighty-seven slaves, valued at $76,238.49. On December 31, 1859, the company owned ninety slaves, worth $80,518.72. The oldest purchase, Anthony, was made in April of 1836; the most recent, Jack, was purchased in November 1859. The value of individual slaves ranged from $400 (paid apiece for Frank and another Jack in February 1845) to $1,500 paid for Jack in November 1859. While it is difficult to determine exactly how this compares to other slave prices, the $1,500 paid for Jack matches the prices that historian Michael Tadman discovered were paid for "No. 1 men" aged nineteen to twenty-five in 1859. Although most slaves appear to have been purchased from individuals (or their estates), forty of the ninety were purchased in April 1852 from J. C. Sproull and Company.[18]

While the SCRR was initially hesitant to purchase, other companies were not. In 1836 J. Edgar Thomson purchased twelve slaves for the Georgia Railroad and another sixty the following year. The LCCRR, which was eventually folded into the SCRR, noted that if the railroad purchased its own workers, it would ensure a "certain and steady control of all in the service of the Company, and thus a more regular and efficient system for the management of the Road." This control would allow the company to distribute the "Villages or farms" for the workers along the road at appropriate points as best suited the company. The company argued that "OWNERSHIP IS KNOWN INVARIABLY TO IMPOSE" a paternalistic obligation, and slaves would respond well: "Labor thus tutored, confined to, and growing up with, and on the Road, would create an identity of interest, and feeling between the slaves and the enterprise; the former seeing that on the success of the latter would depend the permanency, and greater comfort of their own situation." Thus, paternalism was indeed possible with corporate ownership, and the LCCRR urged its stockholders to consider purchasing slaves.[19]

The president of the CSCRR noted that slave hiring was both "economical" and "efficient." When contracts were let to men who were also stockholders, the company realized an additional advantage by allowing stockholders to pay for their subscriptions in labor. This, the president argued, "gives to the slave States great advantages over the free in the construction of Rail Roads." The CRRG agreed, pointing out that if planters could use their slaves to build railroads, the continual employment would help protect the South against "periodic panics." The president of the Mississippi Central praised slave labor as "free of strikes, drunkenness and other labor trouble." Advocating the purchase of slaves in 1859, a New Orleans newspaper declared, "This is the way to build railroads." Slaves were also valued after construction was over. The chief engineer of the Western North Carolina Railroad argued that the company should buy its "train hands, and such as are required at the stations." This was important in part because of the "uncertainty of retaining" slaves who were hired but also because the jobs required "experience, industry, sobriety, honesty and good judgment," which were only "occasionally combined in the hands you hire." Like other advocates, this engineer believed that slaves with the qualities he listed would constitute an excellent investment for the company.[20]

Despite the advantages that these railroads and others saw to purchasing labor, most slaves who worked on railroads were hired. Cost was the primary factor in this decision. Purchasing all the manpower required to operate a railroad would have been a larger capital investment than companies could sustain. Hiring slaves spread out the cost over time, whereas purchase forced a substantial financial commitment upfront. Hiring also minimized costs because workers were brought on only when needed. While many slaves were hired for months or a year, hiring could also be as short-term as necessary. Joseph Franklin White hired out his slave, Simon, to a railroad from October 1 to October 6, 1849, for fifty cents a day. He received his $3 for Simon's work on December 1. Thus, the railroad got the labor it needed but saved hundreds of dollars over what would have probably been necessary to purchase Simon. By forcing the owner to absorb the cost of initial purchase, railroads saved money by paying only for wages and care. Some contractors also increased convenience to companies by providing large numbers of workers. In 1859 the president of the Charleston and Savannah Railroad reported on the extensive work done by two hundred slaves provided by the contractors Drane and Singletary. After working on a railroad in Florida, the slaves arrived by ferry, and "*half an hour* after they landed, they were all at work." More remarkably, the contractor efficiently returned all the slaves to North Carolina and Virginia to allow them "to spend their accustomed holidays at home," and then reassembled them in January to begin working on the road once more.[21]

By hiring slaves, railroad companies entered a triangular relationship with the hired slaves and the slaves' masters. One example of a handwritten contract explicitly laid out the duties of each party. In November 1836 James Moore of Glynn County, Georgia, rented a slave named Maurice to B. F. Perham, an agent of the Brunswick and Florida Railroad Company. The contract stipulated that Maurice would be "attached exclusively" to the company, and in return Moore would receive $240, payable at $20 per month. Perham had one week to return Maurice if Perham did not find him acceptable. Perham agreed to furnish all of Maurice's food and clothing while he was in the service of the company. If Maurice was killed, Perham would pay Moore $1,000; if Maurice was injured Perham would pay Moore in proportion to the injury sustained.[22]

Such trade-offs and responsibilities were often enumerated in these documents. The large number of bonds that have survived on the RFPRR yield rich information on this topic. Fairly standard, boilerplate language emerged, with some additional caveats occasionally insisted upon by owners. Preprinted forms helped ease the process of hiring a large number of slaves. Typical language is given on the bond for the slave Randle, who was hired for the year 1836 for $85 and also was "to be returned well clothed also to have a hat and blanket." Other bonds had additional restrictions. One such restriction was a prohibition against working with white workers. Armistead was hired out to the RFPRR with the stipulation that he was "not to work with Irish or dutch labourers." Some bonds specified that the slaves were not to live with Irish or "dutch" (i.e., *deutsch* or German) workers, either. Of the 149 bonds that the RFPRR paid in the first months of 1837 (for slaves working in 1836), 95 had no restriction on white workers, while 54 (or 36%) placed specific restrictions on working with white workers. While the precise reasons for these restrictions are unclear, it may have been that masters feared that white workers could inspire too much desire for freedom or provide unrestricted access to alcohol.[23]

Owners placed other restrictions on the working environments of their slaves. Nat and Sam were not to work with the Irish or Germans, and it was further stipulated that they were not to work "in unhealthy situations." The bond for Bob and Henry specified that they were only "to work in the Depot at Richmond or in the vicinity of the same." Ben, Lewis, and Braxton were not allowed "to work in water." Although most slaves were supplied with a hat and blanket, George and Tarleton received a great coat instead. On other railroads, masters made even larger demands. When Samuel Smith Downey hired slaves out to a railroad in 1836, he insisted that the slaves receive "plenty of good and wholesome food." He also required that the company "provide good and comfortable houses for said negroes to live and sleep in" and also "pay strict attention to keep them comfortable."[24]

The railroad and the master also made agreements as to who would pay for sickness or desertion. These agreements focused on the value of the hired slave's time. Because railroad companies had no interest in paying for time—and labor—that they did not receive, contracts laid out specifics for different types of absences. When William Boyd hired four slaves to work on the CSCRR in 1849, he signed a contract stipulating that he "agrees to lose all time by absconding," and the contractor would "lose the time of sickness." A contract issued to Edward Lucas in 1847 also set out that the owner would be charged with any time lost by the slave's escape and also that the slaves were "allowed one week to go & see their families & the time so lost is also charged to the owner, but no deduction to be made for time lost from any other cause."[25]

Such deductions and stipulations meant that railroads kept close records of slave labor. Railroad companies were keen to know what they were and were not paying for. John McRae informed a colleague: "The Company do *not* pay for clothing for the negroes the owner has to do that. If you pay any such bill you will have to trust the owner. The best way is to pay for nothing on the Company's account unless *absolutely* necessary & charge nothing to Company as I cannot be responsible for the Company's reimbursing you."[26] McRae's chastisement of another contractor indicates the commitment to accuracy that was expected:

> Please also send me a report of the hands under your direction showing the number of hands on the work each day from the commencement, what those were doing, who were not on the work, the *names* of those who were *sick* & *runaway;* the exact *time* they were *absent* from the work & distinguishing between those who were hurt by the accident & those who were otherwise sick; also the *names* of those who were home on *leave* & the *time*. Also the date of arrival of new hands, their names and the names of the owners. . . . I must insist upon such a return being made to this office every week & if it cannot be done the hands must be discharged. My accounts are much embarrassed by want of these columns & your not signing the vouchers for your pay.[27]

While hiring may have held real cost advantages over purchasing, it also increased the management headaches of engineers and managers who supervised the work. Contractors were required to keep close tabs on who was working, to guarantee that the corporation was paying only for the time that it actually received.

Some owners evidently tried to use the dangerous nature of railroads to exact more money from the companies that hired their slaves. McRae argued against such logic to one owner: "I think too that hands hired on the road do not suffer as many hardships as you seem to suppose, or more than hands usually endure when working away from home." This particular owner may have made reference to a particular accident on the railroad, because McRae then adopted a defensive posture: "The recent disastrous accident can scarcely be claimed as a *hardship* but as an accident which will not probably happen again." The level of danger led some owners to purchase life insurance policies for their slaves. Although by no means universally used, antebellum insurance companies offered policies for slaves hired out to railroads. Sometimes the policies proved necessary. John Buford had a $600 policy on his slave Colonel, through the Richmond Fire Association, which he claimed after Colonel's death on the Virginia and Tennessee Railroad. Other owners may have valued the hard work and danger inherent in railroad work; one planter hired out some recaptured runaway slaves to a railroad contractor as punishment.[28]

Hired slaves were not simply brought into the company randomly but emerged as valued parts of the work force. McRae requested slaves by name when hiring for the SCRR, writing to Thomas Lang in 1846, "I should like to hire two of the boys we had last spring if they are not otherwise engaged. Limerick & Cyrus would suit me I think better than any of the others." McRae preferred to pay only $12 per month because he would be hiring them for an extended time and promised the owner that the work "will be similar to that they did last spring only they would not be as much exposed," thus indicating that the slaves would not be in much danger. Evidently McRae's offer was accepted, for in December he sent payment for the rental of Limerick and Cyrus. Later, McRae requested other slaves specifically from Lang, writing that "Richmond and Bob would suit as well or perhaps better than any of the others."[29]

When Lang asked for his slaves to be returned in June 1847, McRae replied that "two of them Richmond and Limerick are willing to stay but that the other two wish to go home." McRae stressed that the slaves were receiving good care: "I have made arrangements to move the hands of all the Contracts who desire it from the swamp up to this place at night." A few weeks later, McRae reported that Bob and Limerick were "willing to remain." McRae also accommodated Lang as far as the care of the slaves was concerned. When Richmond left the service of the company, McRae wrote to Lang that "I presume that there is a Doctor's bill against Richmond which has not been rendered do you wish that I should pay it." In short, McRae recognized the value that these particular slaves held and was willing to accommodate the owner—and by extension the slaves themselves—in order to ensure their service. We cannot know if Bob and Limerick themselves put pressure on the owner to ensure their good treatment, but their expression to McRae that they were "willing to remain" and the lengths to which McRae ensured the owner of their good care demonstrate that slave hiring was not a simple relationship between the railroad company and the owner.[30]

While the total number of slaves hired by railroads was never high enough to affect the overall market for slaves, it could clearly have an effect on the prices in the areas where railroads were being constructed. McRae's correspondence with another slave owner shows how far he was willing to go to obtain skills that slaves possessed. Attempting to obtain carpenters for one of his contractors, McRae offered L. H. Deas "$50 per month for the two best & $20 for the other if you can send them soon." When Deas informed him that the carpenters could not come before the first of March, McRae responded "better late than never." He directed the slaves to report to Craven's shanty, which was "on the bank of the Wateree just outside of high water mark that is just outside the swamp." McRae understood that this might not be appealing to some of the workers and assured Deas that "if his hands object to remain after the weather gets hot," McRae would "have a locomo-

tive to move them out here every night or to any other place that may be considered more healthy than his present camp, if that will induce them to remain until
the work is completed." Again, McRae had to promise accommodation to get the
slaves he wanted.[31]

In the triangular relationship of corporation, master, and slave, the slave clearly
held the least amount of power. Yet slaves could exhibit some control over their
working environment, dependent on the attentiveness of the master. As we have
already seen, slaves such as Limerick expressed their willingness to remain directly to the civil engineer. Other slaves may have voiced concerns to their masters,
which then found their way into contracts. Still other slaves were more direct in
making requests. In 1835 one slave of the SCRR appealed to the company that he
"desired to be sold as he wished to accompany his mother who had been bought
by a person about to leave the state." A Georgia slave, Cyrus, approached a railroad
superintendent on the Macon and Western and asked to be transferred to another
division of the work in 1854. Naturally, slaves could also demonstrate their displeasure by running away. In 1850 the Montgomery and West Point Railroad
reported that of the eighty-four slaves purchased five years prior, about ten ran
away; those who were found had escaped to Kentucky, Indiana, and Georgia. While
working on the SCRR, McRae reported to their owner that seven slaves had run
away "without any provocation just after getting their allowance. They have not
been flogged nor maltreated in any way. There are now 7 of them about." McRae
recommended that they be punished once caught. Slaves' escape resulted in the
expense of bringing them back. In 1845 a slave on the RFPRR evidently ran away,
because the railroad paid out a sum to a man for collecting the slave. In another
such case in 1855, F. M. Lawson presented a bill to the RDRR's board of directors
"for expenses &c in recovering a runaway Slave," but the bill was rejected by the
board.[32]

White Workers

White workers can seem oddly invisible in the historical record given the widespread presence of slaves. Yet white workers—both skilled and unskilled—were a
crucial part of the labor structure in southern railroads. Skilled whites were prized
for their abilities: the scientific knowledge of an engineer or a conductor's talent
for handling the public. Unskilled whites, sometimes imported from the North,
were also valued as they provided the brawn necessary for construction. While
skilled whites usually feared competition only from other whites, unskilled workers were constantly compared to the vast slave labor force in the South and, as
such, were part of a debate over what race of men should appropriately take on

these tasks. Just as early railroads were frustrated by a lack of competent civil engineers, so too were they frustrated by a dearth of white laborers. Whites did, however, serve as both skilled and unskilled laborers on railroad projects across the South.

Whites figured prominently in the companies' pursuit of skilled labor. "Of all the accidents, difficulties and delays," the SCRR directors lamented in late 1833, "the chief cause is the want of experienced, judicious workmen or mechanics." In 1836 the same company reported that it had only three men "who are professedly Engineers" to run all of its engines; other men who were less experienced were pressed into service when illness struck the three professionals. Although the company attempted to train the replacements as best as possible (and, for safety's sake, put them in charge of the freight trains rather than passenger trains), the lack of qualified engineers contributed to serious delays. The dearth of trained personnel led George Mills to travel to London to acquire some experienced engineers for the RFPRR. The railroad agreed to pay "the expenses of sending the men to Richmond, Virginia, which expences are to be deducted from their first wages," but would be paid back to them if they remained for eighteen months. Economic incentives were necessary to build the required work force.[33]

Some companies tried to solve the manpower problem locally. In response to the paucity of qualified engineers, the SCRR began a system of apprenticeships. Such a plan demonstrates the long-term vision of the corporation; the company wanted to secure its future by creating a more reliable supply of skilled workers. On June 27, 1834, the board of directors proposed accepting twelve apprentices who were "natives of Charleston, or so long residents therein as to be inured to the climate." The company agreed to provide room and board, or a stipend for the same at the end of the year. In return, the apprentices would agree to remain with the company until age twenty-one. They were to receive preferential treatment in the assignment of overtime work and would also receive preferential consideration for hiring if their abilities developed well enough while under the tutelage of the company. Through this plan, the directors hoped to promote the interests of the company and the community by addressing the scarcity of workers in the "Mechanic Arts."[34]

Although most apprentices were white, in 1835 the directors authorized the president to "receive as Apprentices two negroes, the property of Mr Horry." The apprentice system seemed to work; twenty-one apprentices were reported in 1840, and the company boasted, "Many of the apprentices as they become of age, remain in the service of the company, either in the shops or running locomotives, and in both capacities give generally more satisfaction and greater promise than those from abroad, and are receiving the highest wages usually paid for such services."

In an area starved for engineers, the advantage of educating them locally was clear. The apprenticeship program was assessed again in 1843. Between July 1, 1834, and January 1, 1843, eighty-three apprentices entered the service. More than half of these left before their apprenticeship was up, but since January 1, 1841, none had chosen to leave. Tragically, there were five deaths: three in 1837 and two in 1839, but none since that time. Of the twenty-four apprentices who had completed their apprenticeship at the time of the report, twenty had remained with the company.[35]

While skilled workers were critical to the enterprise, the majority of whites— like the majority of slaves—performed unskilled work. Because of the sheer numbers of workers required to build railroads, companies had to import unskilled whites. The CRRG, for example, brought "thousands of laborers, primarily Irish," to Savannah in order to build the road. Importing whites was not always successful. A North Carolina contractor brought down 580 laborers, but only 60 agreed to stay. The result of hiring or purchasing slaves and hiring local or imported whites was that railroad corporations in the nineteenth-century South employed a multiracial work force. The enumeration of the 526 people who worked for the SCRR in 1840 demonstrates this fact: the work force included 288 whites and 238 blacks. African Americans found employment in every department of the road except upper management: they worked in Charleston workshops, Aiken workshops, the locomotive crews, the road department, and the transportation department; at the inclined plane; and on repair crews. The Aiken workshops were perfectly split, with one white superintendent, five white laborers, and six African Americans. Blacks dominated the road department, constituting 134 out of 158 employees (85%). African Americans also made up the entire crew for repairing embankments and ditching: 18 workers under one overseer. But slaves also found a place on the locomotive crews (6 out of 49 members) and the transportation department that oversaw the freight handling (48 blacks and 105 whites). In short, African American workers were fully integrated into the process of running the SCRR, and they worked side by side with white laborers.[36]

This was the case on other railroads as well. Von Gerstner noted that the Petersburg Railroad had a multiracial force in 1839 when he outlined who was in the transportation department: "1 superintendent, 10 locomotive engineers, 10 firemen (Negroes), 10 carpenters, 10 blacksmiths, 4 train captains, 10 brakemen (Negroes), and 45 Negroes for loading and unloading the cars, carrying wood, pumping water, transferring freight from the boats and Blakeley, and so forth." The 1852 report of the Virginia Central Railroad noted that blacks were in service in nearly every department. The railroad employed 11 depot agents and 3 clerks as well as "46 Negroes at depots and water stations." In the workshops there were 3 master workmen, 21 journeymen and apprentices, and 14 "Negroes." Eight overseers supervised 69 "Negroes" in maintaining the roadway, and 28 "negro men" were employed in laying new iron. In 1854 the VTRR reported a work force that included 27 whites and 19 slaves at the shops; 20 whites and 50 slaves at depots; 18 whites repairing roads, including 11 "section masters" who oversaw 145 slaves; and 15 slaves who worked as brakemen. Three years later, the company reported that it hired a total of 435 slaves, "including 30 mechanics."[37]

The multicultural nature of the force was not just limited to black and white. White workers brought older ethnic rivalries with them to railroad work. John Glass described one such battle in Georgia: "Coming up from one of the lower camps, I was passing through Camp No. 2, I soon discovered, there was to be a melee among the Irish and Natives, and likely '*somethun*' to be done. I sat on my pony, a short distance from the crowd, awaiting the sport, which soon commenced in earnest. The Shillalah of the Irish was used with powerful force on the heads of some of the natives, whilst here and there an Irishman was stretched at full length in the sand." A few months later, another fight among white workers broke out in Glass's office, resulting in "the infliction of severe blows and ghastly wounds."[38]

Fighting among the Irish erupted in Virginia in 1850, as "Corkonians" fought the "Fardowners." The former group attacked the latter "in their quarters. They beat the men and even the women, broke into boxes, tore up clothing, burnt down the house, and then returned to the mountain. . . . We heard that many persons had been killed, and that human heads were rolling about like pumpkins." In this particular case the militia was called out, which led to some arrests but few convictions. "Other suspected Irish were arrested in Waynesborough and on the road, so that about fifty persons were secured and brought to Staunton. They were examined by several magistrates during two or three days, but it was impossible to identify many of them as rioters. Only two or three were finally convicted and punished." The labor strife possible among nonenslaved workers could only serve to make slave labor more appealing to railroad corporations.[39]

Working Conditions on the Railroad

Black or white, slave or free, all railroad workers toiled long hours under difficult conditions when building the southern railroad network. Unskilled laborers cleared land, prepared the roadbed, and laid track. N. J. Bell recalled working on the railroad at an early age: "The work was very heavy for a boy and the weather hot, so I did not work very long. All the other hands were stout men, some white and some black, and I tried to do as much work as any of them. . . . We carried and cooked our rations, and camped on the road side whenever night overtook us."[40] Other firsthand accounts of such railroad work are rare. We are fortunate to have the

Figure 2. Laborers at work on a railroad cut on the Orange and Alexandria Railroad near Amherst, Virginia, in August 1859, as depicted in this image from the A. B. Peticolas drawing book. Workers strike at the sides of the cut with picks while horses wait to cart the earth away. Courtesy Virginia Historical Society, Richmond, Virginia.

drawings of A. B. Peticolas on the Orange and Alexandria Railroad in 1859 to illustrate the process of railroad work. Figure 2 illustrates the process of creating a railroad cut: railroad workers chop away at the earth by hand, while carts wait to take away the debris. The cut slices deep into the land, its rough-hewn sides the product of hard labor.

Although most railroad work was backbreaking labor done by hand, workers occasionally received mechanical and animal assistance. As we have already seen, engines helped pump water at the Stumphouse Mountain tunnel. Contractor A. W. Craven purchased two steam engines when working on trestlework on the SCRR in 1846. M. B. Pritchard, moving from New York to Cleveland, Tennessee, to work on the East Tennessee and Georgia Railroad, inquired after the purchase of a steam engine "for the purpose of raising track stuff &c." Once Pritchard arrived in Tennessee, however, he found that buying from the North was not necessary, because he could "get a first rate machine of the kind in Charleston S.C. for $1200."[41]

Far more common, however, were horses. Having played an early role in introducing railroads to the South, horses continued to serve a vital function even after the steam engine became preferred for pulling trains. Civil engineers used horses not only for personal transport but also for construction. To that end, some railroads required that horses be accommodated during the construction process, as a contract on the BRRR indicated: "The Contractor, before commencing his work, and immediately after signing his contract, if required by the Engineer, shall open and maintain a good and safe road for passage on horseback along the whole length of his work." Such paths allowed engineers to travel along the road and inspect the work. When an assistant engineer threatened to resign because of the increased expenditures, John McRae attempted to smooth over the matter by commenting that "wherever the new duties required the Assistants to travel more than a certain number or miles per day," the assistants "should be allowed to hire horses & buggies at the expense of the Company." Likewise, the three resident engineers on the Western North Carolina Railroad each received an "allowance for horse and buggy" in their salary.[42]

Considerable horsepower was required to accomplish the construction of the railroad. Commenting on a contractor who had failed to undertake his work with the necessary vigor, the president of the RDRR noted that completing the work in the contracted time would require a force "amounting to as many as a thousand hands, and seven or eight hundred horses." The Southside Railroad reported that in late 1851 "on contracts between Petersburg and Farmville" it employed 1,500 laborers and 400 horses. As the work went on, the number of laborers was reduced; in 1852 the railroad reported that it employed about 1,000 laborers and 150 horses. In 1859 the Southwestern Railroad employed 115 horse carts and 19 wagon teams. Clearly, animals were an important part of any construction effort, and the move to modern transportation did not end the dependence on older forms of transportation.[43]

But companies worried most about their human workers and the costs associated with them. Because of the extent of the enterprise, companies had to provide a place to live for their workers while they built the railroad and also for the road workers who maintained the way after the road was complete. Railroad employees—both those who constructed the railroad and those who operated it—stayed in a wide range of housing. In 1837 the SCRR reported that it had nearly completed the process of building or purchasing houses for white carpenters and slave laborers at all eighteen divisions along the road. Franz Anton Ritter von Gerstner viewed these houses in 1839. "To accommodate the Negroes," he wrote, "wooden houses have been built along the line. In addition, some of them live in old, out-of-service passenger coaches, which can be moved onto the tracks and run from one location to another if the need arises." Other railroads also used train cars as housing. The Southwestern Railroad reported in 1857 that it "had built in our shops . . . nineteen tent-cars for moveable gangs on repairs of Road, at a cost of $2,000, to be used instead of canvas tents." The company believed that the slaves would be "much more comfortable in cold and wet weather" in these new accommodations. The CRRG devised a similar solution that same year. It built thirty-six "shantee cars of portable huts for repair gangs," which were "valuable substitutes for the cloth tents formerly used for the migratory gangs which it is necessary to keep up for ditching, bridge repairs, &c."[44]

The references to tents indicate that most housing consisted of temporary, low-quality structures. Peticolas has left us drawings of railroad encampments along the Orange and Alexandria Railroad in 1859. In figure 3, tents are pitched along the side of the roadbed. Some labored under worse conditions. Traveling abolitionist James Redpath documented conditions for slave workers on the Wilmington and Manchester Railroad in late 1854: "The railroad hands sleep in miserable shanties along the line. Their bed is an inclined pine board—nothing better, softer, or warmer, as I can testify from my personal experience. Their covering is a blanket. The fireplaces in these cabins are often so clumsily constructed that all the heat ascends the chimney, instead of diffusing itself throughout the miserable hut, and warming its still more miserable tenants. In such cases, the temperature of the cabin, at this season of the year (November), is bitterly cold and uncomfortable. I frequently awoke, at all hours, shivering with cold, and found shivering slaves huddled up near the fire." On other railroads, men may not have had housing at all. Frederick Law Olmsted encountered a labor camp which had "a large, subdued fire, around which, upon the ground, there were a considerable number of men, stretched out asleep."[45]

Some companies built permanent houses for workers. The VTRR, for example, listed wooden "Negro Houses" at its stations to accommodate laborers. Other companies attempted to get out of the expense of building houses by boarding their workers locally. The Farmers Hotel in Fredericksburg, Virginia, appears to have kept a running tab for employees of the RFPRR. Room, board, and fires were provided for the various engineers and train hands of both races; notations included meals for "Black men mail train" and "Black men freight train." When the area in which contractor William Birnie was working on the RDRR was insufficient for "shantees" because of the lack of nearby water, Birnie was not worried because his workers would be able to board themselves in Manchester. Even white employees found their housing targeted as a cost-cutting measure. In 1855 the RDRR began paying one of its station agents an additional $5 per month instead of supplying him with housing. Two years later the company required unmarried station agents to sleep at their stations; married agents were required to have their assistants stay at the stations. In 1859 the East Tennessee and Virginia Railroad required its engineers to begin boarding themselves.[46]

Corporations exerted some social control over their laborers. Workers were under strict prohibitions against alcohol, although some white workers resisted

Figure 3. An "encampment" for laborers on the Orange and Alexandria Railroad near Amherst, Virginia, in August 1859, as depicted in this image from the A. B. Peticolas drawing book. These workers stayed in tents near the track; a table with a pot is visible to the right of the tents. Courtesy Virginia Historical Society, Richmond, Virginia.

this prohibition. Workers on a Florida railroad, for example, broke every day at ten and four "until each man was supplied with a stein of beer." With this demand, these men appealed to a "right" to drink that had considerable history among manual laborers. It also linked free laborers on southern works with their compatriots elsewhere in the Union, and contractors were not afraid to use alcohol as a "cheap alternative to cash wages." By and large, though, the companies worked to restrict access. The SCRR's contracts for contractors prohibited them from supplying their workers with alcohol. Contracts on the BRRR also stipulated that contractors "shall neither give nor sell" alcohol to employees, "shall not knowingly employ any man, either as overseer or laborer, who shall have been dismissed from any other work for bad workmanship, intemperance, or disorderly conduct," or employ "any man who shall be declared, by the Engineer . . . to be either disorderly, habitually intemperate, or a bad workman." The NCRR had a "blanket rule" prohibiting alcohol use among employees. Of course, the rules were not always successful. After giving workers their pay in 1851, John Glass wrote that some of them would be "drunk as Lords tonight (as there is a grocery in the neighborhood)." William Johnson reported on Christmas Day in 1836 in Mississippi, "To Day I Saw fully 20 Drunken men, the most of wich were Dutch men and belonged to the Rail Road." The railroad had an interest in keeping its workers sober, but workers did not always acquiesce.[47]

Railroad work was dangerous, but companies did not always care for their work-ers' well-being.[48] Many companies accepted no responsibility for the conditions that led to injury. Courts generally sided with railroads by applying the fellow ser-vant rule, "which held that an employer was not responsible for injuries to an em-ployee occasioned by the negligence of a fellow employee engaged in common em-ployment."[49] In this vein, the East Tennessee and Virginia Railroad decreed in 1858 that "hereafter all the employees of the Company are employed at their own risk as to life and limb, from accident, or otherwise," and moreover that no person would be employed by the company "unless he shall distinctly accept these conditions."[50]

Other railroads provided some benefit payments to injured workers but with important caveats. In 1840 the SCRR agreed to pay a "gratuity" to James Gros, who had been injured while working for the company, with the understanding that "the same not to be taken as a precedent for the future."[51] In 1852 the RDRR paid a bill for a man whose leg was broken while in the service of the company. However, the bill was paid "with the understanding that the Superintendant be informed that bills of this character will not hereafter be allowed."[52] True to its word, the board of directors denied a bill for $50 a few months later.[53] Clearly, railroads did not want to set a pattern for paying out to injured employees. Some directors, however, could be inspired to sympathy if the employee was a particularly devoted one. The RDRR agreed to provide Thomas Hendrick with a cork leg to replace his ampu-tated leg, "provided it does not cost more than one hundred dollars." The board noted that Hendrick had always been "attentive and faithful in the discharge of his duties both before and since the calamity above named."[54]

When the railroad perceived that the injury was the fault of the employee, how-ever, it was not willing to budge. In 1854 a white fireman, James Lyons, lost his hat and "in thoughtlessly jumping off to recover it, had his leg cut off by the cars." When the doctor who performed the surgery sent his bill to the board of directors of the Virginia and Tennessee Railroad, the board responded that its employees had "full knowledge . . . of the risk, and the wages or hire fully covers all, and the company is not liable in any way for accidents incident to the employment." In this case, no assistance would be forthcoming.[55]

While injuries could certainly spell disaster for workers, railroad work could also result in death. In 1859 the RDRR mourned the loss of engineer L. D. Thomas, whose conduct while in the employ of the company "has been such as to commend him to the favorable consideration of the officers of the Company." Therefore, the board of directors resolved to pay Thomas's funeral expenses and give his widow $100 in addition to his salary. Similarly, the Southside Railroad granted the wife of a laborer $50 after her husband's death.[56]

Not all widows received such instantaneous support. A white section master on the Virginia and Tennessee Railroad, Martin Driscoll, was killed when his handcar ran over a dog. Driscoll's widow appealed to the board of directors in May 1855 for assistance, but the board held firm in its resolve that the company was not financially liable for the deaths of or injuries to its employees. The widow's "appeals draw largely on the sympathies of the Board as individuals," the board claimed, but "stern duty" prevented its members from acting "as officers, having no authority to make donations on account of the Co. and thereby assume undue responsibility in future." The board did note that, "as individuals, their hearts are awakened to the sufferings of the afflicted, and particularly the wants of the widow and the orphans, and therefore, in common with their fellow citizens, will, most cheerfully add their mite to the amelioration of the condition of . . . Mrs Driscoll when called on." But as a *company*, the board claimed no responsibility and could offer no relief.[57]

If companies were uninterested in providing medical care for whites, injuries to slaves could be somewhat more complex because of the involvement of the owner. Railroad companies would deduct costs for the care of slaves, as stipulated in the hiring contracts for them. Accounting for such expenses was routine. For example, when the bond of Robert, hired to the RFPRR, was paid off, $4.17 was deducted for doctor's bills. Another slave, Major, evidently suffered some injuries during 1836, because $12.00 was deducted for doctor's bills during the year. When a hired slave died, the result of death was not just tragedy for the slave's family but also a lawsuit from the owner. As with white workers, companies attempted to disclaim responsibility. The SCRR established early that it would not consider itself responsible for the deaths of slaves. When the slave Stephen died in Edisto Swamp in 1835, the company pointed out that it was not "liable for the loss of Negroes in any way, who are employed on the Road." Two decades later, the Memphis and Charleston Railroad pointed out to stockholders that, after a "negro boy" had been killed, contracts for hired slaves were written "expressly stipulating against any assumption of responsibility on the part of the Company for accidents from any cause whatever."[58] Courts took a different view. The fellow servant rule was "premised on the notion that an employee could bargain for increased compensation for dangerous work, or could quit an unsafe job." Slaves were obviously unable to make such choices, and slave states—with the exception of North Carolina—held railroads liable for deadly accidents to slaves. As a result, owners "invariably received damages for the market value of slaves killed through the carelessness of common-carrier defendants."[59]

Despite wide recognition that the fellow servant rule did not apply to slaves, companies still attempted to evade responsibility. When a slave was killed while digging earth for an embankment on the SCRR, the board of directors noted that that particular slave had been "frequently employed on the same spot at the same business" with the full knowledge of the owner, who had made no objection. Therefore, the company did not feel that it was "responsible for the melancholy casualty which befel the negroe." The slave's owner felt otherwise, and SCRR president Thomas Tupper reported to the company that he was served with a writ in Edgefield District over the slave's death. In 1839 Mrs. Du Pont demanded compensation for her slave, Scipio, killed while working for the same company. As usual, the board declared that it did not "hold itself liable" for the slave's death. Du Pont and her lawyer were persistent, however, and in 1840 the company relented and offered to pay $200, "provided she accept the same in full of all demands whatsoever against the company on account of the said negro."[60]

The Cost of Labor

Wages were a constant concern for railroad companies. One way of saving money was to hire workers on an extremely temporary basis, as demonstrated by the activities of the RFPRR. The company hired men for as little as a few hours, according to its immediate needs. In January 1837 the company paid two men for "cleaning the tracks of snow & removing a rock which fell across the track." James Cavanagh and three of his slaves were paid for similar service: "for the hire of three negro men & himself to clean the snow off the track to the summit & back fifty cents ea. . . . Also one dollar for which I agreed to pay him for setting up & watching the cars &c &c when out exposed all night (*could not be gotten into the house*[?])." Six men received $.25 apiece for unloading cars on the RFPRR on February 1, 1837, with no indication that they were hired on a long-term basis. Thus, short-term employment could meet emergency needs and keep costs down.[61]

Stockholders encouraged companies to keep labor costs down. LCCRR president Thomas Tupper noted that the road saved money by laying off workers when they were not needed. Nearly a decade later, stockholders on the SCRR complained that too many workers were on the road, although the president argued that the railroad's work force was comparable to that of northern railroads. Moreover, he pointed out that the workshops and transportation departments did not employ a fixed number of men but were generally free to hire and fire as the business of the road demanded. The East Tennessee and Virginia Railroad also looked to cut costs by having depot agents double as watchmen and also allowing them to hire assistants only "by the job, or by the hour" so that there were no "permanent hands" at the depot.[62]

Slaveholders could demand varying rates for their slaves. For the year 1836, the RFPRR hired Randle for $85, Daniel for $90, Robert for $110, Henry for $120, and Smith for $150. It was in the area of wages that slave hiring presented one of its clearest advantages: hired slave labor was less expensive than white labor. While working on the Charleston and Savannah Railroad in December 1854, John McRae wrote that he could probably get all the slaves they needed at the rate of $15 per month. In contrast, he offered white worker C. S. Gadsden "$30 thirty dollars per month and expenses in camp" to work as a rodman. Companies sometimes had to make concessions to keep whites from leaving. A "white force" working on the SCRR in 1833 demanded advance wages. Given that much of the slave labor pool was engaged in retrieving the year's cotton crop, the company felt "compelled" to submit to the demands. The demands of the white laborers had a ripple effect: as soon as slaveholders renting their hands to the railroad heard of the advance paid to the white workers, they demanded their money in advance as well.[63]

Extensive documentation is available for salaries and wages paid by the RFPRR.[64] Using the relatively complete figures for 1845, we can reach some general conclusions about white and black labor. In 1845, white salaried workers commanded various amounts for their work, and salaries varied even among workers in the same job. On March 31, for example, A. Omohondro earned $35 for serving as a baggage agent for the month of March; G. W. Derracott earned the same amount as a "train captain." But others in Derracott's position did better: Thomas Chandler earned $40, and B. F. Derracott earned $50. Salaries for engine drivers also varied: N. Simms earned $45, William Teller earned $50, and H. Rollins earned $55. Over twelve months, Rollins would have earned $660 at that rate.

Far more workers were paid a daily wage, and here the amounts exhibit a much greater variety. The RFPRR paid more than $10,400 in daily wages to 127 black and white employees working in three departments (depot, train, and repairs) in 1845.[65] The weekly totals varied, usually settling between $200 and $230 (see figure 4). After a dip in July, weekly payouts increased through the fall, probably a result of increased agricultural production handled by the railroad. This is borne out by the additional workers listed on the depot payrolls during mid-November. Such workers were employed to load and unload trains. New workers like James Weaver and London Chappell arrived the week ending November 14 and earned between $3.60 and $4.00 per week for the rest of the year.

The detailed information allows for a comparison of wages for white and black workers doing similar work. The train department had the most week-to-week work force stability. R. Evans, a white man, earned between $3.00 and $3.50 per week for a year-end total of $166.50. He worked with thirteen hired slaves over the course of the year.[66] Some of these slaves—Isaac Taylor, Ajax, Granville (also called Grandville), and Curtis—were identified as "fireman" on their bonds (see figure 5). Not all of the slaves were hired for the same amount: Isaac Taylor was hired for $80, Granville for $75, Ajax for $65, and Curtis for $50. But the yearly amount paid

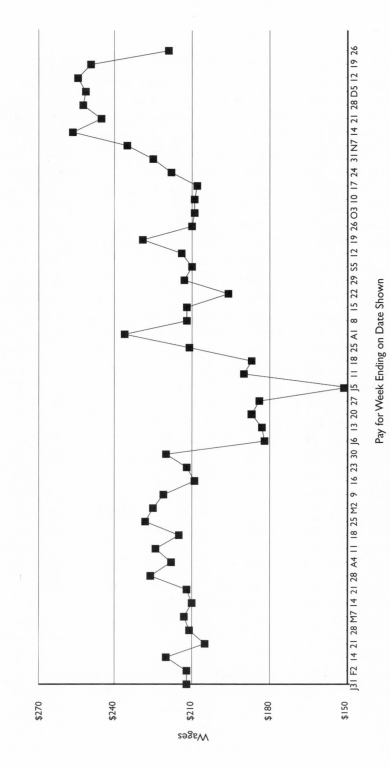

Figure 4. Weekly payroll on the Richmond, Fredericksburg and Potomac Railroad for nonsalaried workers, 1845.

Source: Folders 43–45, box 8, and folders 46–48, box 9, RFPRR-VHS.

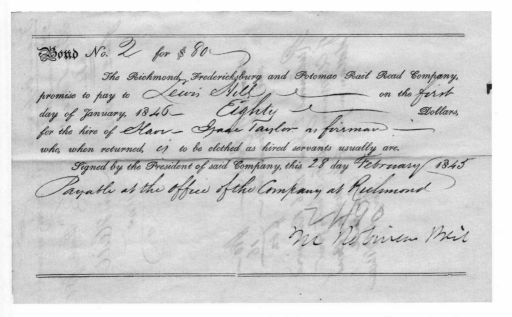

Figure 5. Bond issued for the slave Isaac Taylor (who belonged to Lewis Hill) to work on the Richmond, Fredericksburg and Potomac Railroad during the year 1845. The bond notes that he was explicitly hired to work "as fireman." Courtesy Virginia Historical Society, Richmond, Virginia.

to owners did not constitute the only expense for companies. The payroll sheets demonstrate that slaves themselves received a small weekly allowance for their labors, even earning additional money for overwork. For example, for the week ending February 7, several slaves who worked at the depot—Page, Oliver Carter, William, Captain, and Tony—had $.30 or $.40 marked as "extra" in addition to their normal $.80 allowance. The slaves may have taken additional work beyond their normal duties or worked for an extra period of time. Extra work was undoubtedly available for those willing to work, and the ability to earn an amount equivalent to half of one's normal weekly pay may have proved a strong incentive.

The records do not indicate if slaves were expected to cover some of their own expenses with this money. The bonds specified only that slaves were to be returned "clothed as hired servants usually are," and thus slaves' meals may have been drawn from their weekly allowance. Yet the fact remains that slaves were being compensated for work, and the potential for earning extra money may have allowed slaves to save some cash beyond what was required for covering expenses. In the train department, the slave who collected the most was Waller, who accumulated $55.15 over the course of the year. This made him one of the best-compensated slaves in the company; the highest, James Thompson, earned $59.90. Waller's income did not threaten that of Mr. Evans, but Waller also took advantage of opportunities for overwork, earning up to $2.00 a week (as he did for the week ending May 16), more than double his usual income of $.80.[67]

In other departments there was a wide range of incomes as well. Among white workers, E. Harrison, a machinist, was one of the best-compensated wage workers in the company. By the end of the year he was earning $1.66 $^{2}/_{3}$ a day, six days per week. That brought him $447.00 in 1845, a sum that made him nearly level with the bottom end of the salaried workers. Other skilled white workers in the repairs department—machinists, blacksmiths, carpenters, and painters—also earned more than a dollar per day. Earning slightly less were the car inspectors: M. Tucker, for example, earned $315.63 in 1845. Lowest on the scale of white workers was the apprentice, R. Anderson. His highest wage, $.50 per day, brought him a total of $138.50 that year. Slaves in the repairs department also did fairly well: Warner Burress, Mat, Johnson, and Jefferson all personally accumulated more than $40.00 over the course of the year.[68] Some of the lowest white wages were for employees at the depot, in part because many were hired for only a few days or weeks at a time rather than the year-long employment found by most in the repairs or train departments. J. W. Smith earned $376.81 for the year, but James Roff earned only $152.42, receiving his first pay on July 5. Frank Sullivan, marked as "Colord" on the April 4 payroll, routinely earned $3.00 or $3.60 per week to claim a yearly amount of $160.10. Transient whites earned far less on a yearly basis. W. W. Watts worked at the depot from March to May and September to November; his year-end take was $88.70 based on daily earnings of $.70 or $.75 per day.

There were slaves who worked in the depot for the entire year who did quite well: Page, Oliver, Moses, Warner Lindsey, Roy, and Tony all earned more than $50.00 in 1845. Interestingly, these slaves personally accumulated amounts close to the amounts that owners received for their hire. The RFPRR paid Tony's owner $50.00 for his hire, and Tony himself earned $50.10; Page was hired for $60.00 and earned $55.57; Oliver was hired for $60.00 and earned $53.60. The low amount the company paid to slave owners for hiring these slaves (compared to firemen like Isaac Taylor) suggests that they were not valued for specific skills, as slaves working in the repair shops or serving on the running crew of a train might be. But the vast amount of work to be done around the depot loading freight could work to the slaves' advantage. Working in the depot clearly presented opportunities for overwork, and slaves there routinely claimed extra money in the final months of the year. During the week ending November 14, for example, each slave in the repair department save one earned $.90, and only two of thirteen slaves in the train department earned more than $1.00. Yet among slaves at the depot only two earned less than $1.00, with four of them earning $1.20. The following week all slaves were well compensated (each repairs department slave earned $1.00 save one, who earned $1.25; three train department slaves earned $1.30 and five earned $1.00), but depot slaves again bettered most of their peers: four earned $1.50, two $1.30, and two $1.00. Depot slaves, although performing unskilled tasks, were in an excellent position to earn additional money for themselves.

Slaves were an integral part of the RFPRR's work force. The extant payroll sheets and bonds illustrate the two portions of the company's expenses to hire them: the money paid directly to the master for the slave's hire, and the money paid to the slave. While this latter amount may have been channeled into provisions, the regular amount of "extra" pay earned by slaves suggests they had plentiful opportunities to earn additional money and rarely hesitated to exercise that option. Additional evidence comes from the fact that twenty-eight slaves in 1845 preferred to receive money rather than their winter clothing allowance: twenty-six of the slaves received $7.50, and two slaves received $9.00.[69] Even with the obligation of making dual payments to masters and slaves, slave labor proved cheaper than white. The RFPRR successfully navigated this dual system and took advantage of the savings it presented.

The high cost of white labor persisted throughout the antebellum period. When the general superintendent of the Virginia and Tennessee Railroad reported that slaves had to be replaced by white laborers on a certain portion of the line because the slaves were needed elsewhere, he also noted that employing whites would result in "an increased expense of about fifty per cent." When that same company needed to dispose of one of five train crews in 1857, it eliminated the one white crew. That same year the Petersburg Railroad reported that it had hired 109 slaves (as general train hands) "at an average of $142 each" for a year's labor, but its nineteen white laborers employed in the same capacity were paid $1 per day.[70]

The price of white labor could lead some contractors to attempt to extricate themselves from the work. In 1850 the contractor Robert Harvey and Company attempted to get out of its contract with the RDRR. It made three different proposals to the board of directors, including one that it be given additional funds "in proportion to the difference between white and slave labour." The company refused to release the contractor from its obligation. Five years later, contractors again asked the board of directors for additional funds to cover the higher cost of white labor, at the rate of $.25 per day.[71]

In addition, white workers could strike for higher wages, an option not available to slaves. After Irish workers struck on Virginia's Blue Ridge Railroad in 1855, engineer Claudius Crozet complained that "I regret that we did not hire Negroes at Christmas." When white laborers on the Northeastern Railroad in South Carolina struck for higher wages in 1855, they were jailed, fined, fired, and replaced by slaves. Charles Hardee described another strike in Georgia, in which 150 laborers, "almost all Irish" marched toward Savannah to claim their pay. The militia was called in, "with two cannon planted on the bridge over the canal so as to completely command the approach of the strikers from that direction." Although Hardee did not recall exactly how the strike was decided, he remembered that it was settled by "Father O'Neill, a Catholic priest, who stood high in the esteem of both Protestants and Catholics."[72]

Of course, labor problems also broke out on northern railroads. The Pennsylvania Railroad found in its Western Division that "unhappy feuds" broke out among laborers, even resulting in death. To address the problem, the company arranged for the work areas to be better policed, noting that the military could be called out if necessary. The following year, the company reported that "riotious" laborers remained a problem, but the company helped quell the dispute by offering free transportation to the workers. On the Western Railroad in Massachusetts, there was a "turnout to suppres riot among labourers near the summit" on July 7, 1841. North and South, railroads attempted to keep their costs down in wages while simultaneously fending off complaints from the workers themselves. Southern corporations that could turn to enslaved workers saw advantages in hiring slaves to operate their railroads.[73]

Although the work was constant on southern railroads, there were some light-hearted moments. N. J. Bell reminisced about a particularly good contractor: "The boss was a good one; he would let his hands, in work hours, play leap-frog, and see who could jump the farthest, and take a hand with us himself; also play marbles, and swing on grapevines, and have a jolly old boyhood time—at ninety-five cents a day." William Orr recalled that he worked as a boy on the Mobile and Ohio Railroad, driving "a dump cart" while working with "several little negrows." He also recalled that he would "rasselle with them while the mule ate at noon." The camp at the end of the day also afforded workers the opportunity to relax, as John Glass noted to his son about African American workers in Georgia: "*We* have minstrels *too*, in our camp, and of the veritable Ethiopian stamp. The toils of the day, do not in the least interfere with their amusements at night and I assure you, I frequently envy, when their clean merry laugh rings in my ears, whilst I am plodding over and examining the business of the day."[74]

Such moments of relaxation aside, blacks and whites toiled long hours under difficult conditions. White southerners were determined to use what they saw as a superior labor solution on the railroad. Companies embraced slave labor and as a result, slaves built and serviced the bulk of the southern railroad network. By and large, the experiment worked. Although railroad companies never purchased slaves in large numbers, slave hiring was critical to railroad construction. Purchasing may have had advantages, as civil engineers and company presidents recognized, but required a larger capital outlay than many roads could afford. In contrast, hiring slaves let companies take advantage of a work force more easily controlled than one of rowdy Irish navvies, yet the owners shouldered the ultimate burden of paying the slave's complete value. By employing slaves, railroad companies demonstrated that they were intent on pursing the most modern form of transportation available while retaining the South's preferred form of labor. For corporations, slavery, far from standing in the way of southern progress, facilitated progress.

CHAPTER FOUR

Structure

Railroad companies began operations as soon as they possessed enough track and rolling stock to do so. Building a railroad was a costly enterprise, and whatever revenue might be obtained from operating, even over a short distance, was welcome indeed. The SCRR did not complete its initial route to Hamburg until 1833, but trains began operating much earlier than that. On January 6, 1831, the railroad announced that trains would depart Charleston at four times a day, traveling along the available track. Once operation began, railroads had to develop managerial structures to get the most out of their substantial investments. On the whole, nineteenth-century railroad corporations were highly successful in developing the managerial techniques and structure necessary to do business efficiently. As historian Alfred Chandler has argued, the "swift victory of the railway over the waterway resulted from organizational as well as technological innovation." Chandler saw railroad corporations as integral to the "managerial revolution" in American business. Railroads were critical because they were responsible for some important "firsts," including "the first to have a central office operated by middle managers and commanded by top managers who reported to a board of directors" and the first "to build a large organizational structure with carefully defined lines of responsibility, authority, and communication between the central office, departmental headquarters, and field units." Although Chandler drew most of his examples from the North, southern railroads were also involved in the managerial and organizational innovations that secured the railroad's "victory."[1]

Examining the organization and control of southern railroads reveals several key insights. First, just as southern railroads were part of a national engineering community, southern railroads were also as interested in efficient operation as their northern counterparts. Second, many of the decisions and rules laid down by railroads centered around time. Time management was crucial to the work that railroads performed, and employees were inculcated with the attitude of time's importance. Finally, railroads developed the appropriate equipment to meet their needs. As the business of the companies expanded, so too did the quantity and quality of their rolling stock.

Organization

Chandler posited that an 1841 railroad accident in Massachusetts helped spark organizational improvement in railroads. The Western Railroad's schedule required trains to meet twelve times each day. This happened "on a single track, without the benefit of telegraphic signals, through mountainous terrain," and on October 5, 1841, seventeen people were injured and one conductor killed after a head-on collision. "The resulting outcry," Chandler argued, "helped bring into being the first modern, carefully defined, internal organizational structure used by an American business enterprise." The solution was to break down the road into geographic divisions, "and then the creation of a headquarters at Springfield to monitor and coordinate the activities of the three sets of managers."[2]

Well in advance of the changes made by the Western Railroad, the SCRR introduced a similar structure: the road was broken down geographically in 1834. Workers called overseers ensured that the road was in good condition. Each overseer supervised between one and four other men and maintained eight or nine miles of road. The overseers themselves were under the charge of the two division superintendents (the road was split into eastern and western divisions at the Edisto River). Superintendents also oversaw a single "master carpenter" who could be called upon to do more extensive work than what fell under the purview of the overseers and their gangs. The SCRR's early system meets some of Chandler's requirements. The superintendents on the SCRR were clearly managers who were expected to organize and delegate duties to those underneath them. The SCRR's chief engineer instructed both superintendents that their "indispensable duty" was to see that the overseers under their charge "attend faithfully, and at all times, to their respective duties." The chief engineer expected that the superintendents would have to turn a fair amount of personal attention to this task: "It will not be sufficient that confidence is placed in the Overseer, but it will be necessary, personally to know, that his station is well attended to." This could be accomplished only "by your constant presence on the line, going daily from station to station." The chief engineer thus spared himself the task of caring for the entire length of the line by creating an intermediate layer of management between himself and the overseers who repaired the road. Superintendents were expected to "personally" track the progress of construction and report back to the chief engineer. The two superintendents were required to submit reports on the last Monday of every month. The report was to include the superintendent's assessment of the "state of the road," as well as the dates that they visited each section, directions given to the overseer, and an accounting of money spent since the last report (including a full listing of men employed and the amount they were paid). Clearly, the SCRR was interested at an early date in establishing accountability and tracking the progress of the road. Breaking the road down into manageable units and requiring the superintendents to make regular reports seemed the best way to do so.[3]

Railroads were also interested in accurate information from other parts of the company. The SCRR designated one officer to serve as the "agent of transportation," and that official was responsible for "the Depositories and Agencies along the whole line." The officer was to "make the arrangements for the proper receipt, transportation & delivery of all freight, passengers & Baggage." Like the superintendent, he was responsible for those beneath him and had to "see that every Agent and Keeper of a Depository be diligent and attentive and keep proper accounts." Nearly twenty years later, the Spartanburg and Union Railroad laid out similar duties for its superintendent of transportation. To ensure an accurate accounting of information, the company ordered him to "procure the necessary books & have them properly kept, . . . receive returns from the conductors & agents as often as may be necessary." He was also responsible for keeping the rolling stock of the company in good condition and for reporting "the quantity & condition of the machinery & rolling stock of the Company, the number of men & hands employed in his department & the salaries & wages paid them." Finally, he was to design "such rules & regulations as may conduce to the cheap & easy arrangement of the Transportation & the accommodation of the public," and he held authority over all conductors, agents, engineers, the roadmaster, and their subordinates.[4]

To be sure, such systems of accounting were not flawless. John McRae of the SCRR expressed his exasperation on the difficulty of obtaining information in a series of letters in 1849. In one instance, McRae was trying to root out the source of a problem. He wrote to one of his subordinates that he was sending printed forms "to enter the information I requested you to collect for me. . . . The object of getting this information so much in detail is to show where the detention is and the precise amount of that detention. . . . Send them down as fast as you get them made out."[5] McRae's correspondence also indicated that the process of obtaining information did not always work as clearly as was hoped. "Please send me a list of all the persons in the service of the Company under your charge, state their pay per day, month or year as the case may be, whether permanently or only temporarily employed, with their duties, responsibilities &c," he wrote to N. D. Boxly, agent for the company in Camden. "Let the list be made out as on 1st August. State any changes that may have taken place since. Please also state what you consider to be your own duties and responsibilities & to what officer or officers you consider your-self immediately accountable."[6] Evidently, the chain of command was not func-tioning properly at that time, because McRae had to make a special request to the agent at Hamburg, William Magrath, for routine information: "I enclose a circular which I addressed some time ago to those whom I supposed to be the heads of de-partments. As neither Mr Hacker nor Mr Key claim any jurisdiction over you (ex-cept as regards your money transactions) I send you a copy. Please reply to such facts of it as relate to your division." While such letters indicate that the goal of ob-taining regular information was not always successful, it is clear that from the out-set southern railroads valued the orderly collection of information and a division of labor. They needed neither accidents nor northern example to teach them such things.[7]

Time and the Railroad

Concern with order, division of labor, and accountability all characterized southern railroading. Time was another factor that permeated railroad operations. As histo-rian Mark M. Smith has made clear, the "increasing concern with punctuality" in the antebellum South was "ushered in primarily by the railroads." Smith correctly pointed out the value that southerners placed on punctuality and the ways in which railroads helped them reach these goals. Expanding on Smith's analysis reveals that not all of the railroad's time was governed by the clock, nor did the railroad control all of its own time.[8]

When antebellum southerners discussed time and the railroad, they recognized that different times were important. Clock time itself was described in many different ways. Corporations valued *regularity:* being able to offer services without a break, thereby building public trust that railroad service was dependable. Regu-larity was enshrined in schedules, and the SCRR issued a fine of five dollars to any train that departed too soon from one of six stations that had a clock in 1834, a move made "with the view of attaining the greatest possible regularity in the time of running of the Passenger Engines."[9] In addition to regularity (using the same times from day to day), the fining practice forced employees to be *punctual*—to leave and arrive at specified times. The company also demanded punctuality of passengers: in 1835 the company resolved that each car would have "a large Card" notifying passengers that during meal stops "20 minutes are allowed for Breakfast and 25 minutes for Dinner." An early schedule published in the newspaper also re-minded passengers, "Great punctuality will be observed in the time of starting." Closely related to punctuality was *coordination.* In an era before trackside signals, train schedules had to be coordinated in order to ensure safety when multiple trains moved in opposing directions on a single track. The SCRR and other ante-bellum railroads were surprisingly successful in this regard, with few accidents stemming from coordination problems. Railroads also had to consider nonclock times. *God's time* and the sanctity of Sundays mattered to Sabbatarians. *Nature's time* influenced railroads as well. Agricultural production formed a large portion of the SCRR's business, and the road had to be prepared for the onslaught of cotton during the harvesting season. Time concerns were interwoven throughout railroad work, associated with demands for efficiency, safety, speed, how workers worked, the business cycle, and how passengers interacted with the corporation.[10]

Time was critical in the railroad's drive for efficiency. Railroads prided them-selves on punctual and regular traveling, their much-proclaimed advantage over river transport. Timeliness was at once a necessity for safety and a wonderful piece of advertising. The dual purposes of time awareness for railroads—safety and efficiency—show up repeatedly in antebellum railroad materials. To facilitate these goals, the SCRR put up clocks at six points along the road ("at the Depositories at Charleston, Summerville, Branchville, Blackville, Aiken and Hamburg") in April 1834, six months after the entire road was completed. In a report shortly after the clocks were installed, the company noted that the lack of a "uniform standard of time" at the different places along the road had made it necessary for the company to intervene. Given the early adoption of clocks, the company *anticipated* the prob-lems of multiple trains. Although we do not know exactly how the clocks were syn-chronized, the railroad expressed satisfaction that the system was reliable. Just as the supervisors kept close record of the work done by their subordinates, so too did the company keep close tabs on the timed performance of its trains. Agents at the six stations with clocks were required to submit a return to the main office in Charleston as to when the trains arrived and departed. Individuals who operated trains had a responsibility for keeping them on time as well. Indeed, nowhere was time more enshrined than in the operating rules of the railroad, which gov-erned the actions of engineers and conductors. In 1839 engineers on the SCRR were required to run their trains "as nearly according to the regulations as possible, and to arrive and depart at and from the stopping places upon the line of the Road, except in cases of unavoidable delay at the times set forth on the printed card."[11]

In order to ensure that such adherence took place, both conductors and engi-neers had a key role to play. The first two rules that the SCRR set for conductors in 1853 immediately established the importance of time management and the con-ductor's role in it: "1. Each Conductor will keep a watch, which must be regulated by the time-piece at the depot in Charleston; no excuse will be received for any ac-cident caused by his time being wrong. 2. The Conductor will always have the cur-rent schedule book in his possession, and it is his duty to see that his train runs in accordance with its rules." The first rule made conductors individually responsible, because each had to carry a watch, and the second rule gave them ultimate authority over the train. The rules for enginemen also started with time: "1. Each engine man will keep a watch, which must be regulated by the time of his conductor at the commencement of each trip; and he will always have in his possession the current schedule book. 2. No train will leave a station before the schedule time, or without an order or signal from the conductor, and notice must be always given by the engine man before starting, by the sound of his bell or whistle." Both employees, then, had a responsibility to keep the trains running on time.[12]

Of course, the demands of timely transport required the attention of more than just the operating crew. Stations along the road had to be prepared for the arrival of each train in order to provide wood and water. When this did not happen, time was lost. "As it is now it takes an engine 15 minutes at most of the stations to take wood and water when with proper conveniences five would be sufficient," McRae complained to the agent at Branchville on the SCRR. "The saving of 10 minutes at six stations would be one hour." When recommending that the Spartanburg and Union Railroad erect more water tanks, the chief engineer noted that this would have the effect of "avoid[ing] delays to the passengers." The general superintendent of the Virginia and Tennessee Railroad argued in 1855 that the lack of a water supply at Wytheville meant that the "running of the trains on several occasions has not been as prompt and regular as heretofore."[13]

The combined efforts of running crews and workers at stations to keep trains running efficiently brought substantial advantages to the SCRR and other southern railroads because it meant that multiple trains could be operated on a single track. Railroads could spend their funds improving that track instead of the much higher cost of building a second track. The SCRR reported with satisfaction in 1837 that running multiple trains on a single track required only "a system of arrangement, which a little experience will suggest." As the company grew in knowledge, so too it grew in confidence. In 1837 the company could declare, "Less interruption in passing has occurred the present season, when 5 or 6 trains are met on the road, than with half the number, in years past, when the subject was not so well understood, and the accommodations incomplete." The system of clocks and a trained, accountable staff had proved sufficient for the company to run multiple trains nearly without incident.[14]

Indeed, the success of railroads at time management helps explain why railroads were so slow in adopting another technology—the telegraph. Railroads across the country were slow to adopt the telegraph, and southern railroads fit this pattern. In 1849 the SCRR paid $1,224 to the Washington and New Orleans Telegraph Company for "keeping the line in order." Few details about the relationship are extant.

Not until 1856 did the SCRR announce that it was opening a telegraph office in Branchville, the junction where tracks departed for Hamburg and Columbia. Other railroads also developed relationships with telegraph companies in the late antebellum era: the Norfolk and Petersburg Railroad in 1857; the East Tennessee and Virginia Railroad, NCRR, and East Tennessee and Georgia Railroad in 1858; and the RDRR in 1859. The telegraph was slowly adopted because at the time of its introduction it solved a problem that railroads did not have—or did not solve any problem significantly enough to be worth the cost. Time management was firmly under their control.[15]

Northern railroads also valued time and efficiency. When the Pennsylvania Railroad eliminated some of the inclined planes over the Allegheny Mountains, it calculated the savings in terms of time: it would take three hours to pass through the mountains instead of six or seven. When comparing the ability of the railroad to canals, the company emphasized that, "in this age of steam and electricity, time is too important an element in its effects upon the cost of transportation, to be overlooked." The speed with which the railroad could carry passengers and freight, particularly perishable freight, meant that these sorts of items would be obtained by the railroad "without competition."[16]

Like southern railroads, northern railroads prided themselves on establishing regular service at an early date. In June 1837 the Boston and Worcester Railroad reported that during the past year "passenger trains continue to run with regularity twice a day," and in the winter "there were only three days, on which the cars did not run through the road." Freight trains were also sent regularly—"as many trains daily as are necessary for taking promptly all the freight which is offered." Northern employees also paid the consequences for not making the schedule. Mr. Gleason, an engineer with the Boston and Worcester Railroad, was suspended in 1850 for several reasons, but "his tardiness in starting" was listed first. Thus, in their pursuit of efficiency, southern railroads were fully in step with their northern counterparts. Southern and northern companies alike understood that operating in the manner that saved time would be important to their constituents, be they Pennsylvania merchants or South Carolina planters.[17]

Admonitions about obeying time rules were successful in preventing deadly accidents. Time mismanagement caused few major, deadly accidents in the antebellum era. In his analysis of major accidents nationwide, historian Ian Bartky counted only three (out of forty-two accidents from 1840 to 1859) that could be attributed to time mismanagement.[18] Part of this success stemmed from the fact that railroad companies managed the speed at which railroads operated. Although railroads promised speedier travel than other forms of transit, they did not simply pursue high speed for its own sake. As civil engineer Horatio Allen noted in 1831, the rate of fuel consumption (and its concomitant expense) would rise with the speed of a train. That same year, the SCRR notified the public that trains would travel at fifteen miles per hour when carrying one freight car and passengers, twelve miles per hour when carrying two freight cars and passengers, and ten miles per hour when carrying three freight cars and passengers.[19]

While these speeds may not seem spectacular to modern-day readers, contemporary passengers were impressed. "A Country Stockholder" wrote about his experiences on the SCRR in 1831:

> We got into the Cars, and really Mr. Editor, we were "rattled" along at such a rate I did not think to time her in going up; but in coming down, I timed to a second. We had three cars, and upwards of fifty passengers, and were 20 minutes 37 seconds, exclusive of stoppages, in performing the 5 miles. This is an average of nearly 15 miles per hour—on some of the strait parts of the road, the speed must have been at the rate of 20 miles. As I got off the Car at the Lines, I said to myself, our Horses that are "worn out," don't work in this way.[20]

Interestingly, the "Country Stockholder" brought a watch to time the journey; he came well prepared to judge for himself the effectiveness of this new form of transportation. Time management was important to southerners, and the "Country Stockholder" knew that measuring the time of travel would be a way to measure the railroad's success.[21]

Speeds were increased by 1839, when the SCRR updated its rules, but safety was still paramount; the passenger train was not allowed to run "over 25 miles per hour." The freight engine was limited to sixteen miles per hour. Engineers were also instructed "not to run past turnouts and gates at a speed exceeding four miles per hour"; to reinforce this stricture, engineers were reminded they were "liable for all property injured or destroyed by their Engines."[22] But engineers sometimes broke the rules on speed in order to make up time. When the SCRR's trestlework over the Wateree Swamp collapsed in 1850, McRae claimed that engineers had gone too fast over the trestlework, which damaged the joints and led to collapse. "The freight engines on the SoCa Road were constructed to run an average speed of six miles an hour & I know that they have been running for six or seven months past at more than double this speed," fumed McRae. He charged that engineers took advantage of the fact that the trestlework "was a nice smooth level piece of road" in order to make up their time. In this case, they did so at their peril.[23]

Speeds were adjusted for other reasons, such as regulations of cities in which the railroads operated. The RFPRR encountered this problem in Richmond. The company was not required to switch to horses while operating within the city, but the city limited the train speed to four miles per hour. Railroad companies could also alter speeds for trains running at night. Finally, railroad engineers also had to take care when they went around areas of the road that were being repaired. Not everyone did so, as noted in John McRae's 1850 complaint that a "passenger train was running at great speed over the place where the contractor was laying new rails when one of the cars was thrown off the track & broken probably on account of the speed."[24]

Although railroads grew in technological sophistication as the antebellum era progressed, that did not mean that speeds necessarily increased. This lack of speed was not a reflection of the South's slower-paced society; rather, companies continued to limit the speeds of their trains to prevent damaging equipment and to lessen the risks to passengers. The East Tennessee and Virginia Railroad reported to its stockholders in 1859 that it felt that its schedule was too tight: it allotted seven and a half hours to travel 130 miles, which, allowing for stops for the mail, wood and water, and a meal for passengers, meant that trains had to average more than twenty-one miles per hour. But this rate could not be maintained on the "heavy grades and short curves" of the road, and so engineers made up the time by going faster when they could, as high as thirty-five miles an hour, and also by running at night. The company sensed the danger of risking too fast a speed.[25]

Limiting speed for the sake of safety was also a concern for northern companies. The Pennsylvania Railroad noted that "regularity and certainty of connections are of more importance than high speed in the transportation of passengers." Moreover, running trains at unnecessarily high speeds would lead to damage to the machinery and increase the danger to passengers. Northern railroads also cautioned their drivers against going too fast. Superintendent George Stark of the Nashua and Lowell Railroad informed one of his agents that he had heard that trains leaving his station were "often obliged to move at a dangerous speed. 'The run between Tewsbury and Wilmington was made this morning in 9$^{1}/_{2}$ minutes.'" Stark reproached the agent: "I trust that you will give the necessary directions for correcting these inequalities." Engineers on the Boston and Lowell Railroad were instructed to "*Remember in all cases of doubt or uncertainty, to take the safe course, and run no risks.*" Safety trumped pure speed.[26]

But the railroad was a victim of its own success; by the end of the antebellum era, passengers had higher expectations of speed. Engineers recognized that people expected to move quickly but had to balance these desires with the need for safety. The balance was summed up perfectly by the surveyor of the Savannah and Albany Railroad in 1853:

> It is proverbial that we live in a fast age, (until that term has degenerated, to convey
> the idea of recklessness, it is true,) a speed with which we were astonished a few years

since, now falling far short of the public momentum. This growing feeling in favor of higher speeds in Rail-Road travelling, should, as far as possible, however, be checked. But a rival line, whose success will, by the public, be made to depend upon taking the passenger to his destination twenty minutes earlier than a competing line, will ever foster this feeling and this tendency. . . . Nevertheless, I would not be understood as advancing the opinion that our Roads have reached the maximum speed compatible with safety. Machinery is becoming more perfect, as well as roads, and these improvements may and will advance together, until a speed, the highest yet attained, will doubtless be found compatible with safety.[27]

The engineer had a paramount concern for safety but also conceded that the "maximum speed compatible with safety" had not yet been reached, and that technology would provide a way for higher—and safe—speed.

Scheduling was central to the railroad's efforts to provide punctuality and regularity, and it made possible the coordination necessary for railroad work. Scheduling also helps us understand power relationships between the railroad and its customers. Schedules gave a stark answer to the question: "Am I important enough that the train will stop for me?" Railroads set up schedules quickly after they opened to the public. The SCRR announced such a schedule for trains leaving Charleston in January 1831: "The times of leaving the stations in Line-street, will be 8 o'clock and 10 A. M. at 1 and at half past 3 o'clock P. M. Parties may be accommodated at the intermediate hours by agreeing with the Engineer. Great punctuality will be observed in the time of starting."[28] This early schedule showed some flexibility: while there were definite times of departure, passengers were also "accommodated" if they talked to the engineer. Other railroads set stricter schedules at an early date. The Tuscumbia, Courtland and Decatur Railroad declared in 1833 that "the cars shall leave Tuscumbia for the river with cotton at precisely the following hours 6–8–10–12–2&4 O'Clock, and return at 7–9–11–1–3&5 O'Clock." Other early railroads had multiple trips in one day. The Richmond and Petersburg Railroad ran "three trips . . . in each direction" every day. The Ponchartrain Railroad also ran its cars quite often: "Traffic on this railroad is significant," reported Franz Anton Ritter von Gerstner. "In the summer, trains depart every hour from each end, beginning at 4:30 A.M. and continuing until 9:30 P.M. In the winter the same frequency of service is maintained between 6:30 A.M. and 7:30 P.M. On Sundays, extra trips are also made." Southern railroads created ambitious schedules for themselves and took enough care when scheduling to maintain safety.[29]

In addition to the scheduled stops at established stations, unscheduled stops were possible but did not always work out. Blair Bolling complained in 1836 that, "in spite of a promise to stop for me and my loudest effort to make them hear me," the train passed by "at a rate of about 20 miles per hour," forcing him to secure the loan of a horse. Bolling faced similar problems a few days later and described himself as "disappointed and somewhat vexed as we had been promised by one of the managers that he would certainly stop for us."[30]

In order to secure freight from planters along the road, trains of the SCRR stopped at planters' residences "as a matter of *accommodation*." People or businesses could also get the train to meet them on their time by building a siding for the railroad's use. The Memphis and Charleston Railroad laid out some general principles for the building of such turnouts in 1854. Individuals or companies who wanted these sidings were required to build and maintain them at their own expense. Cars had to be moved onto the private turnout "within ten minutes after they are dropped." Holders of the private turnout could be fined twenty-five dollars for any violation of the rules, and repeated violation would lead to discontinuing the siding. In this way, the railroad could increase its business, but the individual would bear the costs of building the turnout.[31]

Time regulations on railroads did not just apply to those who were operating the trains. Indeed, the effects rippled outward to other portions of the company. The SCRR's board of directors even applied the standards of punctuality to its own actions, establishing rules for its members whereby "The Directors shall meet at the Rail Road Office every Thursday Evening at the first Bell Ring." Moreover, the treasurer and president of the company were required to be in the office between 9:00 a.m. and 2:00 p.m. in order to oversee the business of the company. The secretary was required to be present from 9:00 a.m. to 2:00 p.m. and from 4:00 p.m. to sunset. The Georgia Railroad also demanded in 1841 that members of the board set their watches to that of the president. After a ten-minute grace period, they were fined two cents per minute for tardiness, up to one dollar. For corporations that demanded workers place a high value on time, management practiced what it preached.[32]

Workers in the shops also found their time tightly regulated. The SCRR promulgated the following rules in 1839: "The working hours in the shops and yard shall commence at 7 A.M. from March 1st to Oct 1st and shall be reckoned at 10 hours, and from October 1st to March 1st at half past 7 A.M. and end at sunset, one hour to be allowed for dinner, and to be fixed by the Master of the Workshops. There shall be 8 minutes recess at 10 A.M. and 8 minutes recess at 3 P.M. The calling on and off to and from work, shall be indicated by the ringing of the bell." Stipulations were also in place for tardiness. "Persons not at work five minutes after bell ring in the morning and after dinner will be considered as absent until notice be given the clerk of the workshops. No time allowed after recesses."[33]

As important as schedules were, railroads operated with multiple times, and not all of them were governed solely by the clock. Hardworking train crews learned that their jobs were governed by the length of the task at hand, not simply by the clock. As noted, some workers in the SCRR shops ended their work "at sunset." Other workers labored long hours regardless of the time. McRae informed one stationmaster that engineers were already overworked and could not be held responsible for shuttling trains around the station: "A moments reflection will convince you that after being on duty for 15 to 30 hours on a stretch it is not reasonable to expect more from them." His sympathy also went out to another engineer, who he commended to the SCRR's president for delivering timber and working "early and late in good and bad weather and two nights out of the three that he was here I know he did not get home until 9 P.M. after discharging the last load of the day." The presence of schedules did not absolve these railroad employees from some task-oriented labor.[34]

Overwork could also affect the clerks of the railroad. One detailed complaint of overwork came from S. D. Watkins, the secretary and treasurer of the Southside Railroad, who resigned from his position in 1855. Watkins was frustrated that when he asked for an assistant "the Board postponed action on the subject. I toiled all day and *many* nights until 1 or 2 o'clock in the morning, but found it utterly impossible for me or any other person to keep the Books up with the hindrances I had." In October Watkins was bedridden for two weeks and made up the annual account while lying in bed. He took little time off for personal business, and when he did so, he "travelled day and night (including Sunday) in order that I might not do injustice to the Company." Although Watkins had evidently been offered a lesser position, he wrote that he could not in any case handle "the night labor necessary to keep the business up, although I should be willing to work diligently during the day." But the straw that broke the camel's back was that Watkins now found himself accused of being "*too slow.*"[35]

Other clerks worked long hours; bookkeeper John Glass noted that he had time to write to his family only at night, because there was so much business to deal with during the day. But some clerks were rewarded for their long hours. John Gros, the clerk in the SCRR's workshops, received a $200 increase in salary after pointing out to the board of directors his extensive time in the office: "from 5 oclock a.m. to 6 p.m. and some times to 9 p.m. also frequently on Sundays."[36]

If clerks toiled through the night, operating crews also found themselves running trains after the sun went down. Obviously, the chief challenge to night operations was getting a reliable light source. Horatio Allen recalled that the ability to operate trains at night was desired by early railroad companies: "That the locomotive was to be used in the night, and during the whole night, was plainly to be anticipated." Therefore, he undertook a trial, probably around 1829 or 1830, to place a primitive headlight in front of the engine:

For such trial two platform cars were placed in front of the locomotive. On the forward platform was placed an inclosure of sand, and on the sand a structure of iron rods somewhat of urn shape. In this structure was to be kept up a fire of pine-wood knots. Suitable signals as to the rate of speed, etc., were provided. The day preceding the evening of the trial closed in with as heavy a fog as I have ever seen, and I have seen a first-class London fog. But the fog did not prevent the trial when the appointed time came. The country to be run through was a dead level, and on the surface rested this heavy fog; but just before we were ready to start, the fog began to lift and continued to rise slowly and as uniformly as ever curtain left surface of stage, until about eighteen feet high; there it remained stationary, with an under surface as uniform as the surface it had risen from. This under surface was lit up with radiating lines in all directions with prismatic colors, presenting a scene of remarkable brilliancy and beauty. Under this canopy, lit on its under surface, the locomotive moved onward with a clearly illuminated road before it; the run was continued for some five miles, with no untoward occurrence, and I had reason to exclaim, "The very atmosphere of Carolina says, 'Welcome to the locomotive.'"[37]

While we know little about the technology that followed Allen's early experiment, it is clear that railroads were running trains at night in the earliest years of railroad development. Sometimes it was not by choice: going after a car with a broken axle in 1837, some laborers on the RFPRR left at 10:30 one evening and returned at 5:15 the next morning. Workers could not avoid night work when emergencies struck.[38]

Some railroads scheduled night service in the first decade of southern railroading, the RFPRR doing so in 1839. Traveling observer von Gerstner reported that "the company began to schedule a train departure from Fredericksburg between 1 and 2 o'clock in the morning, so that passengers and mail sacks arriving from the north would be expedited farther without delay." He also noted that the Georgia Railroad instituted night trains by 1839 in order to accommodate the Post Office Department. The CRRG ran night trains at least by 1842, when the company noted that it had "been remarkably successful in our night running; no accident of any importance to the trains has occurred; and their regularity has been fully equal to that of the day trains." Stockholders approved of night running as a way to increase the return on their investment. A resolution offered at a LCCRR stockholders meeting in 1842 declared that "the large amount of capital invested in our Roads ought not to be idle if employment offers, and that freight trains ought to be run by night as well as day, if adequate freight can be had."[39]

Railroad employees expressed some concern about running night trains. John McRae fretted in 1850 that night travel not only required double pay for the workers but kept the engines under constant use, wearing them out. His arguments were to no avail. In August 1851 the SCRR introduced a night express freight and passenger train. The train was introduced "to expedite freight . . . and the result has more than realised expectations." The superintendent of transportation and motive power noted that shipments out of Charleston were regularly handled at night and in the morning: "The up Freights have been uniformly and regularly sent off night and morning for the whole season just as fast as the goods have been received." The demand for making full use of the tracks was clearly met by solutions allowing trains to operate at night.[40]

In later years, the SCRR reported that its night service had been curtailed but not eliminated: in 1856 freight was removed from the night train in order to allow it to reach its destination faster. The annual report also noted that passengers on the night train, which arrived in Charleston at 2:00 a.m., had no services available to them at the depot there, although the company was attempting to rectify this situation. The following year, the company announced that "both day and night trains have preserved their respective schedule with perfect safety to the traveling public, and with almost undeviating punctuality in their terminal connections," meaning that the night service continued through that year.[41]

Governing Railroad Work

Southern corporations governed a wide range of employee behavior. Rulebooks proscribed certain activities and demanded others. Although all railroads clearly had some sort of rules for employees to follow, not many complete books have survived to the present day. Enough evidence is present, however, to help us understand how the companies expected their employees to act and the different jobs that made up railroad work.

One common rule was a prohibition against alcohol. The East Tennessee and Virginia Railroad attributed its success to the "strict temperance" demanded of all employees. "Touch not, taste not, handle not, ardent spirits, is our motto, whilst in the employ of the company," the ETVRR declared. The Spartanburg and Union informed potential employees that only men of "sober steady habits" would work as engineers. Likewise, the East Tennessee and Georgia Railroad claimed that the "simple rule" prohibiting railroad employees from using liquor "proves to be more salutary in protecting life and property than whole volumes of Company by laws or legislative enactments." Such admonitions were not always effective. Henry Bird, a civil engineer working in Petersburg, found on one occasion that after taking a ride on the train the engineer was "in the 5th Heaven, that is to say drunk as the very devil." Other passengers turned to Bird for assistance. Bird obliged and "threw the drunken villain from the engine bestowing a kick or two on him by way of finishing his frolic; and after administering every one to take care of his neck I took charge myself. I blistered and blackened my hands but gained immortal honour among the innocents who fancy an engine as little less than the devil."[42]

Northern and southern railroads alike tried to prevent their employees from using alcohol. The Boston and Worcester Railroad required in 1834 that "no person be employed to take charge of the Engines, or of the cars or to act in any other situation in the service of this corporation, who shall not wholly abstain from the use of ardent spirits." In 1839 the Western Railroad decreed that "no intoxicating liquor shall be kept for sale or consumption, at any of the depots or stopping-places for passengers." The rulebook of the Nashua and Lowell Railroad stated: "No one will be employed or continued in employment, who is known to be in the habit of drinking ardent spirits. The sale of liquors of any sort at the Refreshment Rooms at the Station Houses, is strictly prohibited." Thus, in making the demand for sobriety, southern railroads took the Whiggish position that employers had the right to enforce temperance on their employees. Northern temperance advocates applied pressure to bosses to "assert moral authority over men," and they received a positive response from "those merchants and masters who considered themselves respectable." Southern railroad companies also felt that they could establish such moral authority over their employees.[43]

The position of conductor carried with it a special prestige and responsibility. Conductors were responsible for order on the train. They served as the public face of the railroad corporation and had daily contact with hundreds of passengers and shippers. Thus, a conductor on the Spartanburg and Union Railroad was to "be affable & kind to his passengers, paying strict attention to their comfort & accommodation & be held responsible for the safe keeping & proper delivery of their baggage, use all proper efforts to instruct passengers not to stand upon the platforms between the cars while in motion & observe the utmost care & attention for their safety & the presentation of the property of the Company." Some conductors acquitted themselves quite well in this regard. Traveler Solon Robinson commented that conductors on the SCRR in 1850 were "among the most gentlemanly, well-bred, kind and accommodating officers of my acquaintance." In addition to their public duties, conductors also performed other tasks. The SURR instructions provided that the conductor held authority over brakemen and was to examine the train at each station to "see that it is properly oiled & in good condition." Finally, he was to make a daily report "keeping the time of all hands employed on the train & of the business done & of any matter that may be of interest to the company." Conductors were to report any "mail failures," reasons that the train was detained, or dead stock.[44]

Such rules compared to those used in the North. In 1857 the Boston and Lowell Railroad declared that conductors had the duty to "be in possession of correct time, carefully regulated by the Standard Clock at Boston, and it will be the duty of the conductor of the first way passenger train out of Boston, each morning, to give the correct time to each station that he stops at upon the line." Conductors were also responsible for telling "the Engineman where to stop and when to start, and will see that the train is run as near the Table-time as possible. Each Conductor must report daily, according to the form furnished, the number of passengers carried on his train, each way, the number entering at each station, the amount of fare collected, and the number and names of persons passed free. . . . He must also report the name of the Engine, Engineman, fireman, Baggage and Brakemen, together with the time of starting and arrival, and in case of accident or delay, must state the cause thereof and the injury to cars or persons, if any." In the North and the South, conductors held authority over their trains and were expected to supply their corporations with data on train operations.[45]

Just as they were scrutinized by the railroads and the public, conductors themselves had to have careful eyes: they routinely had to deal with counterfeit money. The SCRR decided in 1835 that "whenever counterfeit Bills are received, hereafter, that they be laid before the Board for their decision." Two years later, the board agreed to reimburse conductor William Bartlett for a counterfeit twenty-dollar bill. Other conductors were not as fortunate. Thomas E. Sims, a conductor on the Richmond and Danville Railroad, asked the board of directors to allow him twenty dollars to make up for a counterfeit note that he had accepted. The board did so but also passed a resolution declaring that "this action is not to be considered a precedent for the future action of the Board." Conductors were evidently supposed to learn that they were on their own when it came to counterfeit money.[46]

Because of their close contact with travelers, conductors left an impression—positive and negative—with the traveling public. Sometimes conductors were quite helpful. Mary Boyce found out from a train conductor that "our cousin Phil had gone on his way to Alabama." But conductors could also seem capricious in their authority. Anna Calhoun Clemson reported that a conductor had refused to let passengers out at a platform near a plantation and forced them to get off at a depot. The passengers then attempted to hire a ride, but when that took them only partway, they were forced to walk and arrived at their destination "muddied up to their knees." The conductor's authority, when abused, could produce poor results for travelers.[47]

Equipping the Company

Railroads required a variety of equipment to meet the needs of their business. Companies needed multiple engines in order to make multiple concurrent trips on the same line and to keep operations going in the event of engines breaking down. When the SCRR opened, it owned nine engines with the expectation of receiving more. Three engines were passenger engines that operated between Charleston and the inclined plane at Aiken. The engines rested in Charleston every third day after making the 230-mile round trip. Each pulled four passenger carriages and some combination of baggage and freight cars. A fourth engine shuttled passengers from Aiken to Hamburg. Five engines were employed to carry freight. Four engines carried freight between Aiken and Charleston; the last took freight from Aiken to Hamburg. Whereas the *Hamburg* could pull twenty-five cars and made the round trip in four days, the other engines pulled between eight and fifteen cars and could make the round trip in three days. Engines appear to have been under the care of specific engineers—the company report listed the name of the engineer after each engine.[48] At the end of 1834, the SCRR owned twelve locomotives: four from New York; two built in the company's own shops; two from Liverpool; and one apiece from Newcastle, Leeds, Philadelphia, and Charleston. The company continued to acquire motive power throughout the antebellum era. In 1851 the company reported that it owned thirty-seven locomotives, ranging in length of service from fifteen years to one month. Three of these engines had been built in the shops of the company.[49]

Railroads had special cars to carry freight. Although the design of American freight cars remained largely unchanged throughout the antebellum period, there were two important innovations during that time: uncovered cars were replaced by covered cars, and double-truck eight-wheeled cars supplanted single-truck four-wheeled cars. The eight-wheeled design allowed the load to be spread over a wider area, and the two trucks meant that the longer cars could still manage curves. Boxcars generally "had an 8- to 10-ton capacity, an arch roof, wood-beam trucks, no truss rods, and a body length of 24 to 28 feet." In 1849 John McRae reported that freight cars used by the SCRR were "30 ft long 9 ft wide & 6 ft high to outside of frame." Freight cars were locked to prevent stealing, but McRae's letters indicate that such efforts were not always effective. "Our cars are so much pilfered," he complained to a colleague. "Who makes your lock? Have you a spare one you could send me as a sample?"[50]

While boxcars changed relatively little, more detail is available on passenger cars. According to historian John White, a "typical car of 1840 had a rectangular body rarely over 30 feet long by 8½ feet wide. Before 1845 the glass windows were generally stationary, with sliding panels for ventilation. The ceiling was low, providing headroom of just over 6 feet. The seats were closely spaced and had narrow cushions and low backs." Contemporaries confirm many parts of this description and also provide some additional detail. S. R. Burford recalled early passenger cars in Alabama, around 1836: they "had no platform at the end. There was only one step about one foot above the ground for getting in and out. The wheels were low and I suppose there were no springs, as the car bed was low down." Traveler Mary Moragne peeked inside some railroad cars in 1838 and declared that "there was nothing which struck me as remarkable— the ladies apartment has curtains for every two or three seats,— & the gentlemen's have bills stuck up [re]questing them *not to smoke.*"[51]

Europeans recognized that American cars were different from the ones they rode in their home countries, so they provided lengthy descriptions. Charles Lyell wrote that a car he rode in North Carolina in 1841 was "according to the usual construction in this country . . . in the shape of a long omnibus, with the seats transverse, and a passage down the middle, where, to the great relief of the traveller, he can stand upright with his hat on, and walk about, warming himself when he pleases at the stove, which is in the centre of the car. There is often a private room fitted up for the ladies, into which no gentleman can intrude, and where they are sometimes supplied with rocking-chairs, so essential to the comfort of the Americans, whether at sea or on land, in a fashionable drawing-room or in the cabin of a ship."[52] German emigrant Louis Heuser described first-class accommodation on railroads in Virginia in 1852: "The first class was extremely well furnished; it is possible to walk from one side to the other, and on the sides are comfortable easy chairs for lounging. Inside the entrance is a place for smoking and across from it is a buffet. At each station a Negro came aboard with water, as well as boys with fruit. In this country women can undertake long journeys alone without the slightest fear."[53]

Figure 6 shows a Delaware-built car on the Charleston and Savannah Railroad dating from 1860. It illustrates the design that caught the eye of European travelers: a central aisle running the length of the car, with seating on either side. The drawing also illustrates how cars progressed from Burford's time. Now, platforms stood at the end of each car, with steps provided to aid the traveler in getting on and off. And a ladies' compartment afforded female travelers some privacy. Although the interior decoration is not revealed by this drawing, some attention was paid to the appearance of the outside, as there is decorative work on the outside corners and the doorframe, as well as paneling on the door and two windows with rounded tops.

Some evidence indicates that passenger cars must have been somewhat colorful or even ornate, but unfortunately few detailed descriptions of color and ornamentation have survived.[54] William Harden reported that the CRRG employed an "ornamental painter," although precisely for how long he was so employed and what work he did are unclear. John McRae ordered two passenger cars from a northern company in 1849, specifying that they should be "perfectly plain painted

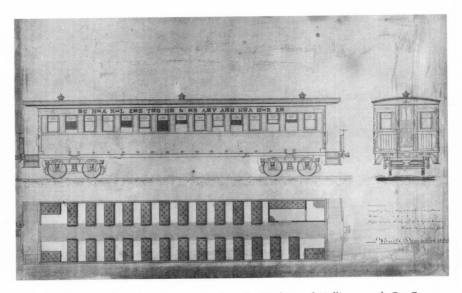

Figure 6. Drawing of a passenger car created by the Harlan and Hollingsworth Car Company of Wilmington, Delaware, for the Charleston and Savannah Railroad, December 1860. The bird's-eye view shows a separate compartment for ladies in the upper right-hand corner. Smithsonian Institution, National Museum of American History, Washington, D.C.

claret color in the best style." McRae also noted, "Curtains do not answer as well as blinds in our climate & it will be desirable that blinds be used in those made for us if done at an additional cost." Evidently the color was not correct, because McRae later complained that the company's president "would have preferred that they had been painted claret colour like one that struck his fancy on your Road."[55]

Beyond color, the SCRR balked at spending too much money on what it viewed as unnecessary ornamentation. In 1850 President James Gadsden noted that although passenger cars were slightly more durable and were not as exposed to the heavy abuse that freight cars were, the ornamental portions of the passenger cars, constructed of wood, were liable to deteriorate quickly. Gadsden charged that cars of "northern importation" were "too gaudy and ornamental in the interior finish and unnecessarily expensive." He urged the company to purchase passenger cars that were "equally as commodious, fully as tasty and neat, [and] without unnecessary ornament." But not everyone agreed with Gadsden. The shops of the Georgia Railroad turned out a sleeping car that included "black-walnut paneling, scarlet plush-covered seats, and drawing room furniture." In short, southern railroads were not necessarily dingy affairs but made an effort to let their passengers travel in style and comfort.[56]

Other specialty cars turn up in the historical literature. As early as 1839, the

RFPRR owned a car "designed for night travel," which featured "swivel chairs in which the passengers can sleep comfortably during the trip." In 1847 Thomas Hobbs wrote that he was able to sleep all night in the train, thanks to "comfortable berths like a steamboat." The CRRG told stockholders that it would need "ten fruit cars" in 1859. Such cars, the company noted, "will answer admirably for other freight, particularly for carrying horses and other stock." The Southside Railroad reported that when its forty-two "rock cars" were finished with their work hauling rock, they could be "converted into box and flat cars for the regular transportation of the road."[57]

Perhaps the most unique item manufactured locally by the SCRR was the patented barrel car (see figure 7). These barrel-shaped cars were used to carry both passengers and freight. The company claimed in 1843 that barrel cars "cost about half the expense of the square car of the same capacity, are much more durable, and require less repairs, and if thrown from the Road are not so liable to be broken." To verify this final claim, the company rolled a car loaded with cotton down a hill, and the car was injured only by the "carelessness" of the workers who attempted to right the car and return it to the rails. When carrying passengers, the design had many advantages: it kept passengers out of the dust and sun, made them "warmer in the winter and cooler in the summer," and provided more room for small baggage. Although the car was never widely adopted, the company still retained at least one, as McRae noted in 1850: "I am sorry we cannot send you the old barrel car just now. We have recently got so many of our passenger and baggage cars broken that we have no other than the barrel to carry negroes."[58]

Contemporary travelers sometimes commented that much of the rolling stock used in the South was of northern origin, explicitly or implicitly criticizing the South's lack of industry and innovation. Frederick Law Olmsted wrote after his travels to the South that the "locomotives that I saw were all made in Philadelphia; the cars were all from Hartford, Conn., and Worcester, Mass., manufactories, and, invariably, elegant and comfortable." Such statements should be treated with caution. Although cars and engines may have been built outside the South, they had doubtless long since been cared for, repaired, or improved locally.[59]

The SCRR relied on its own shops for construction and repair at an early date. In May 1834 the road owned ninety freight cars and was building eight new cars every week. Within a few years, the SCRR noted that although it spent slightly more money on wages, less money was spent on machinery when it was manufactured and repaired in-house. Other companies found similar advantages. Von Gerstner reported that the CRRG had manufactured much of its own rolling stock: "As of the spring of 1840," the company had "6 eight-wheeled passenger coaches; 2 eight-wheeled baggage cars; and about 25 freight cars, some with roofs. All these

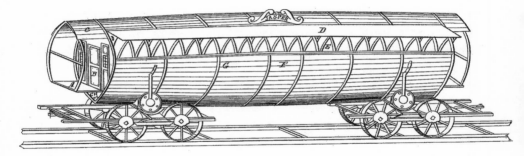

Figure 7. Barrel car patented by George S. Hacker, January 21, 1841. Patent 1937. United States Trade and Patent Office, Washington D.C.

cars were built in the company's shops." The CRRG once "ordered 50 Box Cars built at once in our shops, by Mr. C. C. Millar, Master Carpenter," and bragged that it had built "in our shops, two first class passenger, two second do. and baggage, three express, eight conductors, four stock, two fruit, and one negro sleeping car, at a cost of about $27,500."[60]

Other companies saw that there was an advantage to local work. The East Tennessee and Georgia Railroad resolved in 1851 that the company should start to build its own freight and second-class passenger cars. The Southwestern Railroad expressed the same desires and reported, "We are now equipped . . . so as to be able to perform, with economy and dispatch, all necessary repairs to our motive power and cars, as well as to manufacture our own freight cars." As the company claimed, "The manufacture of freight cars, in our own shops, will be continued to meet the increasing demand of our business." The Southside Railroad went even further, opening its own foundry for the manufacture of locomotives and passenger cars. In sum, southern railroads recognized the cost savings of constructing their own rolling stock, and, while making use of northern manufacturers, were not completely dependent on these companies for their equipment.[61]

Southern railroads went in to their new businesses knowing that structure and organization would be critical in making their ventures a success. To that end, they set up their businesses in ways that would provide for accountability. Conductors and superintendents generated a trail of paperwork that allowed the corporation to track performance.

Contemporary railroaders understood how important time management was to railroad operations. Time infused the rhetoric of railroad promoters, was a guarantor of safety, and was the constant obsession of railroad employees. Southern railroads were quick to master the importance of time, as demonstrated by the

SCRR's adoption of clocks along the route. Corporations also demonstrated control over their employees through rules and regulations, and they used a range of specialized equipment to capture the passenger and freight trade along their routes.

While southern railroads may not have handled the volume of traffic that their northern counterparts did, an examination of their operating practices demonstrates that they were in step when it came to management. Though smaller in scale, southern railroads still valued efficiency, time management, and bureaucratic accountability.

CHAPTER FIVE

Motion

The management structures that railroads put into place soon had to face the challenges of railroad operation. Three important aspects of railroad operations considered in this chapter include the task of delivering freight, dealing with the unpredictability of nature, and averting accidents. Although the standard narrative of southern railroads portrays them as usually carrying cotton to the coast, a close analysis of freight handled by various companies demonstrates that this was not the case. Rather, southern railroads also carried substantial freight to the southern backcountry, suggesting the need to reinterpret our understanding of freight transportation and southern consumerism. Southern railroad promoters felt that railroads offered a clear advantage over river transport. Yet railroads still could not escape nature's grasp and found their operations restricted by nature. Finally, an examination of accidents on southern railroads reveals that corporations were surprisingly successful in preventing major accidents. More important, just as today's car drivers can appreciate the difference between a fender bender and a multicar pileup, antebellum travelers might withstand a minor accident but did not let the unpleasant experience convince them to abandon the railroads.

The Demands of Freight

Although it is convenient to speak of "the railroad," in reality companies dealt with two very different types of service: passenger and freight. Each type of service made its own particular demands on companies. Passenger traffic is, in a sense, freight that loads and unloads itself. Passengers were responsible for ensuring that they were in the correct car at the correct time in order to reach their destination. Certainly railroads had to look out for the needs of their passengers (by employing conductors to take tickets, coordinating with hotel owners to provide meals along the route, and the like), but freight required a larger amount of work on the company's part. The railroad's workers had to load and unload cars at the stations, and its agents had to accept freight, store it, route it properly, and keep paperwork in order. If a passenger overslept and missed his stop, the passenger was to blame. But if freight was delayed or incorrectly handled, the company could expect to hear complaints from shippers and receivers.

Once railroads began operating, people with goods to be shipped quickly flocked to the new service. Indeed, the SCRR discovered to its chagrin that it had not adequately prepared for the amount of business it received. The railroad initially had to refuse freight shipments for want of engines. When new engines were ordered, they were required to be of greater power, "in accordance with the increased business of the Road." In response to this demand, railroad corporations developed regulations to ensure that they would not be caught short again and that freight moved efficiently to its destination.[1]

In 1835 the SCRR established the following rules for freight: "Freight . . . intended for the morning trip must be at the Depository by two o'clock the day previous, in good order, and marked with the name of the station on the line it is to be left at, or it will not be received." Such rules delineated the relationship between the railroad and the shippers. Rules were in place to ensure freight was handled properly, but shippers bore some of the responsibility for seeing that proper care was taken by marking the packages. Other companies established similar rules. The RDRR's 1851 rules dictated that all goods shipped by the railroad were required to be labeled with the names of the persons shipping and receiving the goods as well as its destination station. The company was clear about where it held liability and where it did not. Claims for damage would not be allowed unless the goods were inspected first by the company's agent, and the RDRR "will not hold themselves responsible for the leakage of liquids, the breakage of glass or crockeryware, the decay of perishable articles, nor for goods shipped in bad order." The company made it clear that those receiving goods should claim them promptly, because the corporation would take no responsibility "for pillage, damage by weather or other injury, after goods are delivered at the station to which consigned." Conductors were prohibited from taking packages under their own care; all items had to be entered on the freight list and paid for. A few years later, the company directed that all "articles intended for transportation" were to be weighed "both at the point where it is received, and where it is put off the road, except when in his judgment they are of such a nature or so secured as not to be subject to waste or loss." With such rules, the company hoped to carefully track what it shipped but also make clear that it was responsible for goods only while they were actually in transit.[2]

In addition to regulations, freight soon required its own facilities. Some storage space was needed for the convenience of shippers, and this was also closely regulated. In 1835 the SCRR included one week's storage in its standard rates for freight.[3] As time went on, some policies became less liberal. In 1855 the RDRR decreed that any goods left in the station more than forty-eight hours were left there at the owners' risk.[4] Some railroads required buildings more specialized than simple storehouses. Franz Anton Ritter von Gerstner commented on some of these in 1839. The City Point Railroad created a structure to move freight from ships to the cars. It did so by driving piles to "create a landing where the water depth is 20 feet. Upon this structure a spacious freight depot is built, and two tracks go through it. In addition, side tracks run around the exterior of the building. They connect in a practical manner with each other and with the main line. Cranes are used to remove freight from the ships—two may dock simultaneously—and either put it directly into the railroad cars or into the freight depot." Companies also took advantage of existing structures. In Winchester, Virginia, von Gerstner noted that "the track runs the entirety of the main street so that freight cars can be brought to shippers' warehouses and their contents can be loaded and unloaded right there." Another company won praise in 1857: "Each car of a full train of inward freights can be unloaded on the continuous floor of the delivery side, while each car of a train for outward bound freights can be loaded from the floor of the receiving side, whereas heretofore not more than four cars could be loaded and unloaded at the same time." Freight work required intelligent management of space.[5]

The chief engineer of the Spartanburg and Union Railroad pointed out that the efficient freight movement meant having enough hands on the force. Indeed, freight work required large numbers of laborers. For some types of freight, companies refused to perform the service—the RDRR, for example, required shippers to load and unload cordwood themselves. For most tasks, though, the railroad's employees stepped in and handled the work. Despite the large amount of freight to be moved, workers who loaded and unloaded freight could not depend on regular working hours. Workers did not listen for a bell to signal the end of their working day but labored until the work was completed. Although railroad work is often depicted as strictly clock-oriented in nature, the work of loading and unloading freight before trains could move to their destination demonstrates the persistence of task-orientation in a purportedly clock-driven enterprise. Recognition of the time demands made on workers reinforces the fact that trains used multiple times in their operations. For safety reasons and as a rhetorical device, punctuality and regularity were key. But freight workers saw another aspect of the railroad's time. Their work was complete only when the train was ready, regardless of the time it took to make it so.[6]

Loading goods entailed working through the night. Frederick Law Olmsted described how freight was loaded, at night, in South Carolina:

At midnight I was awakened by loud laughter, and, looking out, saw that the loading gang of negroes had made a fire, and were enjoying a right merry repast. Suddenly, one raised such a sound as I never heard before; a long, loud, musical shout, rising, and falling, and breaking into falsetto, his voice ringing through the woods in the clear, frosty night air, like a bugle-call. As he finished, the melody was caught up by another, and then, another, and then, by several in chorus. When there was silence again, one of them cried out, as if bursting with amusement: "Did yer see de dog?— when I began eeohing, he turn roun' an' look me straight into der face; ha! ha! ha!" and the whole party broke into the loudest peals of laughter, as if it was the very best joke they had ever heard. After a few minutes I could hear one urging the rest to come to work again, and soon he stepped towards the cotton bales, saying, "Come brederen, come; let's go at it; come now, eoho! roll away! eeoho-eeoho-weeioho-i!"—and the rest taking it up as before, in a few moments they all had their shoulders to a bale of cotton, and were rolling it up the embankment.[7]

In this observation, Olmsted captured two aspects of railroad work. First, the work occurred at night, on demand: laboring for the railroad did not occur only when the sun was up. Second, the workers operated with a leader ("one raised such a sound . . . like a bugle-call") who led the group in singing as they moved cotton. As in other work crews across the South, African American laborers timed their work to music.[8]

The vast amount of work done by crews could result in confusion. In Charleston, "empty cars coming up to be loaded, and those already freighted going off, frequently become intermingled, and in the morning before daylight there is a heavy force of men and mules engaged in adjusting the several trains for the engines going to Hamburg or Columbia." Irregularity in the freight service was also reported at the SCRR's inclined plane near Aiken. The company noted that "during the business season" the men who worked at the plane were kept "night after night" until midnight. The stockholders recognized the inclined plane as a literal waste of time and thus a source of monetary loss for the company. Complaints continued in 1848 and a committee convened to look into them discovered that the trip from Hamburg to Charleston (for a freight train) could take more than fifty-three hours. The variance in times recorded by the committee shows that efforts to provide timely freight service were not always successful.[9]

With all the freight being handled, it was inevitable that some things could go awry. Mr. King, in charge of the depository at Charleston, wrote to the SCRR's directors in 1835 that a certain box belonging to John T. Willis was reported missing by the man who was supposed to receive it. Although "enquiry was made up the Line," King was unable to "find, or hear anything of it. I cannot say who it is chargeable to." The problem was compounded by the fact that the conductor of the freight train in question was no longer in the employ of the company. As time progressed, future conductors would not get off that easy. Later that year, the board of directors authorized payment to E. Carson, for his "Keg of Lard lost on 2d or 3d July by theft from Car," and the amount was charged to the conductor. The Virginia and Tennessee Railroad's board of directors complained in 1854 that "frequent errors have occurred, and damage done, in receiving and forwarding freight and baggage," and the company was spending too much money responding to claims of shoddy service. The board of directors moved that any employee who damaged freight would be "held liable" for the damage, "unless it shall appear to the satisfaction of the Board that it was not in the power of the officer, agent or employee to prevent it." Just as conductors held personal responsibility for their trains, now employees who handled freight held personal responsibility for goods on the train.[10]

The World of Goods

What goods did these railroads carry? Southern railroads carried one peculiarly southern item—slaves. Once constructed, railroads become an important link in the trade and transportation of slaves. Slave labor built the network that was then used to move bondsmen, demonstrating how ably white southerners could marry modern transportation with their preferred labor system. Some slaves' recollections of train rides have survived. Slave Robert Glenn recalled that when sold to Kentucky he "traveled by train by way of Nashville, Tenn. My thoughts are not familiar with the happenings of this trip but I remember that we walked a long distance at one place on the trip from one depot to another." Most accounts of slaves being transported en masse, however, come to us from white observers. Although the next chapter shows that some individual slaves could travel with their white masters, slaves being transported in large groups were herded into separate cars. While traveling through Virginia and North Carolina in 1853, Joseph Wharton observed that "two negro traders" on his train "had a car full of human cattle just before ours," which is to say, closer to the uncomfortable, spark-filled air of the engine. "The gang of blacks bound South are numerous but not in the least dejected looking—though a snatch of a mournful hymn came from their car sometimes last night." Wharton later observed chained slaves "surrounded by darkies with baskets of cake and apples which I was surprised to see set among them freely and paid for by the southern bound slaves themselves."[11]

Jacob Stroyer's slave narrative includes the harrowing description of how slaves were taken to a train in South Carolina, having been sold to Louisiana. Slaves who resisted "were handcuffed together and guarded on their way to the cars by white men," while women and children were "driven to the depot in crowds, like so many cattle." The procession attracted more slaves from neighboring plantations as it passed. Slaves "left their masters' fields and joined us as we marched to the cars." At the depot, the train arrived, and "when the noise of the cars had died away, we heard wailing and shrieks from those in the cars." When the conductor called for all to get on board, "the colored people cried out with one voice as though the heavens and earth were coming together." The cries of the slaves continued as the train pulled away, "as far as human voice could be heard." The description Stroyer gives makes it clear that depots could have a painful meaning for slaves.[12]

Frederick Law Olmsted discovered how thoroughly slave traders made railroads part of their normal practice, when he learned that "the negro-dealers had confidential servants always in attendance, on the arrival of the railroad trains and canal packets, to take any negroes, that might have come, consigned to them, and bring them to their marts." Olmsted also observed the process of slave transport firsthand. He rode a train that contained "two freight cars . . . occupied by about forty negroes, most of them belonging to traders who were sending them to the cotton States to be sold." When another train reached a gap in the rails where track was still being laid, some slaves being transported disembarked. "As it stepped on to the platform, its owner asked, 'Are you all here?' 'Yess, massa, we is all heah,' answered one; 'Do dysef no harm, for we 's all heah,' added another, quoting Saint Peter, in an undertone." If the slave trader heard the slave quoting the Bible, one wonders if the trader picked up on the verse's meaning. The slave had quoted from a passage in Acts where Paul was jailed in Philippi. When an earthquake brought down the walls of the jail and loosened Paul's bonds, the jailer rushed in: "But Paul cried with a loud voice," according to the scripture, "saying, Do thyself no harm: for we are all here." When given the opportunity to escape, Paul did not do so. Likewise, the slave claimed that none escaped when released from the train—but also claimed the moral high ground by equating his position with that of Paul. As Olmsted observed, the slaves in this case made good on the promise not to escape. Instead, they prepared for the night by the side of the road: "The negroes immediately gathered some wood, and, taking a brand from the railroad hands, made a fire for themselves; then, all but the woman, opening their bundles, wrapped themselves in their blankets and went to sleep."[13]

Depots could even serve as the location for slave sales. In Chattanooga, slave traders E. A. Parham and A. H. Johnston both sold slaves at a location near the city's depot.[14] European traveler Arthur Cunynghame recounted a slave transaction that took place at a depot:

> Whilst getting into the cars I observed, standing near the door of one of them, a gang,
> as they were termed, of negroes. It consisted of three women and two children. In a

moment the steam-vessel blew a shrill blast as usual, the signal for starting, and commenced its movement. At the same time I observed two gentlemen at the door of the
car, in conversation. One appeared rather well-dressed, the other was a short, stout,
good-natured-looking man. These, it appeared, were slave-dealers. "Come," said the
dandy-dealer, "I'll give you twenty-one hundred and fifty for the lot." "Say twenty-two
hundred," said the stout man, in a huge water-proof, "and they are yours." "Well, done,"
said the first; "I hate not to do business." "Get in, you niggers," exclaimed both; and
the sale was completed between their white masters. These five fellow-creatures were
sold, nor did a compact, embodying the happiness of five fellow beings even take
the same time to ratify that we were employed at our repast, but actually was consum-
mated after the signal had been given for the starting of the cars, and during the time
indeed that they were on the move.[15]

Here, railroads met the needs of the slave trade perfectly—buyer and seller could
be conveniently brought together, and the new purchases easily carted away. The
train's schedule urged both men to complete the transaction in a timely manner.

While railroads were useful for the slave trade, the principal freight on southern railroads was cotton. As Olmsted noted on one of his trips, "The roads seemed
to be doing a heavy freighting business with cotton. We passed at the turnouts half
a dozen trains, with nearly a thousand bales on each, but the number of passengers was always small." He continued, "Plantations were not very often seen along
the road through the sand, but stations, at which cotton was stored and loading,
were comparatively frequent." Railroads had been founded in part to capture the
cotton trade, and clearly planters flocked to this new mode of transport.[16]

Railroads worked hard to attract their business. Because cotton shipments were
crucial to the bottom line of the SCRR, the company did what it could to ensure
that planters brought their goods to the railroad. "As a matter of *accommodation*,"
President James Gadsden reported in 1848, "Passengers and Freight have been received and delivered, at the Turn-outs and Pump Stations along the line of Road:
and at points convenient to gentlemen's residences." Cotton was received at every
turnout on the road, and the loading and unloading were performed by "such hands
as can be obtained from the neighborhood, and from the Carpenter's gangs." (The
"Carpenter" oversaw repair work, thus his "gangs" were slaves employed by the
railroad.) Indeed, the railroad made a practice of hiring hands on an ad hoc basis
as well as having the permanent force along the road. Gadsden indicated that
planters along the route took advantage of the fact that their slaves could be hired
on short notice, using the opportunities to make some additional money.[17]

Historians have been quick to point out that the southern railroads were overly
dependent on cotton for their business. Although southerners themselves recog-

nized the key role that cotton played in their enterprise, the historiographic picture has tended to overstate the lack of balance in freight. The argument is an old one, and goes back to the work of Ulrich Bonnell Phillips. Phillips wrote in 1908 that southerners "had little desire to possess a carrying trade" and were "chiefly concerned in developing a system of internal transportation and commerce, by providing communication between the several staple areas and their gateways." Moreover, dependent on cotton, railroads experienced a "great rush of business in the marketing season and lean months following in spring and summer." Later historians followed where Phillips had gone before. According to Richard Brown, "Southern canals and railroads were not designed to carry people or to promote intra- and interregional communications in general—they were built to bring cotton to market." More recently, Scott Nelson argued that "southern plantations needed outlets for staples like cotton, but they needed few consumer commodities to come in the other direction. . . . With only small crates of violins and suspenders shipped to the interior, locomotives that hauled cotton to eastern ports had little to carry back to plantations." The impression given by historians has been that southern railroads failed in terms of both the quantity (empty cars) and quality (small crates of violins) of trade.[18]

On their face, such arguments have merit. Evaluating the performance of antebellum railroads is a challenging task because of the unevenness of the available documentation. Enough evidence exists, however, to demonstrate that a reconsideration of the railroad's role in southern commercial life is warranted. While plantations were insufficient to stimulate substantial consumer demand, the southern interior was hardly an unbroken landscape of plantations. Small farmers and townsmen benefited from the goods that railroads brought. For example, railroads helped stores in upcountry South Carolina dramatically increase the value of merchandise they held. Year-end inventories in Chester went from $88,950 in 1848 (before the CSCRR began) to more than $137,880 in 1852 (the year it was completed). The benefits reached through the entire South Carolina upcountry: $2,717,776 worth of goods were sold there in 1853; by 1854 the value had risen to more than $4,020,000. While such sums did not match northern states in an absolute sense, it is clear that railroads had a more transformative impact than the Phillips model allows.[19]

Other evidence points to a more diverse "upfreight." Railroads themselves certainly did not intend all of their freight to be cotton, as welcome as cotton was. To that end, railroads established rates for carrying a wide range of goods, and some of these lists have survived. In 1851 the RDRR settled on rates for 26 items (table 3). The next year, the VTRR suggested rates for items in 4 different classes (table 4), and in 1857 reported shipments in 111 categories (in addition to "miscella-

TABLE 3

*Items for Which Freight Rates Were Established on
the Richmond and Danville Railroad, 1851*

Artificial manure	Bacon	Bran	Brandy	Bricks
Butter and lard	Candles	Cord wood	Dry goods	Empty barrels
Fish	Flour	Granite	Groceries	Hardware
Iron/steel/castings	Hay	Lumber	Paving stone	Salt
Oats/corn/corn meal	Shingles	Sugar	Wheat/peas	Molasses
Potatoes/turnips				

Source: Entry for March 10, 1851, RDRR Minutebook.

TABLE 4

Freight Classes on the Virginia and Tennessee Railroad, 1852

First class: Boxes of hats, bonnets, and furniture

Second class: Boxes and bales of dry goods, feathers, shoes, saddlery, glass, paints, oils, drugs, and confectionery

Third class: Virginia domestics, sugar, coffee, liquor, bagging, rope, butter, cheese, manufactured tobacco, leather, hides, cotton yarns, copper, tin, sheet iron, hollowware, queenware, crockery, castings, hardware, marble (dressed), and other heavy articles not enumerated in special or fourth-class rates

Fourth class: Flour (in sacks), rice, pork, beef, fish, lard, tallow and bacon (in casks, boxes, or sacks), beeswax, bales of rags, ginseng and dried fruit, bar iron, marble (undressed), mill and grindstones, and mill gearings

Source: Fourth Annual Report of the President and Directors to the Stockholders of the Virginia and Tennessee Railroad Co. (Richmond: Printed by Richies & Dunnavant, 1852).

neous"). In 1854 the Southside Railroad promulgated rates on 89 different categories, including bricks, figs, pianos, and "tubs in nests" (table 5). Clearly, these companies did not intend for their railroads to become cotton-only enterprises. As the Southwestern Railroad noted, upon opening, "Large quantities of merchandize have been transported to all parts of South-Western Georgia and Alabama," and the company was confident that "still greater quantities will take this route during the current year." Southern consumers soon recognized that the railroad brought them advantages for the goods they could obtain, not just the cotton that was sent away. C. C. Jones wrote to his son, "Yesterday I sent by R. Rd. to Sav[anna]h. for two doors, . . . & they are now today in the house, safe & sound, to be hung tomorrow. This is the convenience of the R. Rd."[20] Upcountry consumers also provided a market for railroads even when the region was not a staple-producing one. Greenville District, South Carolina, had the second-lowest cotton output of any upcountry South Carolina district in 1850 and 1860.[21] Railroad promoter Benjamin Perry acknowledged that freight from Columbia to the upcountry was not initially a priority: "When we were making calculations to induce subscriptions to the Greenville and Columbia Railroad, no one thought of freight on corn to Greenville. All the calculations were made for its transportation the other way." Yet, as Perry went on to

TABLE 5

Items for Which Freight Rates Were Established on the Southside Railroad, 1854

Artificial manures	Bellows	Bricks	Brooms	Bacon
Allspice, ginger, and pepper	Beef and pork	Bran	Barrels	Buggies
Boots and shoes	Bark	Boxes	Cheese	Candles
Chairs	Coal	Coffee	Copper	Carriages
Champagne	Cotton	Crates	Crockery	Cattle
Confectionaries in boxes	Carts	Cement	Corn	Demijohns
Dry goods in boxes	Fish	Flour	Fruit trees	Figs
Furniture in boxes	Feathers	Grain	Grind stones	Gun powder
Guano	Glass	Hay and oats	Hardware	Hives
Hats, caps, and bonnets	Hogsheads	Iron	Lead	Live stock
Horses and mules	Leather	Lumber	Lime	Liquors
Licorice	Molasses	Nuts	Nails	Oil
Oranges	Potatoes	Pianos	Ploughs	Rice
Porter and ale	Sitters	Sieves	Soap	Salt
Slate	Saddletrees	Shot	Sugar	Shoes
Shovels and spades	Shingles	Safes	Sulkies	Tobacco
Staves and hoop poles	Turpentine	Tin	Waggons	Wood
Bark, pitch, and rosin	Tubs in nests	Wheat	Plaster	

Source: Entry for April 19, 1854, Southside Minutebook-2.

point out, the corn shipped from Columbia to the South Carolina upcountry went on to be a major portion of the road's business. Previously, Perry noted, people felt that because Greenville was "not a cotton-growing District, we had no use for a railroad, and could not sustain one." Time, however, had proved otherwise.[22] Farmers had a similar experience in eastern Tennessee. A railroad connection to Chattanooga in 1851 did not automatically spark a transition in production patterns to growing commercial crops in eastern Tennessee; the railroad came nonetheless.[23]

To be sure, there were times when railroads had to transport empty cars—particularly understandable during the months when cotton was taken to market, creating a substantial rush in one direction. In December 1837 the SCRR reported, "For the last three months we have been obliged to send up empty cars to bring down cotton, there not being up freight enough to load them all." In November 1839 the LCCRR was sending about an extra fifty cars per week up the line to handle the additional downfreight. But if the up and down freight were not always equivalent in terms of tons, it did not mean that noncotton freight did not contribute to the company's bottom line.[24]

A more detailed analysis of the income received by the railroads from the different types of freight reveals that freight to the interior region served by the railroad was an important component of these railroads' incomes. Unfortunately, the lack of standardized or complete reporting of statistics from antebellum companies makes it difficult to make sweeping conclusions. Some railroads simply gave yearly aggregate revenue from freight and did not break it down by month. Others gave freight handled by station but did not indicate in which direction the freight

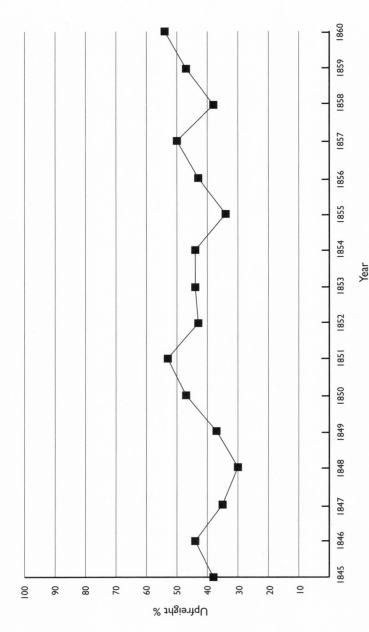

Figure 8. Percentage of freight income received from upfreight on the Central Railroad of Georgia, 1845–60. The 1845 value is for December only. The 1860 value is for January–November only.

Source: Reports of the Presidents, Engineers-in-Chief and Superintendents, of the Central Rail-Road and Banking Company of Georgia, from No. 1 to 19 Inclusive, with Report of Survey by Alfred Cruger, and the Charter of the Company (Savannah, Ga.: John M. Cooper, 1854), and *Reports of the Presidents and Superintendents of the Central Railroad and Banking Co. of Georgia, from No. 20 to 32 Inclusive, and the Amended Charter of the Company* (Savannah, Ga.: G. N. Nichols, 1868).

TABLE 6
Income from Upfreight on the Central Railroad of Georgia, 1845–1860

Year	Value	Year	Value	Year	Value
1845	$7,394.47	1851	$323,880.77	1857	$409,056.77
1846	$109,097.51	1852	$332,040.47	1858	$449,675.63
1847	$143,920.97	1853	$308,989.95	1859	$626,764.50
1848	$142,028.69	1854	$333,801.40	1860	$637,186.52
1849	$211,750.49	1855	$425,284.84		
1850	$269,199.34	1856	$454,622.52		

Source: Reports of the Presidents, Engineers-in-Chief and Superintendents, of the Central Rail-Road and Banking Company of Georgia, from No. 1 to 19 Inclusive, with Report of Survey by Alfred Cruger, and the Charter of the Company (Savannah, Ga.: John M. Cooper, 1854), and *Reports of the Presidents and Superintendents of the Central Railroad and Banking Co. of Georgia, from No. 20 to 32 Inclusive, and the Amended Charter of the Company* (Savannah, Ga.: G. N. Nichols, 1868).
Note: The 1845 value is for December only. The 1860 value is for January–November only.

was moving. Some railroads, however, separated the revenue that they received from freight traveling each direction, and the following analysis is based on these companies.

The CRRG conforms to expectations framed by Phillips. From 1845 to 1860, upfreight (i.e., freight traveling toward the interior) constituted 50 percent or more of the company's freight income on only three occasions: 1851, 1857, and 1860 (figure 8). Every other year, the amount of revenue derived from upfreight hovered between 30 and 50 percent. Yet, as the antebellum era progressed, the absolute value of the upfreight increased dramatically, with some hiccups along the way (table 6). Starting at $109,097.51 in 1846, it more than doubled by 1850, and doubled again by 1859. While freight headed east was clearly the most important for this railroad, the company also received ample business in the other direction.

Other roads were more successful with bidirectional traffic. The SCRR's downfreight was freight brought from Hamburg (and later Camden and Columbia) to Charleston; this is the direction that cotton moved. Upfreight left Charleston toward those destinations. The figures contained in the annual reports suggest that upfreight was a healthy component of the company's bottom line. Figure 9 shows the income from upfreight as a percentage of the total income received from freight each year. This amount never dipped below 40 percent, and indeed it fell to less than 50 percent only six times between 1834 and 1857. But even as the percentages dropped, as they did in the early 1850s, the total value of upfreight continued to climb throughout the period, demonstrated by the figures in table 7. The value of upfreight, after hovering around $100,000 every year from 1836 to 1841, consistently increased from 1844 to 1857. Upfreight increased rapidly before plateauing again in the late 1850s: it reached $200,000 in 1847, $300,000 in 1850, and $400,000 just two years later. As the fortunes of the cotton economy improved, the two-way business of the SCRR improved as well. For the SCRR, noncotton in-

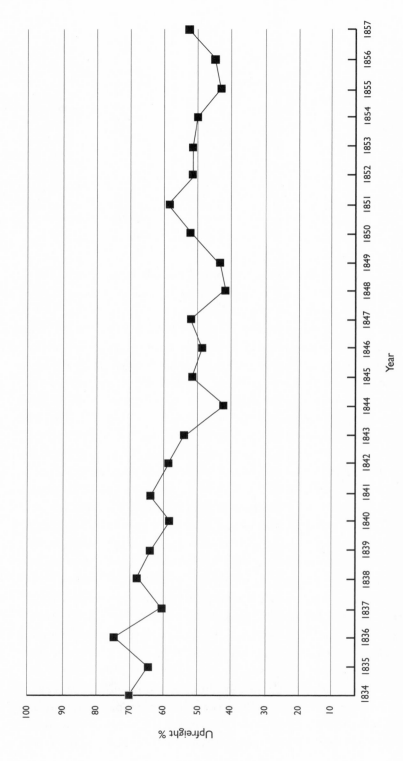

Figure 9. Percentage of freight income received from upfreight on the South Carolina Railroad, 1834–57. The 1834 value is for May–October only. The 1835 value is for July–December only. The 1843 value is for January–June only.

Source: Annual reports of the South Carolina Railroad and its predecessor companies.

TABLE 7
Income from Upfreight on the South Carolina Railroad, 1834–1857

Year	Value	Year	Value	Year	Value
1834	$33,159.20	1842	$114,100.73	1850	$305,611.95
1835	$53,683.90	1843	$49,012.93	1851	$384,872.06
1836	$101,334.96	1844	$157,386.74	1852	$401,026.57
1837	$84,957.67	1845	$168,475.10	1853	$414,932.67
1838	$111,026.72	1846	$172,290.84	1854	$444,796.89
1839	$129,776.41	1847	$201,481.14	1855	$466,428.46
1840	$109,019.98	1848	$217,071.54	1856	$470,114.42
1841	$101,443.29	1849	$268,483.02	1857	$471,739.42

Source: Annual reports of the South Carolina Railroad and its predecessor companies.
Note: The 1834 value is for May–October only. The 1835 value is for July–December only. The 1843 value is for January–June only.

come was a significant portion of the railroad's income, and noncotton income increased nearly every year.

The story was similar for other railroads. In the case of the Southwestern Railroad, the upfreight was that which contained agricultural production and the downfreight was the traffic from Savannah to the interior. In this case, the contribution of the noncotton traffic was not as great as it was on the SCRR but was still present throughout the period 1851–60 (the road began operation in August 1851). Downfreight constituted half or more of the railroad's income for half of the years (see figure 10). As table 8 shows, though, the downfreight increased year after year for this company as well. The CSCRR consistently shipped more upfreight (traveling north from Columbia) than freight in the other direction. Figures are available from end of 1851 through 1858, and income from upfreight dipped below 50 percent on only one occasion. The upfreight's year-over-year growth was not as consistent; income from upfreight climbed sharply until 1855 (see figure 11 and table 9).

Two final railroads offer incomplete figures but similar results. The Virginia and Tennessee Railroad sent goods west from Lynchburg toward Bristol. Its shipment figures (available for all of 1853 and 1854 and parts of 1852, 1858, and 1859) suggest a tight balance between freight traveling west and east. While it is difficult to draw conclusions about the amount of goods traveling west, the income received from western freight was higher at the close of the decade than it was at the beginning (see figure 12 and table 10). Only one year's worth of figures are available for the Greenville and Columbia Railroad, yet the monthly figures still demonstrate the large amount of upfreight (moving from Columbia to Greenville) handled by the railroad. This was the road on which Perry claimed upfreight had not been accounted for, and the figures bear out his analysis. In only two months from June 1854 to May 1855 did income from upfreight constitute less than 50 percent of the company's freight income (see figure 13). For southern railroads, freight in

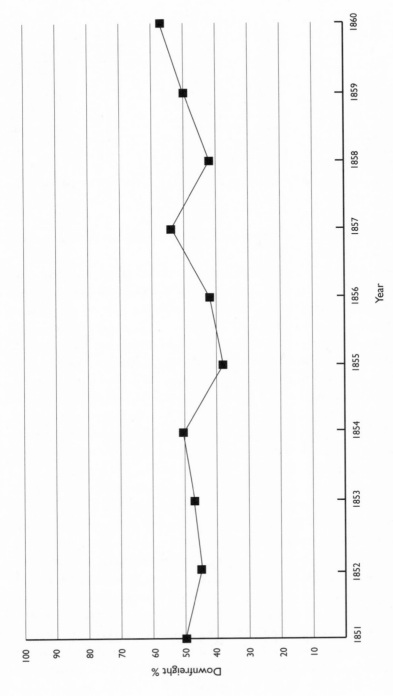

Figure 10. Percentage of freight income received from downfreight on the Southwestern Railroad, 1851–60. The 1851 value is for August–December only. The 1860 value is for January–July only.

Source: Report of the Chief Engineers, Presidents and Superintendents of the South-Western R. R. Co., of Georgia. from No. 1 to 22, Inclusive, with the Charter and Amendments Thereto (Macon, Ga.: J. W. Burke, 1869).

TABLE 8
Income from Downfreight on the Southwestern Railroad, 1851–1860

Year	Value	Year	Value	Year	Value
1851	$17,645.19	1855	$97,477.33	1859	$186,076.54
1852	$37,181.89	1856	$97,874.21	1860	$117,636.05
1853	$47,789.62	1857	$111,217.59		
1854	$68,096.25	1858	$132,668.23		

Source: *Report of the Chief Engineers, Presidents and Superintendents of the South-Western R. R. Co., of Georgia, from No. 1 to 22, Inclusive, with the Charter and Amendments Thereto* (Macon, Ga.: J. W. Burke, 1869).
Note: The 1851 value is for August–December only. The 1860 value is for January–July only.

both directions was of critical importance to the bottom line, even if cotton received most attention as the engine of the southern economy.

Conflict with Nature

Central to the pro-railroad argument was the notion that trains could overcome the reliance on the tides that trapped boats. Such freedom would theoretically allow merchants to take advantage of time savings. But the actual experience of operating railroads proved that the hopes of boosters were overly sanguine. Nature gives us another window to the railroad's time, demonstrating that some critical elements of operation remained beyond the control of the corporations. The railroad's time was challenged and compromised by both weather and disease.

Nature's time influenced railroads when companies confronted the business cycle, particularly in the case of cotton. Perversely, the advantage of the railroad over some parts of nature (it could be more regular than river travel) was matched by the degree to which the train was still subject to nature's forces elsewhere. Namely, because agricultural produce constituted a large portion of many companies' traffic, the growing season dictated the busiest season of the road. Although the year-round availability of the railroad allowed planters to withhold their product until the price was favorable, most cotton appears to have shipped between October and March.[25]

The SCRR attempted to anticipate this business cycle. In July 1835 the president recommended that the company stop running daily trips in order for "power on the Road to be preserved for Fall business as there was a scarcity of Workmen both there & at the North." Despite these attempts, the company reported that in 1835 it was able to adequately handle only the upfreight, not the cotton-laden downfreight. Other railroads felt similar pressure. In 1839 the LCCRR's president noted that the company was simply unable to accommodate the rush of business; as a result, cotton "accumulated at several stations." A few years later, the SCRR reported on its efforts to fully anticipate the waves of business that it could now reasonably expect,

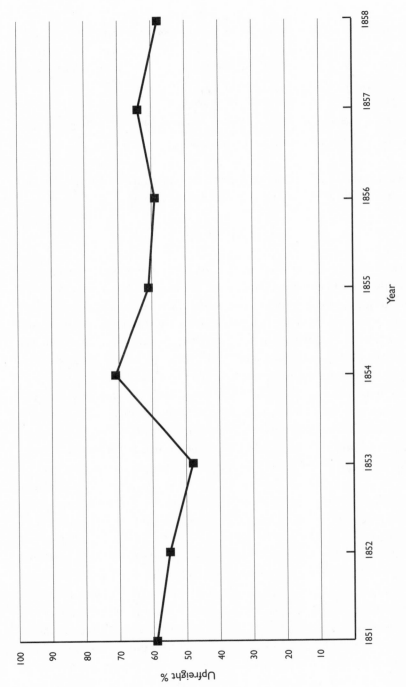

Figure 11. Percentage of freight income received from upfreight on the Charlotte and South Carolina Railroad, 1851–58. The 1851 value is for October–December only.

Source: Annual reports of the Charlotte and South Carolina Railroad.

TABLE 9
Income from Upfreight on the Charlotte and South Carolina Railroad, 1851–1858

Year	Value	Year	Value	Year	Value
1851	$11,487.89	1854	$184,782.39	1857	$147,750.80
1852	$50,159.45	1855	$195,867.35	1858	$156,785.54
1853	$60,038.20	1856	$156,262.28		

Source: Annual reports of the Charlotte and South Carolina Railroad.
Note: The 1851 value is for October–December only. The 1858 value is drawn from the eleventh annual report, p. 13; the column is erroneously labeled "1857."

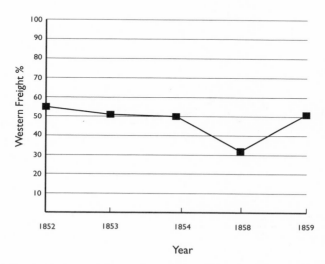

Figure 12. Percentage of freight income received from western freight on the Virginia and Tennessee Railroad, 1852–54, 1858–59. The 1852 value is for October–December only. The 1858 value is for July–December only. The 1859 value is for January–June only.
Source: Annual reports of the Virginia and Tennessee Railroad.

having operated for several years. The company noted that it had enough motive power and freight cars to "be equal to the transportation of 2,000 bales of cotton daily, with the corresponding quantity of up freight." If the business of the company proved greater than that, the company noted that the employees would do "night service, or double duty, if absolutely necessary" in order to make up the difference.[26]

Nature influenced railroads in other ways. Snow could block tracks, as traveler James Davidson found to his dismay in 1836. "I awoke this morning and found the ground covered with snow. Snow is unusual in this climate, and of course the rail road is unprepared for it. The men told [us] that we could not get on. But the Cars started and had not got a mile untill they had to return. They started again and run a little farther, but had to return again." They were finally successful on the follow-

TABLE 10

Income from Western Freight on the Virginia and Tennessee Railroad,
1852–1854, 1858–1859

Year	Value	Year	Value	Year	Value
1852	$6,356.43	1854	$34,859.15	1859	$66,498.50
1853	$38,836.92	1858	$63,452.25		

Source: Annual reports of the Virginia and Tennessee Railroad.
Note: The 1852 value is for October–December only. The 1858 value is for July–December only. The 1859 value is for January–June only.

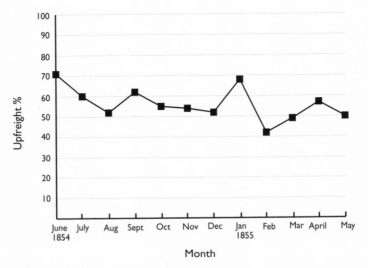

Figure 13. Percentage of freight income received from upfreight on the Greenville and Columbia Railroad, June 1854–May 1855. *Source: Minutes of Proceedings of the Stockholders of the Greenville and Columbia R. R. Co., at Their Annual Meeting, Held at Abbeville C. H., Wednesday and Thursday, the Eleventh and Twelfth of July, 1855* (Columbia, S.C.: R. W. Gibbs, 1855).

ing day. "Our third attempt upon the rail road proved successful. By the time we had dined at Aikin the snow had vanished, and we moved at a rate which set distance at defiance." Snow and freezing weather stopped freight trains on the SCRR for a few days around Christmas 1851. Snow was problematic elsewhere: major snows in 1857 prevented trains from reaching Staunton, Virginia. As a result, the town received no mail or information about the outside world for some time.[27]

The most consistent meteorological challenge, however, was from rain. Newspapers reported considerable damage after a "freshet" (or flood) of May 1840 in South Carolina. On May 27 a newspaper report from Hamburg recounted that the river was "so high as to be running over the rail road, and through all of our depositories. The goods from the stores at Hamburg floated up against the rail road." The report from the 28th noted, "The wheels of the Freight Cars appear to be under water as the Car stands on the track at the depot." An attempt to depart was unsuccessful: "We learn from the conductor of the cars, that an attempt was made on Thursday afternoon, to reach Hamburg with a hand car, propelled by manual labor, and succeeded in getting within a mile or two of that place, when the water was found to be too deep to progress any further, reaching the breast of the negroes, and they were therefore compelled to return to Aiken." No amount of promoter's rhetoric could overcome the reality of a flood.[28]

Fever and freshets damaged the SCRR again in 1852, with the effect of "cut[ting] off all communication entirely between Charleston and the interior." The damage caused by the 1852 freshet was still affecting the road the following year, and the company noted that the road near the Congaree River was "still measurably exposed to damage from the same cause." Bad weather could do serious damage to a railroad's reputation. The credit reporter for R. G. Dun complained about the Greenville and Columbia Railroad in 1855 that every "freshet washes them into a big debt." Such complaints were doubtless embarrassing to the corporations.[29]

In addition to creating problems for freight, weather could delay passenger travel. Robert Habersham reported in 1840 that while riding a railroad in North Carolina sleet forced the cars to stop for five hours on the track and wait for the weather to subside. When the train could finally move on, the passengers got off and waited in a hotel, "it being deemed almost madness to attempt to proceed." As traveler Alexander Mackay reported on one trip, "The bed of the railway resembled that of a canal, which had broken its banks a little beyond, and the water of which was rushing to escape and pour itself with desolating effect upon the adjacent fields." In addition to the rain, a small stream "poured its muddy contents in miniature cataracts. So deeply was the line submerged by this double visitation, that the axles of the wheels were covered, as the train slowly proceeded, groping its way, and following, at a safe distance, enormous pieces of loose timber which were floating before it along the rails." Despite the promises of conquering nature, the experience of Mackay and others demonstrates that the promises were not easily kept.[30]

The other natural threat to the railroad's work was disease, which remained a constant hazard, as it had been during construction. Disease could scare off workers and business alike. In 1839 SCRR president Thomas Tupper noted that while the fever was bad in Charleston during the autumn months, the fear-inducing rumors proved far more damaging. Trade with Augusta was damaged in September and October when the city "suspended" business altogether. Tupper noted that the cause of the sickness was a severe drought, which also dried up the wells used by the engines, and only by "the greatest exertion of nearly the whole force on the line" was the company able to deepen them. Yet, the drought had a positive side because it also damaged the trade on the SCRR's competitor, the Savannah River. Commenting on business in late November and early December 1839, Tupper noted that "the lowness of the Savannah river has given to this Company a heavier business, for the last three weeks, than it has ever before had."[31]

Disease did not just affect the amount of passenger traffic or the willingness of planters to send their freight south; it also damaged the ability of the company to run its trains. Disease laid low company employees, leaving corporations short-handed. This was the case in autumn of 1854, in which the service was so "irregu-lar" that the SCRR found itself besieged with complaints. At the annual meeting, the company reminded stockholders that the majority of their employees lived in Charleston and thus were more liable to succumb to the prevailing sickness. As a result, less experienced men had to take their place, and delays were the inevitable result. Even the end of the disease created more business than the road could handle. "The abatement of the fever," reported the company "was followed by a general rush of absentees homeward, from all quarters, causing such an increase in the travel, that to keep fully up to schedule time, was seldom achieved by any of the Roads, over which the streams passed. . . . At the same time, the Cotton crop came forward in greatly increased quantity, it being the earliest time that a large proportion of the factors could be in Charleston to receive and sell it." The passen-ger figures for October and November 1854 bear out this explanation: 9,270 people rode the SCRR in October 1854, and 16,057 passengers did the following month.[32]

The CRRG faced similar problems in 1842. As a report from the engineering department noted, "Every person attached to this department with a single excep-tion, has had an attack of fever, and several of the assistants are still scarcely able to perform their duties." In 1858 the same company rewarded employees for doing "their duty faithfully and courageously during the sickly season" and presented them with "a gratuity of 20 per centum on the aggregate of their pay for the two months of September and October."[33]

Nature and disease thus proved more difficult to overcome in practice than rail-road promoters had hoped in theory. The importance of agricultural production to these railroads meant that their business was partly keyed to when cotton was har-vested. Disease and drought could withhold business from a company and then unleash it again once the danger passed. Flooding and snow could also provide temporary setbacks or delays. Such realities further complicated the time-oriented promise of railroads. Schedules could be foiled when nature's floods destroyed rails, or when human floods descended upon railroads after a period of disease.

Accidents

If nature could delay travel, man-made obstacles and accidents also created problems. Accidents are a difficult matter on antebellum railroads. There was plenty of contemporary observation—particularly from foreign commentators—about how dangerous American railroads could be in the pre–Civil War era. Grade crossings were often not protected by gates, nor were railroads always effectively fenced in to prevent humans or animals from trespassing. Yet proper perspective should be maintained in assessing the accidents. Major, *deadly* accidents were largely unheard of before the 1850s. Smaller accidents certainly did occur, but on the whole—particularly given the lack of trackside signals—southern railroads had a remarkable safety record for the first two decades of their operation. Contemporaries agreed: "There were not a great many accidents in those days," former railroad operator N. J. Bell recalled of antebellum times, "not so many as one would suppose would be."[34]

Railroads anticipated the need to deal with accidents, as shown by the rules given to conductors and enginemen. In the event that an accident caused a train to block the track of the SCRR, for example, the conductor was to send at least one man a quarter of a mile in each direction from the accident in order to warn any approaching trains. Should such an accident occur at night, the men sent in each direction were to build a fire and wave torches in order to warn oncoming trains. When trains stopped for the night under normal circumstances, conductors had to ensure that the train was stopped at a turnout to prevent collisions and that a "strict watch" was kept over the train.[35]

One accident associated with the antebellum era is the "snakehead," the term for a piece of loose iron rail that would fly up piercing the underside of a car. Such events are prominent in railroad lore. Northerner Henry Whipple gave one such account when traveling in the South in 1844: "The passengers are amused on this road by running off the track, sending rails up through the bottom of the cars and other amusements of the kind calculated to make one's hair stand on end." Whipple's tone paints southern railroads as dangerous compared to their northern counterparts. Yet, in the next sentence, in describing his own travels Whipple admits, "We only ran off the track once and that was in running backwards." Whipple terrifies his readers with the specter of the snakehead but is unable to come up with any explicit examples himself, nor did one occur on his journey. It is easier to find tales such as Whipple's than it is to find actual references to such an event. Indeed, one comprehensive survey of early railroad accidents uncovered only two documented snakeheads in the antebellum era—neither in the South. Given the prevalence of strap-iron construction and the lack of uniformity in reporting accidents, it is likely that it happened more than twice, but it is doubtful that passengers routinely feared rails bursting through the floor of cars.[36]

Accidents had diverse causes. In 1834 civil engineer Horatio Allen noted that the primary cause of accidents on the SCRR was axle breakage. In 1837 axles and wheels were reported as "frequently giving way" under freight cars, and the company responded by importing a large number of higher-quality replacements. Miscommunication between engineers could also lead to accidents. In 1840 SCRR engines collided on two occasions. In the first case the engines were damaged, and in the second the damage extended to the tenders and cars. Another accident illustrated the dangers of modern technology malfunctions. A boiler explosion on the *Reading* led to the death of the fireman, a free black named John Humphrey. The engine detached itself from the train and ran for a quarter of a mile before anyone was able to catch it. Weather could also contribute to serious damage. An accident occurred on the Camden Branch of the SCRR in 1847, when a dense fog prevented two trains from seeing each other. Trains also barreled ahead on the assumption that the road in front of them was safe, but such was not always the case. Virginia farmer Daniel Cobb recorded in his diary in 1859 that a railroad bridge burned down, unbeknownst to trains approaching from both directions. Only "by providence" did a passerby discover the danger and alert the trains by building large fires.[37]

The presence of work crews on tracks while the railroad was operating also led to accidents. In 1854 a work crew was sent to clean up slides on the Virginia and Tennessee Railroad. Although instructed by their foreman not to put dirt cars on the track, a gang of workers did so anyway. When a freight train approached, the workers "fixed" the cars "more securely in their position" instead of moving them out of the way. The engineer of the freight train attempted to stop, but the engine struck the rearmost dirt car. The engine, tender, conductor, and fireman (a free African American) were "thrown from the trestle work & down the high embankment" (the engineer escaped injury by jumping onto the first freight car). The fireman, Henry White, died the next day. None of the employees emerged well in the eyes of the company from this accident: the track workers should not have placed cars on the road, and the conductor had improperly strayed from the rear of the train. Had he been there he would have spared himself injury and more importantly would have been in position to ensure that the brakemen, who "fled from their breaks on seeing the danger" would have retained their positions.[38]

Another accident on the VTRR shows how the actions of employees could lead to or prevent damage. In June 1855 the passenger train headed by the engine *Henry Davis* ran over four head of cattle while heading toward a bridge. The collision threw the baggage car off of the track and tore up sixty feet of iron and cross ties, leaving the second-class passenger car sitting on the "string pieces of the bridge." The engine escaped relatively unscathed, and continued to carry the passengers, albeit in freight cars, about ninety minutes after the accident. The engineer, however, did not escape as easily. Engineer Daniel Jones, after striking the cattle, jumped from the engine before it reached the bridge. The fireman, a free African American named William Cotrell (also spelled Crotwell in these minutes) "remained at his post and stopped the engine after crossing the bridge & no doubt prevented it from receiving serious injury." As a result, Jones was fired the following day and Cotrell presented with fifteen dollars in appreciation for his service. Clearly, while serving as a fireman he had observed enough to know—or had been taught—how to stop the engine. He also showed a bravery that Jones did not and was accordingly rewarded by the company.[39]

Antebellum accidents had multiple causes: equipment failure, miscommunication, weather, direct disobedience of instruction. Although companies reasonably wanted to find out what caused accidents and how to assign blame, investigations did not always guarantee easy answers. In an investigation undertaken by the RDRR after an accident on September 15, 1858, the company appointed a special committee, whose report detailed the multiple factors at play in the accident. The morning of the 15th, the down passenger train left at the "usual hour," but engine problems forced the train to delay its journey at Amelia Court House for two hours and twenty minutes so it could be repaired. After repairs, the train continued on its way, although behind schedule, and met the up passenger train, without incident, at Chula. Near Tomahawk, it met an up freight train, which was bearing a flag signaling that an extra freight train was following it. Accordingly, the down passenger train sat on a siding for one hour and fifteen minutes, waiting for the extra train. The superintendent, who had gone on the regular up freight train to determine what was delaying the down passenger train and then remained with the down passenger train, gave a special order for the down passenger train to proceed. The down passenger train then collided with the extra freight train about one and a half miles south of Tomahawk. The committee noted that the rules were quite "explicit" about how a train must "keep out of the way" when it was as far behind time as the down passenger train was. Therefore, the question hung on whether the superintendent was "under the circumstances justified in suspending for the time, the rules and regulations of running."

The committee then laid out what it felt to be the "circumstances" by which it could make that judgment. First, the extra up freight train had itself been delayed by engine problems, which it took about an hour to repair. Second, the passengers on the down passenger train had "been detained 5 hours and upwards without supper or breakfast, and were exceedingly clamorous" for their train to continue. Third, the superintendent was concerned about the condition of the extra up freight train and wanted to determine why it had become so separated from its flag train—the train bearing the signal that an extra train was following it. Therefore, as he reported to the committee after the accident, he determined that the passenger train should proceed at the rate of about six or seven miles per hour and would "flag all the curves" to ensure a proper lookout, and he stationed himself on the front of the passenger engine as a lookout, where he remained "until the trains were within 30 or 40 feet of each other." The fourth circumstance was the nature of the track where the trains collided. Although the trains could have perhaps avoided collision after they spotted each other, the committee noted that the extra freight train was descending "out of the heaviest and longest grades on the road," and that the passenger train was also going down a slight grade.

The fifth circumstance was beyond the control of the employees operating the train. When the freight train was spotted, the engineer of the passenger train gave the signal for the brakes to be applied, and the conductor and brakeman did so. The conductor and brakeman also removed the brakes on signal so that the engineer could begin to reverse the train. However, unbeknownst to the brakeman, a passenger who understood the first signal applied the brakes in another car, doubtless believing that he was helping. Unfortunately, the passenger was not familiar with the "remove brakes" signal, and so when the engineer attempted to reverse the train, the passenger was still applying the brakes. As a result, the engineer was initially unable to reverse the train, and "when it did move back it would only move at a very slow rate, not greater . . . than about half the speed of a man walking."

Thus, the committee did not lay blame at the feet of the conductors, engineers, or the superintendent. Rather, outside circumstances seemed to rule this case: "If the trains had been approaching where the grades did not so materially aid the collision, or if the brakes had not been applied by the passenger, or been taken off at the signal, it appears that the collision would not probably have occurred, notwithstanding the violation of the rule which the Superintendent had himself committed. There is cause of congratulation that no life was lost and no person seriously hurt. . . . Your committee not only entirely acquit of all blame the conductors and engineers of both trains but have ample testimony that they together with the Sup[erintenden]t made every effort and exertion under the circumstances that could be expected of the best of officers."

The board of directors agreed for the most part with the committee's assessment, and passed resolutions to that effect. First, it reaffirmed the right of the superintendent to change the rules when necessary and likewise reaffirmed that that right was denied to all subordinate officers. Second, it resolved that, while the board was not going to fire the superintendent for his actions, it still deemed his decision "improvident." Third, it reaffirmed that "any such violation of the Rules by a subordinate officer" was a "flagrant violation of duty" and was "deserving instant removal from the office." The board reaffirmed the hierarchy of the road but also acknowledged the difficult circumstances under which trainmen operated. "Clamorous" passengers, timing the length of pauses when adjusting for late trains, the presence of one's superintendent on the train and giving orders: conductors and engineers had to process all of this information and still maintain safe travel.[40]

Although the data are not as complete as we might like, dangerous accidents increased in the 1850s. Before this increase, southern railroads announced their safety records with pride. In 1841 the SCRR noted that 273,362 passengers had passed over the railroad without a single death and with only one serious injury. Railroad employees had been less fortunate during the same time span: "10 or 12" had lost their lives, "most of these during the early and imperfect construction of the Road, and in some cases from the negligence of the sufferers themselves." The first accident fatal to passengers on the SCRR occurred in 1852; when an axle broke, a "German passenger" jumped from the second-class passenger car and died. While obviously any loss of life was tragic for the parties concerned, the track record of the railroad in its first years of operation remained remarkable, given the newness of the enterprise and the fact that railroad safety depended almost entirely upon the accurate timing of trains.[41]

Other railroads also prided themselves on lengthy records of safe travel. In 1860 the Southwestern Railroad reported that since July 1, 1851, 705,320 passengers had traveled on the road, "of which there has been but one killed, and he lost his life by imprudently—under excitement of the moment—jumping off the train when a collision was about to take place; had he kept his seat in the car he would have received no injury, not one of the all others having received a serious injury." The CRRG was pleased with its safety record as well, reporting that for "a period of over ten years, no passenger has lost life or received serious injury." The East Tennessee and Georgia Railroad noted in 1860 that only one passenger had been injured in the previous nine years of operation.[42]

How did passengers respond to these accidents? Joseph Wharton reported back North that he found southern railroads to be safe: "The stories about the hardships and dangers of travelling in the South are humbugs. The rail roads and Steamers are good and safe nearly all the locomotives are built in Philadelphia the cars in Connecticut and the steam boats in New York."[43] Blair Bolling suffered some accidents while traveling in 1838. He was delayed twice, yet "nothing particularly attracted my observation this day," and he remained in wonder at the "magnitude of the work." Bolling seemed to accept that some level of accidents and delay were a part of traveling. This does not mean that travelers were not frustrated by delays—or that they were reckless—but that they saw minor accidents as an acceptable price to pay for the convenience of railroad travel.[44]

One reason for this attitude is that railroads did not introduce accidents into American life; before the railroad, accidents had struck steamboats and carriages. When travelers mentioned railroad accidents, the primary complaint was about time: the accidents caused delays. G. J. Kollock described his 1851 trip as mostly pleasant, despite an accident. "We had a very comfortable trip from Mrs. Nash's to Athens; & reached the cars in time to eat our lunch before starting. . . . We were detained by meeting an Engine off the track 24 miles from Macon, having run over a cow—no damage done however, excepting the detention." Kollock appears to have had mostly a good trip—the weather was "cool" and the cars were "comfortable"—but the only complaint about the accident was "detention."[45]

Indeed, travelers took most minor accidents in stride, seeing them as a necessary accommodation to the modern convenience of railroads. James Dobbin's train ran off the track in 1847, and he complained about the delay of two hours, but noted that "no injury was sustained." Jeremiah Harris also acknowledged an accident, but this did not prompt a major complaint: "The train from Richmond was delayed some two hours, in consequence of an accident to one of the coaches; so it was late before we set off for our several homes." The derailment of Stephen Douglas's train in 1847 definitely inconvenienced him, because the derailment destroyed the baggage car, and passengers had to wait overnight for a new car. This meant that Douglas missed his connection, but he wrote a friend that he had "a jolly set of fellows in company, and besides had a good appetite for breakfast when we got here today. . . . I shall sleep well tonight & be ready for an early start in the morning."[46]

Benjamin Babb found his trip from Virginia to Alabama in 1844 greatly delayed by breaking down on all stages of his journey: coach, steamboat, and railroad. Yet, his main complaint was over the loss of time, not that any of the forms of transport were particularly deadly: "I met with a great many accidents on my way owing to the railroads breaking down and also the stages and steam boats but all without any damage to any of the passengers only detaining them on their journey. I was 12 days on the road when I ought to have performed[?] the journey in 9. I was three nights in succession without any sleep." Though uncomfortable and frustrating, it was not enough to scare him away from future travel. Carolina Seabury was once on a derailed train in North Carolina that struck and killed three mules. As she noted, the animals "were sleeping too soundly to 'look out for the engine when the bell rang,'" quoting the warning that humans heeded when near train tracks. Seabury was grateful, however, that the "slow pace" of the train had prevented further damage. The passengers walked to the next depot while the train was being righted.[47]

To be sure, some passengers were quite angry over lost time, and some attempted to recoup losses. When passenger W. F. Davis failed to make a connection on a passenger train in 1853, he filed a claim against the RDRR. A freight train's accident had prevented his passenger train from getting him to his destination on time. He demanded a refund for his ticket as well as $16.12 for "expenses alledged to have been incurred by him in consequence of the detention." The board of directors refunded the $5.00 for his ticket, but declined to pay Davis's additional expenses.[48]

Southerners were well aware of the railroad's dangers. Contemplating the railroad journey that would take her home to the South from her northern school, Jennie Speer noted, "Railroad and steamboat disasters come to us in almost every paper. You have no doubt seen in the [papers] the account of the terrible accident at Norwalk, Conn., and that was only a part of the many heart-rending scenes which have of late been acted." However, such stories did not prevent Speer from continuing to travel by train, for she believed that "the same strong arm that protected me here can take me safely home." Moreover, passengers recognized when individuals were to blame for injury. T. Campbell Girardeau wrote to his mother in 1859 about a railroad conductor in South Carolina who had broken his leg. But Girardeau reassured his mother about the railroad's safety, noting that he "heard it must have been nothing but imprudence on his part in attempting to get off while the train was in motion." The individual—not the company or mode of transit—was responsible for injury or the loss of life.[49]

By and large, by 1860 travelers appeared to have been inured to the minor breakdowns that characterized railroad travel. Traveling in 1860, Samuel Burges wrote in his journal, "Engine broke down. we changed with them, which delayed our arrival in Ch[arlesto]n till 11:20 whereas we were due at 10:30 P M." Burges was simply stating the facts. The problems he encountered were not worthy of lengthy hand-wringing or future mistrust of the railroad. The southern response was a forthright response to the realities of mid-nineteenth-century travel. After journeying through Connecticut on the railroad in November 1856, northerner Anna Marie Resseguie noted, "No accident occurred on our journey worth noticing. The cars soon after we left N[ew]. London went off the track . . . but the train was immediately stopped, another car substituted and nothing harmed." Rather than protesting or abandoning railroad travel altogether, southern travelers, like their northern counterparts, learned to adapt to the demands of modern traveling.[50]

Despite the standard historiographic image that portrays southern railroads as simply carrying cotton to market, railroad operations were in fact far more complex.

Railroads developed regulations and facilities to handle a wide variety of freight, and southern consumers saw the difference when they visited their local stores. While southern railroads may not have had the same robust two-way trade that northern railroads did, an examination of railroads' income demonstrates that southern railroads still counted on noncotton traffic to provide a meaningful part of their business. Railroad operation also brought surprises and challenges to those working on the railroad. Freight handlers soon discovered that trains needed to be loaded and unloaded regardless of the time of day. A flash flood could drown out the voices of railroad boosters, and with them any hope of punctual service. And accidents— large and small—could prevent railroads from keeping their schedule. But southern travelers learned to accommodate the accidents that sometimes delayed their journeys. This was easy to do, given the dearth of major, deadly accidents in the antebellum era. While travelers may have complained about delay—hardly different from travelers today—they also recognized the advantages railroads brought them.

|||

CHAPTER SIX

Passages

It can be difficult for the modern reader to understand the appeal of railroads to the antebellum traveler. Living among jet planes that promise speed and highways that promise independence, a steam train puffing along at twenty miles per hour and only at certain scheduled times seems unbearable. When reading accounts of railroad travel in the antebellum era, however, it is crucial that we remember the vantage point of the traveler. Overemphasizing the discomfort that travelers experienced on the railroad overlooks the fact that trains presented enormous advantages over other types of travel. Rather than consigning railroads to failure because their speeds do not compare well with what we are accustomed to in the twenty-first century, we can grasp why southerners used them so eagerly by taking a closer examination of the passenger experience. Moreover, a detailed look at individual travel accounts makes it clear that traveling was a rich sensory experience. Travelers wanted a convenient route to their destination, but they experienced and noticed much more than that.[1]

Prerailroad Travel

Although slow to modern sensibilities, antebellum railroads compared favorably to the other types of transportation available to contemporaries: stages, boats, horse riding, and walking. Those who lived through the transition realized the magnitude of the change. Kentuckian Thomas W. Parsons wrote at the close of the nineteenth century: "Now that we have turnpikes and Rail Roads and all of the facilities for travel are so increased, we look back with curiosity at thoughts of the rude pack saddle." Contemporaries also recognized the discomforts of the other types of travel available. M. F. Kollock described one stage trip in the North in 1826 as "one of the most uncomfortable journeys I have ever taken." The journey began at 2:00 a.m. and after traveling only two miles, the driver was lost. The driver groped in the darkness "with his feet and hands" to recover the trail. After going another mile, "he suddenly stopped again, at a place which even appeared worse than the former." When the situation appeared that it could not deteriorate any more, the driver stated that "he was totally unacquainted with the road, as he had not travelled it for several years." Eventually the correct road was found, but as the writer complained, "It had taken us *two hours* to travel *six miles*." Early steamboating also presented its share of challenges: "We left West Point at eight in the evening, in the rain," Kollock reported on another journey in 1829, "and were obliged to get into a small boat, (which was *pretty* well *packed*) in order to reach the Steam Boat, where we found *only 500 persons;* of course no births were to be had and there was no alternative but to sit up all night, accordingly we wrapped ourselves in our cloaks and took our seats on deck where we remained until two in the morning when we landed at New York, and immediately proceeded to the Hotel." When railroad promoters lauded the advantages of railroads over boats, exasperated passengers like Kollock were a receptive audience.[2]

Antebellum travelers could also strike out on their own, either on horseback or on foot. In 1841, for example, Parmenas Turnley walked from Tennessee to New York. The first day was "a pretty hard day's walk, as the sun was intensely hot." After they second day of walking he stopped at a hotel and found his feet to be "much blistered, for I had unfortunately started out with a new pair of shoes instead of wearing my old ones." Turnley did not relish walking after that point, "but my limited cash was a fact I could not ignore." In all, Turnley walked 363 miles in twelve days.[3]

In this context of uncomfortable stage and boat or arduous walking, it is not surprising that some people took to railroad traveling rather quickly. Anna Calhoun Clemson wrote to a friend in 1838 that she was planning a trip from Charleston to Washington, D.C., and expressed relief that the steamboats and railroads would allow her to do "very little stage travelling which is a great advantage, at this season of the year." She described the proposed route as an "excellent one" and believed that the journey would be "both safe and pleasant." James Davidson was also glad to see his stage traveling come to an end. "My stage travelling has terminated here for the present," he wrote when arriving in Augusta, Georgia, in 1836. "I will now take the rail road for Charleston South Carolina. The change will no doubt be an agreeable one."[4]

Railroads were more dependent on freight than passengers for income, but southern railroads never lacked for passengers. From 1834 to 1857, the SCRR experienced a steady increase in the number of passengers it served. Although only six months of figures are available for the early years of operation, thousands of passengers flocked to the railroad as soon as it was introduced (see table 11). And "the facilities of travel beget travel," claimed the president of the Cheraw and Darlington Railroad: "A few months ago a two horse-hack was found sufficient to con-

TABLE II

Passengers Served by the South Carolina Railroad, 1834–1857

Year	Passengers	Year	Passengers	Year	Passengers
1834	13,575	1842	33,924	1850	117,351
1835	15,109	1843	17,370	1851	128,590
1836	39,216	1844	54,146	1852	131,286
1837	44,161	1845	56,785	1853	141,083
1838	44,487	1846	64,036	1854	156,571
1839	37,283	1847	77,429	1855	152,019
1840	29,279	1848	75,149	1856	160,662
1841	35,141	1849	92,713	1857	171,554

Source: Annual reports of the South Carolina Railroad and its predecessor companies.
Note: The 1834 value is for May–October only. The 1835 value is for July–December only. The 1843 value is for January–June only.

vey to and from Cheraw all the travellers who offered themselves on our line of travel. During the past year 9,447 persons have passed over your road." The ability to travel encouraged more people to take part.[5]

New Sensations

As a wholly new mode of transport, the railroad elicited no shortage of commentary from those who rode for the first time. Their rich descriptions of train travel allow us to understand how contemporaries understood the changes around them. The first trip was often worthy of comment. An anonymous contributor to the *Charleston Courier* reported on an early ride taken on the SCRR:

> On Saturday last [January 1, 1831], in company with about *one hundred and forty persons,* we enjoyed the pleasure of a ride on our Rail Road, in cars drawn by the Locomotive Steam Carriage. We say the pleasure of the ride, because it was truly delightful. Although hundreds of our citizens have enjoyed such a ride, there are yet very many who have not done so—to them we would recommend a trip to the six mile house; and if there be any who doubt the advantages of Rail Roads, or still question the utility of applying steam power to carriages, to these we particularly recommend the "Charleston's Best Friend," (such is the significant name of the steam carriage) and we doubt not the "Best Friend" will find a *friend* in every visiter.[6]

As this correspondent's writing makes clear, a trip on the train did not have to be long to make an impact—only six miles, the writer felt, would be enough to convince anyone of the train's utility. Other early riders in South Carolina were also impressed. "Tis quite [a] sensation," wrote a northern minister in 1832, "to be flying over [the] ground at [the] rate of 20 miles an hour without [the] aid of animal or water power. A novel but rather pleasant mode of travelling, had agreeable company." Four years later, Anne Carolina Lesesne recorded that when a friend rode for the first time, "she was highly pleased with it." That same year, James Davidson wrote approvingly of railroad travel after a long ride: "Rail roads are a delightful mode of travelling. It is Whiz—Whiz—Whiz—as the mile Posts recede behind you in rapid succession. . . . Rail roads will in time supersed stages entirely." Davidson's appreciation for railroads did not prevent him from being critical of the SCRR, however: "This railroad is a flimsy affair, and not well managed." But if Davidson did not care for management, he still welcomed the new form of transportation.[7]

Railroad travel was a complete sensory experience. The stirring sight of the engine, the shrill blast of the whistle, the uncomfortable feel of the seat: nothing went unnoticed by antebellum travelers as they embraced this new mode of travel. By exploring the sensory impact of train travel, not only can we better understand how southerners themselves perceived train travel, but we can begin to understand the way in which southerners apprehended modernity.[8]

One of the most commented-on features of the railroad was its arresting visual appearance. The railroad's linear trajectory gave it the appearance of driving resolutely to its destination—a modern juggernaut that could not be stopped. But railroad descriptions could also take on a darker character. Engines "snorted" or "breathed fire," suggesting that this purveyor of progress was actually something hideous. Historians should pay attention to these darker images but be wary of overemphasizing their importance. Americans embraced railroads despite the ambivalence inherent in these descriptions.[9]

Some found the appearance of the train to be truly frightening. Maria Glass commented in 1854 that on one train ride "the curves in the winding of the road, [were] so great, that as you sat, you could see the great puffing, and smoking monster, dragging us on." Former slave Ann May recalled that when she first saw a train in 1858 she "ran and it was a long time before they caught me. I sure was scared." South Carolina slave Al Rosboro remembered that the first engine passed his plantation "puffin' and tootin', lak to scare 'most everybody to death." Sallie Layton Keenan reported that although fellow slaves on her Mississippi plantation were overjoyed about returning to South Carolina, some balked at the prospect of taking the train: "Some o' de ole niggers lowed dat dey wuz feered to ride on dem things, bein as dey was drawed by fire. Dey thought de debbil, he wuz a workin' in de inside of dem." These objections fell away when the plantation mistress noted that anyone who did not want to ride the train could remain in Mississippi. This last story illustrates how easily negative imagery was overcome. Trains may have been driven by the devil, but when faced with the prospect of staying in Mississippi, these slaves were willing to accept the devil they did not know over the one that they did.[10]

Aside from the imagery of the train itself, vision played an important role in other aspects of travel. Travelers often commented on what they saw in the cars. Foreigner Alexander Mackay noted while a "New York railway carriage is a clean affair," in the South, spitting was more prevalent: "The floor is regularly incrusted with its daily succession of abominable deposits; so much so, that one might almost smoke a pipe from its scrapings. It too frequently happens, also, that the seats, the sides of the car, the window hangings, where there are any, and sometimes the windows themselves, are stained with this pestiferous decoction." While the description makes clear that riding on the railroad was not entirely pleasant, it also shows that travelers let little escape their notice.[11]

Traveling allowed passengers to view the world from the convenience of the window. Southern society and landscape both drew the attention of travelers. Many commented on the slaves that they saw en route. Traveling through Virginia, Carolina Seabury noted that "the fences were mounted by smaller woolly heads — with every shade of complexion in their faces, all pictures of happiness, looking as though not a thought or care ever disturbed the even tenor of their way — either with large or small specimens."[12] Lucy Carpenter noticed slavery when traveling in South Carolina as well: "We then passed cotton plantations and villages of Negro houses which looked very comfortable even pleasant with there blowing fires of light wood." Frederick Law Olmsted reported that slaves "almost invariably stop their work while the train is passing."[13]

Travelers also commented on the land, although it was difficult for travelers to assess the landscape when moving at speed. Louis Heuser found that when traveling in Virginia "attractive farms passed on both sides, but we couldn't see much of them on account of the speed." Most other travelers were not impressed with what little they saw. Carolina Seabury found the trip from Acquia Creek to Richmond to be characterized by "miserable sandy country," and the land south of Richmond offered worn-out plantations. Northerner Henry Whipple described Georgia as "one continual succession of swamps and pine barrens, as desolate a looking country as one could wish to see." Foreigner Charles Lyell was more interested in the SCRR than the land it passed through. During his journey on the "excellent railway" he "scarcely saw by the way any town or village, or even a clearing, nor any human habitation except the station houses." Despite the emptiness, Lyell recorded that "the spirit of enterprise displayed in such public works filled me with astonishment which increased the farther I went South." Poor surroundings did not overtake his satisfaction of the train.[14]

The "scenery" could also include fellow passengers in the cars. When a spark from the engine landed in the hair of one of Anna Calhoun Clemson's traveling companions, she noted that she and the rest of the group found his "contortions" rather amusing. Packed into railroad cars, passengers had the opportunity to pick up gossip and rumors. John Herritage Bryan gathered some political gossip in 1840: "I heard by a passenger in the Cars this afternoon that Maine had gone *against* the adm[inistratio]n if so their case is more hopeless than I had supposed," he wrote to a friend. The train could also be a place to unexpectedly meet friends. In 1855 Mary Boyce recorded in her diary that she "met an old acquaintance on the cars." Jeremiah Harris enjoyed the company: "The cars from Charlottesville, to the number of three coaches, were crowded with passengers, and among the number I was pleased to meet and exchange a little conversation with parson Cooke of Hanover, and a Miss Tisdale, a Northern lady, who was on her way home, from Mr. B. Winston's having completed, at his house, in Louisa, a term of tuition service." Trains brought strangers and friends alike together in their cars, allowing passengers to pass time in pleasant conversation.[15]

Travelers took the opportunity to catch up on their reading while taking a train.[16] One correspondent to the *Southern Literary Messenger* noted that, "as the train went whirling through the swamps at the rate of thirty-five miles an hour, we regaled ourselves with the tropical esculent, and read in the diverting pages of Pisistratus Caxton, how Riccabocca espoused Miss Jeminma, and how the Squire made a peace-offering of the stocks."[17] If passengers were uninterested in southern scenery, they could exercise their imaginations through reading. Railroads even accommodated those who left reading material at home by allowing special agents to sell it on the trains. In 1856, for example, the Greenville and Columbia Railroad contracted with W. W. Van Ness, who was granted the "exclusive privilege" of selling "Books, magazines, Periodicals and newspapers of a moral, Religious and useful character," in passenger cars and depots. Each train coming and going would carry one newsboy for this purpose. For the privilege of selling books, Van Ness paid the company $150 per month, agreed that all of his employees would be "polite and attentive to the passengers" and that the newsboys were "not to carry for sale or distribution any work of obscene or improper character."[18]

Visually, then, railroads made an impact on southerners—whether it was through their striking appearance, the way in which travel reduced the landscape or framed scenes of slavery, or the presence of amusing fellow passengers. Historiographically, visual impact has been closely linked with modernity. For theorists such as Marshall McLuhan and Walter Ong, modernity brought a transition from an emphasis on the oral/aural in premodern times to—with the invention of the printing press and the growth of literacy—an emphasis on the eye. Yet close examination of the experience of southern travelers demonstrates that travelers relied on a whole range of senses in order to describe their travel experience. Train travel was eye-catching, but it did not represent the eye's triumph over the putatively "premodern" senses. Other senses also perked up when trains arrived on the southern landscape.[19]

The sound of the train was a new sensation for inhabitants of the South and formed an important component of their interactions with and memories of the railroad. When Samuel McGill went with his father to sell cotton in Charleston, they camped about ten miles away from the city. "Here we heard, for the first time in our lives," he remembered, "the roar and rattle of the railroad cars, which a mile or two away resounded through the woods like a thunderstorm. Jumping to our feet we listened with some fear, as the horses pricked their ears and became somewhat fractious." McGill's reminiscence demonstrates the power that the railroad's sound had, as the first-time listeners leapt up. So new was the experience that McGill's only frame of reference was nature: the "roar" like that of an animal, or comparing the effect to a "thunderstorm."[20]

The sound of the railroad caused a range of emotions. For some, the experience was wholly negative. College professor Frederick Porcher hoped to take his mother on the railroad, but the night before the planned trip a train passed and his mother "was so very unpleasantly excited by the noise it made, that she sent for me and told me that she could not think of travelling in it. She had never, I believe, seen a railway train." Former slave Berry Smith also recalled being frightened by the sound: "De first train I ever seen was in Brandon. . . . Hit sho' was a fine lookin' engine. I was lookin' at it out de upstairs window an' when it whistled I would a jumped out de window, only Cap Harper grabbed me." But the sounds of the railroad attracted as well as repelled. When the first engine ran into Marietta, Georgia, in 1845, the engine whistle drew a crowd: "It looked like everybody in town ran for life," recalled future conductor N. J. Bell, "especially men and boys, to the spot where the engine would stop, myself with the rest." Marietta's horses had the opposite reaction, running in "every direction."[21]

Because sounds were used to warn passengers and employees about emergencies, passengers had to accustom themselves to the sounds if they wished to travel safely. Train employees communicated to each other and to passengers through a series of signals on the whistle (for engineers) and bell cord (for conductors). For example, conductors on the SCRR were to ensure that each train had a cord from which the conductor could ring a bell to alert the engineer of danger. This cord was to be reachable from any point on the train. One contemporary author described this practice as universal by the 1850s: "Every train in the United States can be stopped by the conductor, or even by any passenger who may be ill, by means of a cord carried along the roof of each carriage, and attached to the bell on the locomotive."[22]

Other aural cues were part of railroading, and passengers learned to understand what these whistles and signals meant. "About half an hour before leaving time the engineer would let off his whistle in a blow that lasted for fully a minute," recalled R. H. Whitaker. "That long blast was understood to mean that in thirty minutes the train would leave; so, if one had not yet gotten out of bed, he knew just how many minutes he could devote to his toilet, how many to his breakfast, and how much time would be left to reach the depot. One minute before the train left the whistle blew again, when the tardy fellow would strike a trot, and reach the train just as it began to move." The signals that Whitaker remembered had a long history. In 1835 the rules of the SCRR held that when the bell rang, passengers "will be allowed one minute to take their places." Sound signals enforced time discipline on the passengers.[23]

Railroads also used sound to communicate with other modes of transportation. Frederick Law Olmsted, for example, was traveling once when his train came to a stop because the rails ended (the track was still being constructed). He was informed by the conductor that the gap would be covered by a stage, "which would come as soon as possible after the driver knew that the train had arrived. To inform him of this, the locomotive screamed loud and long." At 3:30 a.m., Olmsted was "awakened again by the whistle of the locomotive, answering, I suppose, the horn of a stage-coach, which in a few minutes drove up, bringing a mail." Although it interrupted Olmsted's sleep, these signals helped coach and train coordinate their efforts to provide timely service.[24]

So unique and recognizable was the railroad whistle that it soon took on symbolic meaning well before the close of the antebellum era. Speaking at a meeting in favor of funding a railroad in South Carolina, J. D. Allen remarked, "When I hear the whistle of that engine, it appeals to my senses, like the voice of God, giving me encouragement and urging me by every tender time to my people, to increasing zeal and energy in its [i.e., the railroad's] support." Such symbolism percolated among other members of society. In 1852 Mississippi farmer Isham Howze wrote in his diary: "I can hear the steam whistle—the railroad cars are moving. What improvements in science and art!" When describing the changes that had come over Charleston since the beginning of the railroad, Charles Fraser noted that on King Street, the "hissing of steam" had replaced the "smack of the cracker's whip." For Jennie Speer, hearing the train's whistle reminded her that she would soon be able to return home. "I love to hear the car's whistle," she wrote to her mother. "I think they will whistle for me some day." For each of these people, the train's whistle held symbolic meaning: as a sign of the changes taking place in southern communities, or as a reminder that a trip home—and all of home's comforts—would soon be at hand.[25]

Railroad travelers also commented on tastes experienced while traveling. Some trips were long enough that railroads accommodated their passengers with meals. The taste of the food—or its blandness—aroused commentary from passengers. Advertisers tried to appeal to the taste buds of their potential customers. W. Wall declared in the *Charleston Courier* that he provided "substantial and savory Breakfasts to travellers." And some passengers found that such promises were not far from the mark. One traveler recounted that a stopping place on the Baltimore and Ohio Railroad in 1859 featured "a delicious Southern supper of fried chicken, corn bread, baked sweet potatoes, fresh biscuit, butter, honey, tea, and coffee." As delicious as the meal may have been, travelers were not allowed to relish their meals. On this particular journey, there was only twenty minutes to complete the meal, so "the conductor's cry of 'all aboard' made us drop the biscuit and honey and hurry to the train." Other travelers brought their own snacks to enjoy on trips. Lucy Carpenter declined to get off at a thirty-minute pause on her trip on the SCRR but treated herself to "an orange and some ginger cake." Clara von Gerstner brought food along on one of her journeys: "A big supply of oranges kept us refreshed, for the heat of the day grew ever more intense."[26]

Perhaps Carpenter and von Gerstner realized that food quality varied and thought it safer to provide their own. William Elliott commented that while traveling "in our southern thoroughfares . . . the fare is generally execrable," consisting largely of "fried bacon, fried lard—fried fat!" Carolina Seabury also found the food unappealing on a particular journey. After stopping at a "log house — in the woods" that served the meal, she was presented with "the most uninviting looking mixture of ham, eggs, chicken & corn bread — all but the bread swimming in gravy." Being informed that the meal was "our only chance for a mouthful until we reached Augusta" Seabury managed to "overlook appearances of food as well as waiters who evidently had been working in the field."[27]

Beyond vision, sound, and taste, the tactile sensations also mattered to travelers, as they noted the comfort of their travels, both in the train itself and at the necessary hotel stops. Anglo-American society had a firm culture of physical comfort that equated comfort with social progress by the time railroads were introduced. Travel involved a great deal of physical movement, of course: jostling in crowds on the platform, elbowing for room with an inconsiderate nearby passenger, sweat on hot days and chills on cold days, feeling wind and dust through the open window, and the rattling of cars down the track. The railroad car offered more protection from the elements than did walking or horse riding, but passengers could still find their bodies and insides jolted by a journey. Indeed, Frederick Douglass credited the physical turbulence of a rail journey with assisting his escape from slavery to freedom. Instead of purchasing a train ticket in the station, he "considered the jos-

tle of the train, and the natural haste of the conductor in a train crowded with passengers," and decided to leap aboard the train after it had started moving. His assumption that the bustle of the train would help his cause proved correct.[28]

Sometimes the bumpiness of a trip could produce amusement: Charles Hentz
commented on "a little boy with a handkerchief full of oranges and apples who
amused us very much, running after his fruit, as the rattling of the car caused them
to roll about the floor." On other occasions, physical proximity to other passengers
was an annoyance. Frederick Law Olmsted complained about one boorish passenger who, after a speech "in most violent, absurd, profane, and meaningless language," proceeded to claim a seat and "immediately lifted his feet upon the back of
the seat before him, resting them upon the shoulders of its occupant. This gentleman turning his head, he begged his pardon; but, hoping it would not occasion
him inconvenience, he said he would prefer to keep them there, and did so; soon
afterwards falling asleep."[29]

Descriptions of physical comfort during travel also included assessments of the
ability to catch sleep on long journeys. Antebellum travelers had certain expectations when they traveled, and the hotel accommodations they found did not always
measure up to this standard. Carolina Seabury reported that the hotel in Wilmington, North Carolina, at which she stayed was permeated with "a close indefinite
odor of all the dinners served up during the past summer." Moreover, sleep was
impossible because of the "permanent residents" of the beds. Hoteliers had an important advantage over travelers because they were often the only available option.
Albrect Koch found this much to his dismay in 1845 while traveling in the South.
At one point "a man of about fifty introduced himself as the innkeeper, but he had
more the appearance of a robber-chief than that of a hotel owner, as he called himself. Discouraged by that, we had not the slightest inclination to put up at his
house, but he described to us a great many advantages we would enjoy in his house
and made also the only-too-true remark that, first, it would be a long way to the
other hotel, and second, nobody was here who could take our heavy suitcases
there."[30]

Indeed, sometimes hotels were not available, and passengers had to sleep where
they could. Arriving late in Mississippi, Lydia Lane found that no rooms were available, and eventually "some men connected with the railroad took compassion on
the poor tired children, and let us go into a baggage-car, filled with mail-bags, over
which we spread some shawls, lay down, and slept soundly until the cars were
ready to leave next morning." While exhaustion may have produced sleep, it was
clearly not preferable to a comfortable bed. For other travelers, sitting for the
length of the journey was what prompted discomfort. At the end of a railroad journey, Ann Maury found that she would have to walk in order to complete her trip,

but "After sitting still in Railroad car & stage for nearly 12 hours I found the walk a relief rather than any additional fatigue." Mary Hering Middleton had a similar experience, noting that the "jolting" of the railroad "kept up my painful sensations—the stoppages were numberless & at length fortunately for *me* the Engine got out of the track, which caused so long a detention that the Cars arrived too late for the others which were to have carried us on at 2 o'clock, for we did not reach the starting point until 5 & then stopped at a house two miles from Weldon. The relief of lying down on a feather bed was great, & I prudently remained on it until the next day at 12."[31]

Sleep on the train itself was also problematic. Although Virginia Clay-Clopton clearly preferred travel in the time she wrote her memoir (1905) than that of a half century prior, she recalled that "Sleeping cars were not yet invented, but the double-action seatbacks of the regular coaches, not then, as now, screwed down inexorably, made it a simple matter to convert two seats into a kind of couch, on which, with the aid of a pillow, one managed very well to secure a half repose as the cars moved soberly along." John Shaw found sleeping on his train through North Carolina impossible because the car was far "too crowded." In Georgia, however, he was supplied with a railroad car that included a bed, much to his gratification. "These cars contained beds, where the weary traveller might rest his limbs and sleep as soundly as if he were on a bed of down at his own home. This is one of the many instances to be found in America, where something new and original will be constantly meeting the eye of the European traveller." Shaw wrote, "I found the bed to conduce much to the sleeping propensity, notwithstanding the constant shaking and agitation of the cars, so much so that I was enabled to pass a very good night, and on waking found that we had travelled 170 miles in twelve hours." Arthur Cunynghame was also "denied" sleep "by reason of the frequent visits of the conductors, who at each small stopping-place constantly made a request to see our tickets, and also in consequence of our being obliged to change four different times from one set of cars into others; once, indeed, during the night we had to leave the railroad and pass through a town in an omnibus."[32]

Of course, the visual, aural, and tactile sensations of travel were experienced together, not in isolation. Writers remarked on the senses in combination as well. Mary Moragne wrote about her first train trip in 1838. She commented on how the increasing speed of the train altered her vision: "At length my head grew dizzy in looking out upon the flying roads." Soon, the physical movement of the train combined with sights and sounds to produce a strong sensation: "The jarring of the wheels combined with the whizzing, & sputtering of the boiler, & the flapping of the curtains & loose sashes in the strong concussion of wind, caused a commotion, which gave me rather a disagreeable sensation of hurry, & alarm—, not that I felt at all frightened,— but it gives one the idea of danger." Moragne was quick to point out that she was not "frightened," but clearly the physical dislocation of the jostling train, the noise of the boiler and curtains, and the visual chaos of the "flying road" created a disconcerting experience. Moragne went on to talk about another visual dislocation—the "singular appearance to be travelling sometimes between two high banks, & the next moment looking down from a precipice on either side." While paused at a station, Moragne found the engine to be "a diabolical image of ugliness," an image facilitated by the "black smoke spouting from its chimney." The aural signal to board the train was met with another episode of physical jostling: "Every person goes in indiscriminately, & gets a seat as he can, & 'tis only after they are seated, that the owner finds how many passengers he has." Clearly, the experience left a deep impression on Moragne. But the dislocation was not limited to Moragne alone: another passenger, a young John Graham, "whose head had become so bewildered by the ride" spent the walk home "constantly running back in the greatest distress crying 'Pa that a'nt the way home'— the poor child was frightened out of his wits, & all that his father & Mother could do would'nt satisfy him that he was going home." The railroad, then, offered much for the senses of antebellum travelers, even if the unfortunate result for some was confusion, as in the case of poor John Graham.[33]

From the dramatic image of the steam engine to the rough accommodations in hotels, railroads excited the travelers' senses. The richness of sensory expression revealed in travelers' writings demonstrates that those who encountered trains had a complex sensory experience. This experience allows us to understand precisely how southerners and others confronted this modern marvel. Modernity here was not simply a visual phenomenon, as important as the engine's visual power was. Rather, multiple senses were animated by the railroad. Understanding the depth of these reactions allows historians to capture the full meaning of the experience and demonstrates that modern developments excited all of the senses.

Time

Railroad corporations placed a high value on time, but southern passengers were also well aware of time's importance. Indeed, southerners possessed a "railroad mentality" in advance of the railroad's arrival.[34] Thus, they were acutely aware of the advantages that railroads possessed in terms of saving time and were quick to be critical when the service did not measure up to par. Ample evidence that passengers were aware of the railroad's temporal advantages is provided by the references to time that litter their descriptions of travel. William Blanding's notes on an early trip on the SCRR are instructive: "Up at day light and at 20 minutes before 7 oC left Charleston by the Rail Road; we stopped twice for wood and three times for water say about 10 minutes each and twenty five minutes for Breakfast and arrived at Branchville 62 miles at 11 oC some short parts or distances of the way we travelled at the rate of 25 to 30 miles an hour, these were the strait parts of the way; we seemed to fly." Clearly armed with a watch, Blanding noted precisely the time of departure, length of travel, and the length of all breaks. Doing so allowed him to calculate the speed of the train, to his approval. Again, while thirty miles per hour may not seem impressive to a modern reader accustomed to much faster speeds, we must remember that given the alternatives Blanding had, thirty miles per hour was quite impressive. The train's speed also yielded unwitting advantages, because the route was not that attractive: "We hardly passed a Plantation all the way. I am told that the whole distance to Augusta is much the same."[35]

James Davidson also marveled at the speed of the train. In the first part of his 1836 journey he had been traveling largely by stage, and so he had ample basis for comparison. Although worried that he would be delayed because of bad weather in Aiken, South Carolina, once the train started, "we moved at a rate which set distance at defiance. If my great grand father had have been told that his great grand son, at this day, would dine at one in the evening and go to bed the same evening 120 miles distant, he would have called the Phrophet a fool." Davidson also came well prepared to judge the speed and efficiency of the train because he brought a watch. "I examined my watch several times," he reported. "The result of the first examination was a mile in 3 minutes—of the second, a mile in $2\frac{3}{4}$ minutes—of the third a mile in $2\frac{1}{2}$ minutes. Being fully satisfied with my speed, I ceased to examine further." As Davidson's story illustrates, antebellum southerners already had a well-developed time awareness and did not need railroads to instruct them in that regard.[36]

Travelers continued to record the times of their travel throughout the antebellum era, sometimes to the exclusion of other details about the trip. For some, time seems to have emerged as the only necessary marker to fully recall the trip's meaning. Mary Boyce's description of a trip in October 1855 demonstrates the recording of both natural and clock-oriented times: "Miss Boatwright, Mr & Mrs McMahan, Mr B, the children and I left this morning just at Sunrise to take the cars at Youngs Crossroads. We got here in good time, we arrived and [sic] Newberry about 10 oclock and had to wait for the other train until 12. . . . We arrived at Alston about 2 oclock and had to wait some time before we started on the Spartanburg train. We arrived at Waters about half past 3 oclock and found no one at home. We rested awhile and then got in a buggy and come over. We got here just at dark." Here, Boyce has mixed the clock times of railroad travel and the natural times of her journey ("just at dark") in order to fully describe her day.[37]

Southern passengers' awareness of time's importance was reinforced by railroads that demanded that the passengers be punctual to keep the trains running on time. Such rules and regulations were settled upon at an early date. In 1835 the SCRR resolved that "Conveniences be erected at the Stations that may be determined upon, [and] That a large Card stating that 20 minutes are allowed for Breakfast and 25 minutes for Dinner be placed in each Car." Passengers were aware that they had to meet the railroad's time requirements. Mary Moragne noted as much when she took her first train ride in 1838. "It was just three oclock, the other car was ready to be off— & soon as the passengers were seated, the bell was rung, & away we went, leaving the rest of our company behind, who, as they afterward informed us, arrived just five minutes too late!— just as well an hour!" The *amount* of time one was late, as Moragne realized, was irrelevant. Mary Boyce and her family made several attempts to take the train in July 1855; on the 10th they found themselves "disappointed by geting (sic) there too late." The following day, having learned their lesson, they went early to meet the train and waited for half an hour before it arrived. In this case, the railroad offered a lesson in punctuality, forcing the Boyce family to modify its behavior in order to travel. James Frederick noted that he arrived at the station one day in 1856 "just as the cars came up," demonstrating that he knew when he was supposed to arrive. But if trains were willing to leave passengers behind, passengers also demanded punctuality from the trains. Their dissatisfaction could lead to demands for reimbursement. In January 1837 the RFPRR paid out $64.73 to passengers for "delays in the arrival, departures & running of trains." Time was literally money to the RFPRR when it was forced to atone for its errors.[38]

Although passengers could be stranded if they were late, most considered this a fair price to pay for the punctuality trains offered. Given a choice between the railroad or a boat when traveling south to Beaufort District, South Carolina, from Philadelphia in 1839, William Elliott noted, "I prefer the rail road to Wilmington." He questioned bitingly "if indeed there is any certainty of the times of sailing by that boat." By the end of the antebellum era, the importance of time was so complete that it replaced distance as the measure of space between two places. "Annapolis is two hours distant by railroad from Baltimore," commented a writer in the *Southern Literary Messenger*. The actual distance, in miles, no longer mattered. Time had replaced distance altogether.[39]

As passengers became ever more sophisticated, they considered the timing of their journeys when making travel plans. Ann Maury attempted to schedule her travel in 1843 in order to avoid layovers. "The communication between Fredg. & Charlottesville is very unpleasant in one respect," she noted. "It cannot be accomplished in one day. You reach the junction, where the Richd. & Fredericksburg

Road joins the Louisa Road, about 2 o'clock in the afternoon & remain there until after breakfast the next day." Maury did not appreciate the enforced layover, finding it "disagreeable to spend the day alone at the junction." If passengers were careful about arranging their schedules, railroad travel also allowed people to change their schedules on short notice. Samuel McGill recalled that as a boy he accompanied his father to Charleston from Williamsburg, South Carolina. When he fell ill in Charleston, it was so arranged that he could "lie over till next morning, take the train twelve miles up to Woodstock, walk out from there to the public road and meet the wagons at a designated place." Thus, he was allowed to rest one additional day in Charleston while his father started the trek back to Williamsburg with the wagons. Indeed, although the train trip "with its velocity, its rockings, jumpings and bumpings produced dizziness" that did little to ease his illness, McGill received additional rest and even arrived at the meeting point before his father's wagons did.[40]

Passengers demanded punctuality and regularity and, in the case of the Boyce family, quickly learned that timeliness was a critical component of railroad travel. Because southerners already had an appreciation that "time is money" when railroads were first built, it is hardly surprising that passengers began to prefer railroads to other forms of travel. Moreover, as the experience of McGill makes clear, passengers soon manipulated these advantages to their own ends. Railroads not only improved the speed of travel but gave travelers flexibility as well.

Who Rode the Trains?

The character of American rail passengers attracted attention from an early date. It appeared to contemporary observers that the railroad did not simply replace the previous modes of travel but actually increased the total number of people who were traveling. "The rapidity and comfort of the Steam car," wrote William Grayson, "have induced multitudes to travel who would never had been tempted to try the horrors of a Stage coach at four miles an hour over deep ruts and heavy sand." Some saw the benefits to specific routes. "Previous to the building of the Charleston and Hamburg Rail Road," noted one observer in 1840, "a triweekly stage was found sufficient for the travel on that line; and to my own knowledge for several years it was rarely full. The same may be said about the route from Augusta to Milledgeville. So soon as the R.Road in question was completed, the increase of travel is well known." Traveling on the railroad could influence where people chose to spend their time. As Emily Sinkler informed Mary Wharton in 1850, "We will probably spend a day with William and Anna and leave from their house as we did last year, they being rather nearer the Railroad and over a better road." According to a contributor to the *Southern Literary Messenger* in 1852, "In these days of steamboats and rail roads every body visits Mount Vernon." Other travelers discovered that railroads could change the dynamic of lodging available to nonrailroad travelers. Frederick Law Olmsted complained that when traveling through the Virginia backcountry he was denied lodging at four successive houses. He asked the fourth homeowner why so many had denied him and received this response: "Well, you see, since the railroad was done, people here don't reckon to take in travellers as they once did. So few come along they don't find their account in being ready for them." More people traveled by railroad than by other means, and thus those traveling by other means found themselves with fewer options for lodging than before.[41]

Why travel? Natural reasons included business and pleasure. Some also felt that travel had beneficial consequences for health.[42] Wilson Lumpkin urged his ill daughter to "not remain stationary" but attempt some "moderate" railroad travel to cure her ailments. "There is scarcely any great Rail road line," he informed his daughter, "that you may not find comfortable staying places, for a few days at a time." As travelers became more experienced, they began offering advice to others making trips. Instructing his sons on traveling from Princeton, New Jersey, to Charleston, South Carolina, Charles Colcock Jones wrote: "Check baggage to Washington, thence to Charleston, at Baltimore, Washington, & Richmond & Petersburg particularly look out as soon as the cars stop, for the omnibus that carries the passengers from one depot to the other, & get in as quick as possible as you'll be left, & have to pay a hack to take yourselfs & dont lose your baggage." Henry Robertson Dickson wrote to Zebulon Vance recommending the "Rutherfordton and Chester route" to Charleston, "unless the object be expedition and in that case I suppose via Greenville &c will be fond a rapid and directer route. But you are so well treated at the Inns and the Outs, and the scenery villages &c &c are all so delightful that I would protest against any friend of mine coming any other way." P. M. Kollock advised George Kollock to select a route that would allow him to sleep. "Owing to the great competition in this country travelling is pretty cheap. Your best route from Philadelphia is down the Chesapeake to Norfolk, thence in Steamboat to Richmond, thence on Rail Road by the Louisa Rail Road to Charlottesville to the Springs. In this way you are enabled to sleep at night. On the other route I understand you lose some rest in stages and Rail Road cars." Antebellum southerners did enough travel that they soon learned to pass on their experience to others, hoping that they could benefit from it.[43]

Historian Eugene Alvarez has argued that American railroads offered "a conspicuous example of the American concept of an egalitarian society. Unlike his European counterpart, the American traveler was almost totally lacking in the type of class consciousness which would have resulted in different cars for each social class." There is some truth to this; as we have seen, Europeans were struck by the openness of

American railroad cars. But numerous distinctions were made on American trains. Alvarez allows that "most Negroes, some immigrants" and women were treated differently in America, and these exceptions are significant enough to undercut his main point. Indeed, it is clear that by the 1850s most southern railroads had at least two different classes of cars. As John White has noted for the national experience, the "much-glorified single-class ideal began to break down at an early date."[44]

Passengers on antebellum trains were categorized both by race and by class. This is clearly expressed from the "classes" of cars that railroads listed in their inventories. Before it officially opened, the RDRR authorized the purchase of "two passenger cars, one a first class & one a second class, of such construction and dimensions, as . . . the wants of the company may require on opening the road." When the Southside Railroad began operating in Virginia, it owned two "passenger cars" and one "2nd Class passenger car." Clearly, southern railroads themselves were prepared to divide customers, despite the popular myth of the egalitarian rail car. While the car itself may have been more open than its European counterpart, passengers were segregated before they even boarded the train.[45]

Yet, as important as it is to recognize that southern railroads separated their passengers, we must also recognize that racial segregation did not occur "naturally" on antebellum railroads. Previously, we have seen how southerners easily used railroads to further slavery's ends: gangs of slaves were transported in low-quality cars. For other slaves, though, the issue of where to board the train was less clear-cut. Although fully committed to racial slavery, southerners still argued about the proper place of slaves in railroad cars. While most slaves rode separately, some did not, which some white southerners found problematic. On trains, slaves were "transported" in a physical sense; the danger was that the fluidity of the passenger car could also "transport" slaves racially, rendering blurry supposedly bright racial distinctions. Railroads did not offer a simple mirror to southern racial attitudes; rather, they were an arena where these attitudes were hashed out.[46]

Slaves traveling with their masters often rode with them, resulting in a mixed-race car. The 1835 passenger regulations on the SCRR noted this different treatment and gave the following rule for "servants": "Servants not admitted, unless having care of Children, without the consent of all the Passengers." Only by securing the permission of other whites would slaves be admitted. Thus, the early rules permitted some flexibility. But for slaves traveling without the benefit of a white companion, segregation was the order of the day. The SCRR's board of directors decreed in 1835 that trains should be "so arranged as to afford Ladies & other passengers accommodation while travelling & that the Negro Car be also similarly arranged." A separate car—the "second class" or "smoking" car—was established for slaves traveling without white company.[47]

The presence of slaves on trains could cause considerable controversy; that same year the SCRR's agent of transportation reported to the board of directors that the "attempt of a Mr Heard to take passage on the Road to Charleston, with Four Coloured Children," had caused a "riot" by "certain individuals" in Hamburg. Unfortunately, we have little other information of what transpired. Therefore, we do not know precisely why the "certain individuals" protested. But the event itself illustrates the tension: Mr. Heard believed he had a right to bring the slaves aboard the car, but others strongly disagreed. The Hamburg residents disturbed by his action may have feared that newfound mobility for slaves threatened to upset racial hierarchies, or they may have sensed the larger implications of allowing slaves to be more broadly mobile than before. Slaves whose skin color pushed the limits of readily apprehensible racial identification might more easily take flight if they could move to an area where "genealogy and history" would not provide the "correct" identification. The resolution to the riot is unfortunately shrouded in mystery. At their next meeting, the board heard another letter from the agent of transportation on the subject but made no mention of the contents.[48]

The importance of local knowledge in determining race was confirmed two decades later by instructions given to conductors and ticket agents on the East Tennessee and Georgia Railroad. Agents were "prohibited from selling Tickets to Slaves under any circumstances" and could sell to free blacks only when given "unmistakable evidence of their freedom." Under such restrictions, whites had to buy tickets for blacks and those whites had to buy tickets "in person, or through some reliable person well known to the Agents." Thus, the personal knowledge of the agent cleared the way for the admittance of African Americans on to the trains. Whether train employees judged the race of blacks or the character of their white ticket purchasers, in each case they held considerable authority.[49]

Contemporary observers saw that different slaves were treated differently. For example, some slaves who traveled alone were sent under the care of conductors. Civil-engineer-turned-planter John McRae sent a slave via railroad from his plantation near Camden, South Carolina, to a friend in Charleston in 1858. He asked that she be placed "under charge of the Conductor from Columbia." McRae later wrote that he was "somewhat anxious to hear of her safe arrival" because she had never ridden a railroad before, and "conductors do not always take much trouble about servants placed under their charge," indicating that McRae was not the only person to make use of this practice.[50] Franz Anton Ritter von Gerstner reported that on the Petersburg Railroad the "negroes pay half fare when they ride in the baggage car; but most of them take seats in the passenger coaches." In this early case (von Gerstner made his tour in 1839), accompaniment of the master does not seem to be have been a prerequisite. In this respect, von Gerstner felt that "slaves in the South have it better than their colored brothers in the North, for the latter are not permitted to ride on the railroad in the same coaches with whites." James Redpath, an abolitionist who traveled in the South around 1854, noted that slaves "and the free persons of color, are put in the first half of the foremost car by themselves, unless they are females travelling with their mistress, when they sit by her side. The other half of the negro car is appropriated for smokers, and is always liberally patronized."[51] In both cases, a racially mixed car resulted: slaves were "promoted" to the white car when they accompanied whites. Whites were "demoted" to the African American car when they chose to smoke. Frederick Law Olmsted offered a detailed description of how slaves could mix in the first-class accommodations:

> The railroad company advertise to take colored people only in second class trains; but servants seem to go with their masters everywhere. Once, to-day, seeing a lady entering the car at a way-station, with a family behind her, and that she was looking about to find a place where they could be seated together, I rose, and offered her my seat, which had several vacancies around it. She accepted it, without thanking me, and immediately installed in it a stout negro woman; took the adjoining seat herself, and seated the rest of her party before her. It consisted of a white girl, probably her daughter, and a bright and very pretty mulatto girl. They all talked and laughed together, and the girls munched confectionery out of the same paper, with a familiarity and closeness of intimacy that would have been noticed with astonishment, if not with manifest displeasure, in almost any chance company at the North.[52]

Olmsted's anecdote demonstrates the ease with which supposedly strict boundaries could be violated when it suited one party. All of these tales illustrate the difficulty of maintaining rigid barriers in an open railroad car, designed for free and easy passage in and out.

The conductor was the arbiter of these boundaries and had to adjudicate when disputes arose. The plight of conductors is exposed by a complaint of Charles Clark to the RDRR, who felt that the conductor, William Gilman, had been too rude when removing Clark's slave from the first-class passenger car. The board passed a resolution to instruct Gilman how to act: "Resolved: That it is the duty of the conductor to remove all *negro men* from the first class car and particularly from the Ladies saloon, but it is equally his duty to do so with as little violence or rudeness as possible, notwithstanding any rudeness or incivility on the part of the owner, and that which the Conductor did no more than his duty in removing the servant of Mr Clark he was and should be censured for entering into any controversy with Mr Clark and being uncivil to him, and his conduct is therefore disapproved." Interestingly this resolution applied only to *men;* slave women acting as nurses may have been seen as less of an affront to white sensibilities. Moreover, conductors clearly had to navigate a fine line between upholding the demands of segregators and not offending those who wished to bend the rules.[53]

The power that conductors held to judge the race of passengers was terrifyingly potent, as Frederick Douglass recounted in the story of his escape from slavery on a train. Douglass dressed as a sailor, carried a friend's "sailor's papers," and hoped that the "kind feeling" Baltimore extended to sailors would protect him. Soon, the conductor approached Douglass, to take his fare and examine his papers. "This was a critical moment in the drama," Douglass recalled. "My whole future depended on the decision of this conductor." Douglass showed the conductor his sailor's protection, and, to his relief, the conductor did not examine it long enough to see that it described someone "much darker" than Douglass. Having judged Douglass's race and status as a free man, the conductor took Douglass's fare and proceeded through the car.[54]

Companies established special rates for slaves. In 1835 the SCRR established that dogs would be allowed on the railroad "at the same rate as negroes." Fares paid by slaves sometimes matched those paid by white children, implicitly reinforcing the dependent and perpetually childlike status of slaves. In 1838 the SCRR recommended that children under the age of twelve be allowed to travel at half price. The same allowance was recommended for slaves, and a special rate of $4.00 was recommended for "gangs of negroes ten or over . . . including children at the same price." Other railroads also offered discounted rates. When it instituted its fares in 1832, the Tuscumbia, Courtland and Decatur Railroad charged $.25 for the full route and $.12^1/$_2$ for partial trips. Children under twelve and slaves were charged at half that rate. As the road expanded, the rates went up, and special regulations were put in place for slaves. In 1835 slaves "with inside seats" were charged "the same price as white persons" and those "with outside seats" were charged half price (evidently some cars on this railroad were either uncovered or more exposed to the weather). The final regulation was "male servants not to be taken inside of the Carrs," suggesting that they could be allowed only on the "outside seats." When the railroad changed its rates later that year, it left the rates for servants and children the same in proportion to the full rates for other passengers.[55]

In 1839 the RFPRR had two "8-wheeled baggage cars [that] have a special compartment for Negroes, who travel at half fare." The Richmond and Petersburg Railroad also had two "eight-wheeled baggage cars with sections for the Negroes, who travel at half fare." Lower rates for slaves continued at the close of the antebellum era, urged on by whites who believed low rates would assist the slave trade. The Southside Railroad established the same rates for slaves and children in 1858: "The Board also determined to increase the rates of Passenger fare on children of 3 years or age and under to 12 years to two thirds of the present full rates and directed the same charge to be made on negroes passing over the roads." The Virginia and Tennessee Railroad general ticket agent explained to a customer in 1859 that tickets to Nashville were $24, but all white children between the ages of three and twelve were half price, as were all African Americans over the age of three. African Americans under the age of three rode on the railroad free of charge.[56]

Railroad companies had to be wary that slaves did not use them as a route to freedom. Arthur Cunynghame, a British traveler in Georgia, noticed that a train conductor made some odd movements on one of his trips: "I observed the conductor constantly opening the door and looking out at the rear of the carriages." After asking the conductor what was the matter, Cunynghame learned that "frequent instances had happened of slaves escaping from the plantations, by getting up behind the cars; and he added, 'We have to be very particular now, because we are held responsible for them by law should they escape in this manner.'" Although the underground railroad has captured the attention of modern Americans, the dangers of the overground railroad were also real to the antebellum master class.[57]

Like Frederick Douglass, some slaves did use the railroad to escape to freedom. Former slave Mary Moriah Anne Susanna James recalled, "Whenever a man wanted to run away he would go with someone else, either from the farm or from some other farm, hiding in the swamps or along the river, making their way to some place where they thought would be safe, sometimes hiding on trains leaving Virginia."[58] Slave owners Abraham Rencher and Charles Manly petitioned the state of North Carolina in 1850 because a slave couple, one person belonging to each petitioner, escaped via the Raleigh and Gaston Railroad. The slaves met with "Nelson, a northern interloper and a journeyman tailor by trade," who pretended to be the slaves' owner and "purchased tickets for himself and for these slaves of the Rail Road agent at Henderson, and thus by means of the Road these slaves were enabled to make their escape out of the state." According to the petitioners, the fact that they requested tickets beyond the terminus of the Raleigh and Gaston Railroad "ought to have aroused suspicions in the breast of the least cautious." Rencher and Manly did not "impute to the agents or any of them improper connivance with these fugitives" but charged that "by the sale of tickets under such circumstances without demanding proper indemnity for the true owners, the Rail Road has become legally responsible for the value of these slaves."[59] More dramatic was the rumored insurrection in Tennessee in 1856. One slave reported to her mistress that on election day slaves at a mill were planning to

> take advantage of the absence of the white men on that day, and while they were all
> from home at the polls voting, to kill all the women and children, get all the money
> and arms, and waylay the men on their return home from the election and murder

them; then make for the railroad cars, take them and go to Memphis, where they could find arms and friends from up the river to carry them off to the Free States if they did not succeed in taking this country.[60]

While southern railroads faced a unique issue due to the presence of slavery, they were not alone in creating different classes of passengers. Northern railroads also offered different pricing tiers based on what people were willing or able to pay. The Western Railroad, for example, resolved in 1842 that it would have a special rate in the second-class section for groups of five or more. Passengers themselves recognized the distinctions among the different classes of passengers. When the Western Railroad established a night train in the summer of 1846, the fare was set lower than the normal second-class fare and had the effect of attracting second-class passengers, "to the discomfort of the regular passengers by the night train," recalled one railroad employee. When the train did not make money, it was canceled by the railroad after just over three months of operation. Northern railroads also established "emigrant" cars: low-quality transportation aimed at new immigrants to the country. During the year 1852, the Pennsylvania Railroad owned twenty-six eight-wheeled cars for this purpose. Six years later, it was using thirty-one cars for this service. Other railroads pursued the immigrant business. The Western Railroad decreed in 1841 that "emigrants destined for Albany in parties of Twenty" were to be given a special rate for taking the freight train and riding in "second class or merchandize cars." Further, it was decided that this arrangement would be advertised abroad in Liverpool and Havre as well as in Boston.[61]

North and South, women received different treatment from men on railroads. Separate cars for women "were relatively common by the mid-1840s, and they usually had a comfortable sofa where a woman could stretch out for a nap." Husbands could accompany wives. Southern railroads were less likely to have separate cars for women, but their passenger cars still featured a "ladies' compartment" that allowed women to separate themselves from the rest of the car. Women commonly traveled in the company of a companion. Lucy Carpenter wrote that in 1848 she traveled under the "care of the President of the road" when traveling from Charleston to Camden, South Carolina. When unable to travel with a male companion, women were sometimes under the care of the conductor. "Left Macon at midnight very lonely," Dolly Lunt Burge recorded in her diary. "Doctor put me aboard the cars introduced the conductor to me. I went in found not a single lady aboard. A few gentlemen all sleeping." Elizabeth Lomax wrote that a friend experienced a similar situation when traveling from Washington, D.C., to Richmond: "Bob Crawford escorted her as far as Alexandria and put her under the care of a reliable conductor." Women who traveled with a companion or under a conductor's care did so as dependents. Similarly, when McRae had asked for his slave to be sent under the care of a conductor, the dependent status was clear as well. Yet the dependencies sprung from different sources and would not have been confused by contemporaries: white womanhood in need of protection versus a perpetually childlike slave. Although ironically afforded the same protection, they were two very different cases in the antebellum white mind.[62]

Ann Maury began a trip in 1843 without a companion but soon found one. "Both Mr. Richardson's sons accompanied me in order to see if they either of them were acquainted with any passenger under whose care to place me," she wrote. "They knew no one going to Charlottesville, but were able to introduce me to a gentleman going to the Eastward who would be my companion as far as the junction." Maury wrote that the chaperone was "very attentive to my comfort until we parted after breakfasting at the junction." Deprived of a companion, Maury found that in the next train she was "the only female passenger & therefore considered I had better ensconce myself in the ladies compartment of the car, & there I travelled in solitary grandeur all the way to Gordonsville, without exchanging a solitary word with any one except when the Captain asked for my ticket." Maury described herself as "very lonely," but was able to complete her trip safely. Other women may have been more confident in their ability to get along. Thomas Hobbs, who was traveling with a group of women, left them at Knoxville: "Here I parted with the ladies, who, having their baggage checked through and tickets provided, think they can get along without me."[63]

Anna Calhoun Clemson passed along advice to her daughter about traveling on the train: "If you young ladies come along, do be prudent, & careful, & behave with dignity, & propriety. No giggling, loud talking, &c &c. You had better bring something to eat on the road & if you are delayed in Baltimore go to the cars, & remain there, rather than go alone to a hotel, or wander about the city in the hot sun." Women were given protection by the corporation in that a separate compartment was provided for their use, but Clemson's instructions make clear that ladies still needed to govern their behavior aboard the train.[64]

From Reality to Metaphor

Railroads and the experience of travel assaulted all of the traveler's senses. Shrill whistles, frightening engines, and uncomfortable hotel beds all formed a part of the antebellum travel experience. Travelers also found themselves surrounded by others, black and white. Railroad corporations allowed for racial segregation by creating separate cars, but such rules could be bent or broken under the proper circumstances. For travelers, the impact of the rail journey extended far beyond the time of travel itself. "Trains," "engines," and "railroads" began to occupy important metaphorical positions in southern language—an instant shorthand for speed, progress, and modernity. The general presence of this language—not just among railroad promoters—illustrates the antebellum South's embrace of the modern mind-set epitomized by railroads.

Southerners used steam engines as a metaphor in their writing. When James Henry Hammond noted that he had been forever shunned from politics and society because his sexual escapades had damaged his reputation, he wondered if a man could do anything without politics and society by asking, "Can a Steam Engine work without fuel or water?" When arguing that teatime was important for society, Charlestonian Charles Fraser called the tea table a "great engine, by whose well-regulated steam, more has been done for the humanizing of modern society, than all the contrivances of art, or than all the ceremonious courtesies of chivalry." North Carolinian William Woods Holden informed a correspondent in 1850, "I write you in haste, under high steam pressure." Whether as a prime mover or indicator of pressure, engines assumed an important metaphorical role.[65]

Unsurprisingly, the railroad became the symbol for speed for the southern literary audience. "Vagueness is the idol of the hour; incomprehensibility the railroad to popularity," the *Southern Literary Messenger* complained in 1842. Another writer in that magazine, who protested against the "premature use of books in the education of children," termed the system that supplied children "in almost every department of literature with text-books" the "rail-road system of education." Considering speed in a more positive light, civil engineer Henry Bird expressed his disappointment that he was so far away from his fiancée (he in Virginia and she in Pennsylvania). "I wish there was a railroad between us," he wrote in 1833. "I could see you in 12 hours."[66]

Railroads also earned a place in the fictional literature read by southerners. In the didactic story "The Happy Child" by Harriet E. B. Stowe, the lives of two children were contrasted: a sick, poor child who had no worldly belongings but contented himself with reading the Bible, and a rich child with a wide range of toys, including "rail-road cars, with a little rail-road for them to go on, and steam-engine and all," in the excited words of one of the other children.[67] In olden times, another writer commented in the same magazine, "the traveller thought himself a fortunate being, if he, in a stage-coach or mail-cart, could get shaken over thirty miles of mire and dust in half as many hours. But mark the change. See the traveller now breakfasting in one state, dining in another, supping in a third and sleeping in a fourth."[68] Clearly the railroad had brought about this change. Former slave Fannie Berry recalled that steam engines figured into a tall tale that her master had told her: "My Master tole us dat de niggers started the railroad, an' dat a nigger lookin' at a boiling coffee pot on a stove one day got the idea dat he could cause it to run by putting wheels on it. Dis nigger being a blacksmith put his thoughts into action by makin' wheels an' put coffee on it, an' by some kinder means he made it run an' the idea was stole from him an' day built de steamengine."[69]

The railroad was a positive symbol of progress. This is demonstrated by an advertisement for a shoe and boot salesman (figure 14), in which a "shoe train" attracts a variety of customers, black and white. "Shoes are low—O how fast they go," announced the advertisement, linking the speed of the train (demonstrated by the pedestrians' elongated bodies) to the quickly moving merchandise. Railroads equated progress in literature as well. In one 1840 story the prerailroad era was already referred to as "old times": "I will not decry rail-road-cars; those of England are capital; but there are other things besides speed which a traveller needs," chimed one character. "In old times we did not use to think it a matter of such moment to fly over the country at a pigeon's pace. . . . A rail-road or post-coach, will whirl you by the seat of your nearest and dearest friend, without the chance of a *how d'ye.*" Here, the fictional character warned against the dangers of speed, that it would destroy the bonds of friendship. A railroad partisan, by contrast, might have claimed that railroads made it easier to visit distant friends.[70]

Eleven years later, in the story "Recollections of Sully," railroads were again used to illustrate the difference between "the old generation and the new." The older character in the story, Frank, was pontificating one evening "on his usual topic 'how all the old honor had from Virginia gone.' His reasons for this state of things were manifold—among other causes, yankee immigration, and—could the truth be got at—*railroads.*" When young Sully remarked that a railroad tunneling through the Blue Ridge would be a positive development, Frank retorted "You are wrong, doubly wrong! Your railroads will perhaps carry off a few barrels of flour, a few bushels of wheat, but beyond that?" Frank projected the influx of tourists in the region: "Well, some day you will see in a *Northern* book, ' *Scene from the Blue Ridge; the fine old house in the back-ground is Inglewood, the residence of Francis Sully, Esq.*'—so goodbye the country life and privacy which my father bequeathed me; they have made me and my house the public property." Frank did not object to all industry, but did object to that which brought in outsiders: "Sully, I may be old-fashioned in these notions, but I wish this valley to remain in its ancient obscurity, with nothing to disturb the mountain air but the clack of the old mill yonder, and the cheering blast of the stage's horn as it rattles along in the morning sun." We have seen how railroad travel excited multiple senses in travelers. Here, Frank recruited those same senses to protest the incursion of railroads; the only sounds he wanted were the "clack" of the mill and the "cheering blast" of the stage horn.[71]

Come one—come all—come give me a call,
And I'll be sure to sell you your shoes this Fall;
For Shoes are low—O how fast they go,
At the Steam Boot and Shoe Store in "Mayblu's Row."—SHAKESPEARE.

STAND BACK AND MAKE ROOM FOR THIS TRAIN, B'HOYS AND ALL!!

Figure 14. South Carolina Temperance Advocate, October 8, 1846, p. 53. Reproduced from the collections of the South Caroliniana Library, Columbia, South Carolina.

Other writers used railroad travel as a device to draw the reader into the story. "If the reader is not a resident of that vicinity," opened a story about a haunted house, "but has passed along the rail-road between Richmond and Petersburg, before the fire occurred by which the wood work of the building was a few months since destroyed, he may have been struck by the lone and desolate appearance of the house, and been led to make some inquiry respecting it." The reader in 1855 could easily imagine himself on that railroad and then have a picture in his mind about how such a house might look. In this way, the author drew the reader in to what was to follow.[72]

Some felt that the presence of railroad symbolism in literature went too far. A critic in the *Southern Literary Messenger* remarked on a poem of Elizabeth Barrett Browning: "We confess ourselves of the uninitiated, but, by dint of hard study, have made out the 'resonant steam-eagles' to be, in plain prose, locomotives." Overwrought metaphor was clearly not welcome here. Critics also assessed literature by determining whether it was appropriate for reading on the train. Regarding the work of William Gilmore Simms, one reviewer remarked, "Mr. Simms' poetry is for the closet, the bower, the forest aisles, and grand cathedral of Nature; for the solitary muser, the companionship of thinking minds and deep hearts, the quiet circle of intelligence and love; but not for the steamboat and railroad, and laughing drawing-room, and half-thoughtless party, wanting something light, and pretty, and amusing." Other literature was just fine for that purpose: a review of

You Have Heard of Them declared it "will wile away the hours in a railroad car or steamboat as well as any work lately published."[73]

As boosters knew, railroads offered clear advantages to travelers, and antebellum travelers agreed. No longer were they dependent on a stagecoach driver who could lose the way—railroad tracks offered an undeviating path to the destination. Railroads also had time advantages over steamboat travel, even if trains were not as free from nature's influence as people hoped. And any railroad journey, no matter how jolting, was preferable to walking. It can be difficult for us to understand the advantages of rail travel because of our own available transit choices, but for antebellum southerners, the choice was clear. Paying too much attention to the difficulties that travelers faced—as real as those difficulties were—obscures the fact that travelers were willing to accommodate some discomfort in order to accept modernity's advantages.

The South's accommodation of railroads required attention to race and gender. Railroads constituted a danger as an escape route, and some slaves were able to take advantage of that fact. Railroads also required southerners to determine acceptable behavior for slaves. Some slaves could ride with their masters, others were consigned to lower-class accommodation. As the early riot in Hamburg demonstrated, the presence of slaves on trains could be contentious. While separate "ladies cars" were not as common in the South as they were in the North, women were still accommodated in separate compartments.

All passengers appreciated the benefits of rail travel, even if they grumbled about delays. But the impact of railroad travel was far deeper than simply speeding travelers to their destinations. Traveling was not simply about getting from point A to point B. Railroads were social arenas where travelers could meet old friends or be entertained by peculiar passengers. Railroads fascinated southerners, and antebellum travel was a complete sensory experience. A whistle's blast to signal arrival, the rushed hotel meal, and the feel of upholstery all featured in descriptions of travel. Travelers had a rich experience that could excite, terrify, or frustrate them. Modernity, then, was not simply an economic change from one mode of transportation to another. By understanding the sensory impact of travel, we can understand precisely how antebellum southerners experienced the change to modernity. Some adjustments, such as the time savings, were worth making. Others, such as whistles that interrupted sleep or a meal rushed by the conductor, were less pleasant. But southerners continued to embrace railroads as a form of travel. Passengers were willing to tolerate some discomfort for the modern advantages of railroads.

|||||||||||||||||||||||||||||||||||||||||||||||||||||||||||||||||||||||||||||||||||||||||||||||||||||||||||||||||||||||||||||||||||||||||||||||||||||||||||||||||||||||||||||||||||||||

CHAPTER SEVEN

Communities

Passengers and shippers were obviously important constituencies for railroads, because they supplied revenue. But railroad corporations dealt with a wide range of groups, including some that never boarded the trains. As soon as railroads entered communities, different interest groups interacted with and placed their demands on railroads. Thus, the railroad did not possess a single meaning for southerners. Rather, different groups and individuals invested the railroad with multiple, even contradictory meanings, according to their own expectations and needs. Communities welcoming a new railroad saw it as a cause for celebration. Sabbatarians saw the railroad as an affront to their religious sensibilities. Landowners saw tracks as a boon to their real estate values or as a source of constant damage and frustration. Pedestrians saw the track as a pathway for their own travel. This chapter examines some of the different southern communities that interacted with railroad corporations and the meanings that the communities attached to the railroad.[1]

Celebrating the Railroad's Arrival

Given the excitement surrounding railroad projects, driving the last spike was always a cause for celebration. This was true when railroads were new to the South and remained true when they were accepted parts of the landscape. When the RFPRR began operating in 1836, Blair Bolling recorded that "the scene was novel here, and attracted a great concourse of spectators, who filled every door, window and ally as well as much of the street itself, to the extremity of the town anxious spectators evinced their astonishment and gratification at the stupendous specimin of internal improvement." A few weeks later he attended the official opening of the road. It began with a ride "in fine stile amidst the shouts and applaus of an admiring and (in part) an amazed multitude." At the end of the journey there was "a table spread, and loaded with the finest wines and liquors, to partake of which a cordial invitation was extended to the whole company, who accepted and partook apparently with enthusiastic delight, and very soon began to exhibit its exhilarating affects in their animated countenances, voices and movement." After drinks, they proceeded to "a most sumptuous dinner" which was followed by "a profusion of champaigne." At the end, they "reembarked and returned home, with speed, without accident and with grateful acknowledgments."[2]

The celebrations could be large and multiracial events that engaged the entire community. James Henry Hammond organized a celebration when the railroad was completed to Columbia, South Carolina. Hammond reported that three thousand persons were probably in attendance, including "women, children and negroes." He estimated that eight hundred people alone made the trip up from Charleston. Former slaves also remembered railroad celebrations as large affairs. Nelson Buck of Mississippi recalled that when the Mobile and Ohio Railroad got "as far as Macon, they celebrated with a big picnic at the bend of the river. Folks cum from near and far and they wuz all havin' a grand time when a big rain cum along and washed away mo' good grub; I sho' hated to see it go, too." Former slave Andrew Jackson Jarnagin remembered that the Mobile and Ohio was "celebrated with a big picnic." A large crowd turned out in 1857 for the Memphis and Charleston Railroad. The celebration reportedly drew thirty thousand spectators and featured a parade, speeches, dinner, fireworks, and the "Marriage of the Mississippi and Atlantic," whereby a barrel of saltwater from the Atlantic was poured into the Mississippi River.[3]

Ceremonies could be quite elaborate, as the Memphis and Charleston demonstrated. At the completion of the East Tennessee and Virginia Railroad in 1858, a crowd of fifteen hundred gathered to hear a speech about the history of the road, as well as an ode composed by Reverend Andrew Shell. When the last spike was set, each director took a blow with the hammer, with the final blow being taken by the president. Three "hearty & enthusiastic" cheers were given by the crowd, and the road was complete. Even more pageantry accompanied the SCRR's entrance into Orangeburg, South Carolina, in 1840. The train "passed through an arch, garlanded with roses and evergreens, and bearing a number of suitable inscriptions." After the dinner, toasts, and thirteen-gun salute, the "ladies . . . were taken into the cars, and gratified with an excursion of several miles on the road."[4]

Celebrations appealed to many senses: the arresting sight of the engine and the prospect of tasty food. They were also noisy events. The Wilmington and Raleigh Railroad celebrated its completion with "161 cannon rounds, one salvo for every mile of railroad." When completed in 1856, the North Carolina Railroad held a celebration and a newspaper noted that "such hallooing, singing and cheering by the negroes—commingled with the bellowing of two engines, perhaps was never heard before by our citizens." Although Basil Thomasson did not see the railroad celebration in Statesville, North Carolina, in 1858, he certainly heard the thundering of "the cannon over at Statesville, a distance of about twenty miles. The cars rolled up to that place, I'm told, to-day, and a large crowd gathered in to see the sight, hear the music, help eat the dinner, ride the Rail Road, etc." Throughout the antebellum era, celebrations such as the one Thomasson heard marked the general public's acceptance of the railroad's entrance into the community. As we saw in the first chapter, plenty of southerners were active boosters and agitated for railroad development. But even those who never harangued a crowd or penned a pro-railroad pamphlet could still express their approbation by attending such celebrations. Railroads were not just a matter for self-appointed boosters but met with general approval.[5]

Sabbatarians

After celebrations died down, railroads had to deal with the numerous interest groups they served. Early in southern railroading, companies had to respond to Sabbatarians: groups that were interested in prohibiting trains from operating during the Sabbath. Although Sabbatarians never had a formal regional organization in the South, diverse groups worked in concert to present their demands to railroad companies. The presence of Sabbatarians demonstrates that southerners already possessed ideas about the importance of time before the introduction of the train, because Sabbatarians placed a specific value on Sunday and wanted to enforce that vision on the new technology. The Sabbatarian efforts also show that railroad companies did not have complete control but had to negotiate their time with community groups.[6]

Religious leaders did not seem to object to trains *qua* trains or technology in general. Rather, they objected to behavior that was facilitated by the railroad. Indeed, the Alabama Bible Society declared in 1857 that "if the men of the world go by steam, the men of the church must go by steam." Thus, the technology itself may have been unobjectionable, but ministers in the South railed against the practice of running trains on Sundays. In Richmond, Reverend A. D. Pollock complained in 1837 that the "private consumer" could not get wares "untouched by the sacrilegious hand of the Sabbath breaker." Ministers also warned about the dangers of not observing the Sabbath from the pulpit. One railroad employee, after having heard a sermon in Wilmington, North Carolina, against Sabbath breaking, "was convinced by his sermon that I could not consistently continue in this business, and so resigned my situation." Not all railroad employees were swayed as easily as this man, however, and the antebellum era saw numerous debates between Sabbatarians and railroad companies.[7]

The first recorded Sabbatarian effort to influence the SCRR occurred in 1835. At the June meeting of the stockholders, one Rev. Dr. Gadsden introduced a resolution to end Sunday service. The company's existing policy on Sunday travel is not entirely clear from the extant documentation, but debate over the resolution suggests that some limitations were already in place. On these grounds, railroad superintendent Alexander Black argued against the resolution. He pointed out that "every thing which could be done to promote the views of the mover had been tried," and passengers were carried on Sundays only "in cases of extreme necessity." Moreover, the necessity of delivering goods "in regular order" meant that trips could not be cut back any further. To lose a day's work would damage the business that the assembled stockholders had worked so hard to promote. Another stockholder, Mr. Masyck, moved that the directors be asked to make a report to the stockholders on the topic. After the second, Mr. McBeth argued that "he thought the Direction were already pursuing the object of the Motion," thus rendering any activity by the stockholders unnecessary. Before the vote to table Gadsden's motion could be taken, yet another stockholder, J. Harleston Read, asked if the motion had been demanded by "any religious Society, as there had been movements of a similar nature under similar circumstances and he was extremely adverse to any such intervention." Gadsden stated that he was not under the influence of a "religious Society," but considered the matter to be an important one for the stockholders. With that, the resolution was tabled.[8]

While the precise policy of the SCRR is unclear, it appears that the company already had a policy against general passenger travel on the Sabbath. In 1836 the company rejected an offer from "several persons, among whom were large Stockholders," for a special train to be sent to Hamburg on a Sunday, for which they were willing to pay $200. It was moved and adopted that "the rule prohibiting the transportation of passengers on the sabbath be strictly adhered to." Thus, it appears that the railroad had some policy against Sunday trains—outside of the trains required by the U.S. Post Office Department—at this early date. But even the Post Office faced a challenge from the company. When battling with the Post Office about the mail contract at the beginning of 1836, the SCRR considered two different compensation packages to propose to the government. In both, it specified that it had no desire to do Sunday delivery. When contemplating coordinating with stagecoaches to run the mail north of Aiken the following year, the company likewise proposed that the mail not run on Sundays.[9]

These efforts did not satisfy all Sabbatarians, however, because opposition to the trains running on Sundays resurfaced again in 1842. Five petitions and memorials were received at the stockholders' meeting that year of the LCCRR, which would soon form a part of the SCRR: one from the clergy of Charleston, one from the citizens of Charleston, one each from the members of St. Paul's and St. Peter's churches in Charleston, and one from the citizens of Columbia. The committee formed to address these complaints reported back with something of a compromise. It recommended two resolutions to the stockholders: first, that the no freight trains should run on Sunday; second, that the board look into the possibility of not running the mail and passenger trains on Sunday and report back to the stockholders.[10] The following year, the SCRR reported that it had attempted to accede to the wishes of Sabbatarians. In either late 1842 or early 1843, the company attempted to obtain a mail contract without having to deliver the mails on Sundays, offering to give up one-seventh of the compensation that it received from the Post Office. This was refused by the government.[11]

Sabbatarian petitions reappeared in the 1850s. Now, the petitioners had a much more complex argument, and perhaps one intended to gain the ear of money-minded stockholders.[12] Led by Dr. Whitford Smith, the memorialists adopted a multipronged approach. First, they argued that Sabbath operations prevented morally inclined investors from purchasing stock. Second, the petitioners appealed to the stockholders' sympathy for the workers on the road. Every man deserved the opportunity to rest one day out of seven. "Is it right, is it justice," the petitioners asked, "to any class of men in any station or pursuit in life, to cut them off from every opportunity of domestic tranquility and enjoyment, and doom them to toil without intermission, and to labor without rest?" Although they accepted that not all men would avail themselves of the opportunity to go to church if they were not required to work, the Sabbatarians argued that at the very least the workingmen would benefit from time spent with their families. To force the worker to toil every day of the week and prevent them from receiving the benefits of domestic life would serve only to promote "carelessness and recklessness, eminently dangerous in such a service." Third, the petitioners argued that Sunday operation violated city and state ordinances. Finally, the petitioners noted that Sunday operation violated the law of God.[13]

The petitioners were all preachers, and they were concerned about the moral effect that the railroad was having on their charges. They were finding it increasingly difficult to hold the attention of the public on the day appointed for religious instruction: "Often are the villages and hamlets in which they preach roused from their Sabbath stillness by the shrill shriek of the whistle. The children they would train and educate to virtue are frequently led off to see the passing cars. And the slaves whom they would teach their duty to God, and those in his providence placed over them, too often find employment where the temptation of a pecuniary reward is too powerful for their weak resistance." Ministers feared that the sounds of modernity would pull their flock away from their sacred obligations. Children and (childlike) slaves were believed by these ministers to be the most susceptible to the influence of the railroads and the distraction from the Sabbath. Children would want to see the trains, and slaves would not want to pass up an opportunity to earn more money. Like previous memorials, this one was to little avail and garnered only a resolution that the "sense of the meeting" was that Sunday service should be stopped "as soon as practicable." It was not rhetoric that matched the urgency the petitioners felt.[14]

In 1854 the petitioners tried again, when a committee from the Protestant Episcopal Church showed up at the stockholders' meeting. The arguments were largely the same, and the public nature of railroads and their capacity for distraction again elicited special notice. "The trains traverse great distances, pass through many localities, attracting great attention and exciting curiosity wherever they go. Thus the evils are not confined either to operatives or travelers; they extend to the eyewitnesses and observers, to the idlers who are attracted to the stations, and to others whose attention, whether voluntarily or otherwise, is disturbed and distracted by the noise and bustle and business, properly belonging only to the working days of the week." The railroad was acknowledged as a great distraction, which was too powerful for people to overcome on the Sabbath. Despite their efforts, the petitioners were yet again unable to earn unequivocal victory. The stockholders, however, did pass three resolutions: one thanking the directors for ending the loading and unloading of merchandise on Sundays at the depots of the road during the previous year, another acknowledging that the directors were considering ending all Sunday service, and a third asking the directors to follow through on previous requests and stop all Sunday operations, except what was necessary to transmit the mail. The stockholders stopped short of an outright ban but continued to move toward the vision sponsored by the memorialists.[15]

In 1856 the company finally gave in and reduced its Sunday service. In the annual report for that year, the company reported that ending some Sunday trains had been a success. Although there was some "additional expense in minor points," the savings in labor cost "has worked rather advantageously than otherwise in many respects." Importantly to those who valued the regularity of service, "The transit of freight has not experienced any delay from the discontinuance of the practice." The following year, the general superintendent was able to report that Sunday trains had been reduced yet again, with the permission of the Post Office. Only two trains operated on that day, one from Columbia and one from Charleston, both of which left in the mid to late afternoon. After two decades of effort, the various Sabbatarians—mainly churchmen and sympathetic stockholders—were able to get the company to reduce its Sunday service to a bare minimum. The movement was not uncontested, as stockholders demanded that they get maximum value out of their investment, but it was ultimately successful nonetheless.[16]

The Sabbatarian experience on the SCRR was repeated on other roads throughout the South during the antebellum period. In some cases, Sunday travel was tolerated. In 1839 Franz Anton Ritter von Gerstner reported that the City Point Railroad in Virginia ran on Sundays for passengers only, "and mostly those belonging to the poorer classes," charging half fare. In May 1858 an ETVRR board member offered a resolution that demanded that the railroad refuse to operate on Sundays if it was awarded the mail. This was rejected by a vote of eleven to six. The company, however, was not entirely opposed to the goals of Sabbatarians. At a meeting that September, the board of directors unanimously passed a resolution that applauded the Memphis and Charleston Railroad's "movement" to dispense with Sunday service and expressed the "ardent desire of this Board that our Road shall cordially co-operate with the different Roads in the South to effect this great object." Although the resolution did not commit the road to cutting its own service— and profits—it was a harmless expression that it supported the basic goals of the Sabbatarian movement.[17]

Sabbatarianism was not a purely southern phenomenon, of course. The Pennsylvania Railroad discussed the prospect of closing on Sunday soon after it began operation. In 1849 the railroad adopted a resolution asking the superintendent to end all Sunday operations and to not require any employee of the company to work on that day. Explaining the decision, the company noted that it felt it had a responsibility to its workers to spare their laboring on Sunday: "Universal experience shows the necessity of occasional rest." By letting its workers devote one day per week to rest, it hoped to obtain better-quality work from them. Albrect Koch reported that in Connecticut "even the steam carriages and the steamboats rest [on Sunday], and anybody who should by chance take it into his head to travel can be punished by law."[18]

The Boston and Worcester Railroad received petitions from the citizens of West Needham and Natick regarding the running of trains on Sunday in 1835. The Western Railroad ordered that none of its trains would run on Sundays in 1841, except to carry the mail or on the special order of the engineer. John Rockwell, a director of the Norwich and Worcester Railroad, carried his opposition to the extreme, putting forth a successful resolution whereby the express freight train on the Norwich and Worcester Railroad left Norwich at 1:00 a.m. on Mondays instead of Sunday evenings. Although employee reaction went unrecorded, it is doubtful that being required to be at work for a 1:00 a.m. departure allowed the employees the repose on Sunday that Rockwell professed to value.[19]

Railroad companies were highly clock- and schedule-oriented enterprises. And railroads were praised by promoters as offering reliable and regular transportation. Yet, as the example of the Sabbatarians demonstrates, companies were never completely in control of their own time. Rather, the railroad's time was negotiated with community groups, who were sometimes able to exert moral pressure on companies to change their behavior.

The Post Office

Another important entity with which the railroad had to negotiate was the federal Post Office Department. As with Sabbatarians, debates with the Post Office revolved around time. The mail was clearly important to southerners. The status of the mail, especially a breakdown in its delivery, was a regular feature of the newspaper in Charleston. A typical newspaper report read: "All the Mails due from the North, came to hand yesterday—no mail was received last night from West of Milledgeville, (Ga.) The mails, both from the North and West, were not received until about two hours after they were due." The timeliness of the mail, as well as its origin, was of substantial interest to readers.[20]

In an 1832 report, the SCRR indicated that it had been carrying the mail since around February of that year on the completed railroad immediately outside Charleston.[21] But the company reported difficulty securing a contract once the entire track was near completion. The SCRR approached the Post Office Department about a possible contract in August 1833. Not receiving an answer, it tried again in April 1834, and the next month SCRR president Elias Horry promised stockholders that every effort was being made to secure the mails. The mail contract must have been settled shortly thereafter, because in 1835 the SCRR's board of directors received two letters from the Post Office regarding "failures between Aiken and Hamburg." It was the beginning of a tumultuous relationship between the company and the government.[22] The SCRR's board of directors decided to cease carrying the mail in 1835 when it thought that the Post Office was not handling its account correctly. The community's response was swift. Fifteen "merchants of Charleston . . . feeling seriously the embarrassment occasioned to the trade of this city and the interior by the suspension in the transportation of the mails on the Rail Road," petitioned the company and asked that the company "tender to the Post Master of this city the use of the Rail Road for carrying the mails until the first of next month free of expense, unless before that time some satisfactory arrangement is made between the company and the Department." The mail was important enough that the merchants demanded that the company acquiesce. The company listened to its customers and agreed to carry the mails for free until January 1, 1836.[23]

In 1838 the company communicated to the stockholders the terms of the new mail contract for the following year. It was clear that the main debate was over who would control the railroad's time. There were nine places to deliver the mails between Charleston and Hamburg, and the company estimated that ninety minutes would be lost in stopping the train, delivering and taking on mail, and restarting. Thus, the company announced its plans to attempt to get the Post Office to agree to a schedule that would allow the railroad to deliver and take on mail at way stations without stopping and place an additional charge if they were required to stop. The company also balked at having to transport the mails on Sundays, given that on Sundays "in many instances extra wages are required to be paid."[24]

In the next report, the company could address the status of the offer that it made to the Post Office. The Post Office was happy to allow the railroads to set the time of operation—if they would in turn accept a reduction in pay. The Post Office offered $237.50 per mile if the railroads would accept the Post Office's timetable, and $200.00 otherwise. The company was not pleased with the times offered by the Post Office, because it made no allowance for delays, and if delayed "as the cars frequently are by unavoidable accidents," the trains would have to operate "at night, to the great discomfort and some risk of the passengers."[25] Ever mindful of its own time requirements, the company did not hesitate to comment on how the process might be more efficient. In 1839 the mail arrived in Charleston from Wilmington via boat between five and six each morning. Mail for Charleston was separated from other mail at the post office and then taken to the railroad. The railroad suggested that the sorting process take place on the boat, thus eliminating the hour or so that was wasted by conveying the mail to the local post office and sorting it there. Without this change, the company worried that the trains would start too late. Because the freight trains left after the passenger trains, any delay to passengers invariably delayed the freight. Finally, the company was disgruntled with the necessity of running the mails on Sundays, which it estimated would lead to an additional expense of $20,000 per year. The company finally agreed to take the mail from Charleston at 7:10 a.m. and deliver it at 4:50 p.m., and it expressed the hope that improved transportation between Raleigh and Wilmington would eventually allow the mail to arrive sooner than that from the Wilmington steamboats. Despite the difficulty in the contract, the company still reported with pride that it was able to transfer the mail in an efficient manner, reporting in 1839 that since the aforementioned contract was signed, there was only one day of delay, and even on that day the mail was still delivered by the evening of the day it was supposed to be delivered.[26]

If time coordination was one source of frustration, the level of pay was another. The SCRR was paid per mile per year, regardless of volume, and the company found that by 1842 the volume of mail had increased "say three fold," yet it received no additional compensation. The increased volume of mail meant that the company had to turn away freight, which it would have carried at a higher profit than what was allowed for the mail. The company argued that it lost more than $80,000 per year by carrying the mail because it lost money by not carrying additional freight.[27]

Disputes with the Post Office Department were not private battles but very public ones. Southerners noticed when the Post Office and railroad were in dispute. "The South Carolina Rail Road & Mr. Secretary [James] Campbell have got into a dispute about carrying the mails & the R. Road has discontinued service the 1st," Charles Colcock Jones informed his son, who was studying in Massachusetts. "Consequently we are some 6 or 8 mails behind, and, in this we may account for our not receiving letters from yourself or your Broth last week." Likewise, the credit reporters from R. G. Dun were not particularly impressed by the SCRR's attempt to play hardball with the Post Office. The reporter in 1855 characterized the road as "A miserable monopoly. Now under the Presidency of 'John Caldwell.' . . . All his operations indicate a narrow mind." Referring to the mail dispute, the reporter commented that Caldwell "endeavd to extort a higher price altho' getting enormous pay before. The conseq[uenc]e was a return for some weeks to the old stage system." While the company may not have felt it was being treated fairly, its negotiations with the Post Office demonstrated again that the railroad was not a full master of its own time.[28]

The struggle between the demands of private enterprise and of the federal government was not one that was solved during the antebellum era. It was an issue that centered around who properly controlled the railroad's time. Railroad companies were quick to defend their claims. Charles Ashby of the Orange and Alexandria Railroad argued that "the responsibility of Sunday mails rests mainly with the President & Directors of the company & not with the Government, as is generally believed." Yet the government possessed a powerful trump card—it was dispensing a popular and necessary service. As the experience of the SCRR demonstrated, even fellow merchants who may have otherwise been sympathetic to the rights of private capital demanded that the mails move on time, and the general public's patience wore thin in the face of delayed mail. The railroad could pitch its arguments to the Post Office, but it was not fully in charge of the process.[29]

Links to Local Communities

In addition to interest groups such as Sabbatarians or the federal government, railroads dealt with the communities that were immediately around the railroad. Many enjoyed watching the work being done on the road or watching the trains fly past.

The view from Mary Boyce's plantation in Fairfield District, South Carolina, was "greatly *enlivened* by the passing to and fro of the cars" after a railroad was constructed there in 1855: "We have enjoyed [a] view of the cars passing today." Living near a railroad did more than just provide an entertaining view for Boyce and her family; it also made it easier to travel. As Boyce noted, "It is quite convenient to only have a short walk to the cars." Likewise, Anna Calhoun Clemson noted that her husband was considering moving based in part on the proximity of the proposed land to the railroad. Clemson also appreciated the view. When a platform near her home moved to the crossroads, she admitted that it was much easier to get to but that she could no longer view the cars coming and going. But not everyone enjoyed the privilege of watching the train. South Carolina slave Anne Broome remembered that her master would "whip me just for runnin' to de gate for to see de train run by." Here, trains were a distraction too great to be tolerated by the master.[30]

Some corporations believed that they brought moral—as well as economic—benefit to the regions that they served. "It is gratifying too," the SCRR noted in its 1839 report, "to find that the state of society is improved through the country which we pass. The opportunity for improvement has not been lost to the good citizens who have been realising from our large expenditures. Schools, Churches, and Temperance Societies, have been, to a certain extent, substituted for dissipation and idleness." Locals recognized these benefits as well—the Knoxville Sons of Temperance, for example, held stock in the East Tennessee and Georgia Railroad. Not only did the railroad represent technological progress; it represented moral progress as well.[31]

Despite the acclaimed moral benefits of railroads, many of the community links centered around economics. Railroad companies believed that the construction of railroads would have a positive economic impact on the community. In 1837 the SCRR estimated that the company's constant need for wood, for both fuel and construction, "gives employment to hundreds of inhabitants on the line." Moreover, the company claimed, expanding that calculation to the families of all employees of the railroad in every department "would increase the number to thousands who have their support from this institution." Indeed, railroads did purchase a wide range of goods from locals and even slaves; some slaves in Barnwell District, South Carolina, for example, "sold timber to the South Carolina Railroad." Railroads also eased the ability of common whites to market cash crops. In North Carolina, railroads expanded the markets for wheat, tobacco, and cotton, effectively "diminish[ing] the apparent risks of entering the cash-economy."[32]

As railroads expanded in the South, companies turned from trumpeting potential to actual benefits. After it had operated in South Carolina for more than a decade, the SCRR argued that railroads improved the state's economic condition, which helped to discourage people from leaving the state. Railroads also encouraged capital to remain in the state, and capital spent on railroads had a range of benefits: "Rail-Roads have introduced a new element of civilization, which is making wonderful revolutions in all the pursuits of peace, and the affairs of war, and even in intellectual improvements, affecting physically and morally, as well all the relations of business and property, as the extension of human knowledge, and the diffusion of social sympathies and connexions." What was more, these public benefits could also lead to private gains: "While they do all these things and more for the public good, they bring business and increased profits to our Road."[33]

Even those who did not share such high-flying ideals could still applaud it for raising the value of their land. "As the Rail Road from Walterboro is to pass through my Land in time I will inform you of the number of shares that I will take in it as I like the plan of construction and the power to be used on said road," Thomas Napier wrote to his agent in 1847. "From the various movements in respect to Rail Roads connected with Charleston SC I feel much confidence in the advance of real property there." Napier was fully convinced that the railroad would have positive benefits for the cities in which it terminated. In fact, railroads did have a positive impact on land value. Land in Barnwell County along the SCRR saw the price increase from less than fifty cents per acre to one to five dollars per acre after the road had been completed to Hamburg. Land along Virginia's Louisa Railroad "increased in value nearly one-third between 1830 and 1850." Andrew O'Brien noted that, when subscriptions were being pushed for a railroad project, he was sure that his property "would be worth 100 pr. ct. more." The railroad brought real economic benefits to those located along the route.[34]

In addition to the rise in land values, railroads brought economic benefits to businesses located on land near the road. Passengers needed nourishment and a place to rest their heads during long trips. At the end of 1835 the board of directors of the SCRR asked the president to "confer with Mr Rice who now keeps the public house at Summerville, and state to him that if a good house be built and well kept at George's Station that shall be the down dinner house." If Rice was unable to meet the company's needs, the president was authorized to "establish" such a house "by advertisement or otherwise." John Cassford Kerr was granted free passage on the SCRR in 1840 so long as his place was the "breakfast house" at Woodstock.[35]

Traveler James Davidson noted that there was significant competition for the business of travelers along the SCRR. He described Aiken, South Carolina, as "a cluster of Taverns. The railroad cars stop here during the night, and the principal business of this place is of course to entertain the passengers. There is always a scuffle amongst these landlords for the patronage of the passengers. When the Cars arrived there was a fellow present bowing and inviting the passengers—'Gentlemen will you walk to Major Marshe's?' It was the first time I had ever seen strangers invited to a tavern in that manner." Some railroads ran their own hotels in an attempt to capitalize on the opportunity. The Pontchartrain Railroad spent $11,000 in 1831 to build a "hotel and three bath houses" to entice travelers to relax at Lake Pontchartrain, and two years later it built separate facilities for free blacks. The SCRR owned its own hotel in Aiken. The Petersburg Railroad listed in its 1845 annual report a hotel as an expense item as well as $800 in income from "rent of hotel." In 1860 the CRRG reported that it held $2,300 worth of stock in the Milledgeville Hotel Company.[36]

The railroad's economic power could cause damage as well as create benefits. Railroads closed inefficient stations. A special committee of stockholders of the Southside Railroad felt that "Wilson's Depot in Dinwiddie and Wellville in Nolloway are too near together (being only about three miles apart) and that it would be to the interest of the Company to discontinue the former." By doing so, "a saving of some seven or eight hundred dollars would result to the company annually and no serious inconvenience to any one." Railroads could also force businesses to close. Elliott Story worked at a store in Virginia owned by a man named Trezvant, but after a railroad "intersected a part of the country near Jerusalem the trade of that place was much curtailed," although Trezvant's "cool treatments towards his customers" probably did not help matters. Charles Lyell observed that the railroad's introduction could lead travelers to neglect previous routes and businesses declined as a result. Traveling by carriage from Augusta to Savannah in 1841, less than a decade after the SCRR's completion, Lyell noted that he was "pursuing a line of road not much frequented of late, since the establishment of the railway from Augusta to Charleston. Our arrival, therefore, at the inns was usually a surprise, and instead of being welcomed, we were invariably recommended to go on farther. When once admitted, we were made very comfortable, having our meals with the family, and being treated more like guests than customers."[37]

Most impressive, though, railroads created towns where none existed before. The SCRR caused new communities to blossom, as the company noted in 1833: "Twelve months ago, Branchville, 62 miles from Charleston, and Midway, 72 miles distant, were in the midst of pine woods, unseen and unknown." Now, communities formed to serve the needs of passengers and freight. The development of Atlanta was even more spectacular. Thomas Hobbs noted in 1855 that Atlanta had blossomed to "a population of 10,000 souls, with extensive workshops, and factories, and is a beautiful instance of the development of a country by railroads." The railroad transformed the landscape in an obvious way when it cut deep in the land to lay tracks, but it also transformed lives as it recentered populations in the South.[38]

Just as railroads could create communities, they could also destroy them. The decision of a railroad to put its line in one area instead of another could damage a town's chances for survival. The town of Amsterdam, Mississippi, lost a great deal of its population to cholera and then went completely extinct when it was bypassed by the railroad. Likewise, Tatumsville, Mississippi, suffered with the construction of the Mississippi and Tennessee Railroad. The population of Tatumsville—led by Tatum, the founder of the town—moved to nearby Senatobia, which was on the line of the railroad. Even towns with considerable commercial prosperity could be done in by the lack of railroad connection. Preston, Mississippi, had, at its height, about six stores and "an excellent school." But around 1858 the population began to move to Garner, on the Mississippi and Tennessee Railroad, and a few years after the Civil War the town was completely "abandoned." Towns in other states had similar problems. Monck's Corner, South Carolina, suffered from being neglected by transportation. Although it was a healthy community before the Revolutionary War, the opening of the Santee Canal initially damaged the community's business prospects, and the Northeastern Railroad, which bypassed the old town, finished it off. The same name was given to a new settlement on the line of the railroad. Macon, Georgia, also suffered when new railroad towns diverted trade away from it.[39]

Some businesses worried that railroads would destroy their commercial prosperity. Thus, they were reluctant to allow "union" depots that would allow passengers to make easy transfers and forgo the necessity of spending their money in a city. As von Gerstner noted, "Neither in Richmond nor in Petersburg did the termini of the various railroads meet, and no connecting railroads unite them. Therefore passengers must be transported by omnibuses from one depot to the other. . . . In this way passengers and freight do not simply pass through, and innkeepers, forwarders, cart drivers, and so forth, are not deprived of their livelihoods." Of course, the lack of union stations was not purely the creation of the antebellum South: Chicago, for example, still drew complaints from travelers in the 1890s annoyed at having to travel crosstown when attempting to travel from the eastern to western United States. A similar desire drove Chicago businessmen: the hope of capturing business from travelers as they passed through. Philadelphia also forced passengers to disembark from trains and board omnibuses. The City Railroad's monopoly on downtown travel meant that other companies had to build their stations to connect with the City Railroad. Moreover, the ban on downtown steam travel meant that "additional terminals were constructed outside the original city limits to take advantage of the absence there of any restrictions on steam power."[40]

If some businessmen feared business would bypass them, southern planters eagerly cashed in on the opportunities presented by the railroad. Indeed, slaves noticed the difference in how hard they worked as the result of a railroad's appearance on the landscape. Abolitionist James Redpath, traveling near Weldon, North Carolina, in the mid-1850s, reported that slaves were increasingly unhappy with their working conditions. "De times is getting worse and worse wid us 'specially," one slave commented, "since dese engines have come in here." Redpath misunderstood the slave to say "Indians," but the slave's explanation made it clear that he was referring to the iron horse: "You see it is so much easier to carry off the produce and sell it now; 'cause they take it away so easy; and so the slaves are druv more and more to raise it." Railroads did not seem to harm slavery's prospects. As the Western and Atlantic Railroad moved into northwestern Georgia, the white population increased by "nearly 89 percent from 1840 to 1860," but the slave population increased 165 percent during the same time. Likewise, the booming economy of the 1850s South Carolina upcountry was assisted by the expansion of railroads. Slave values and populations increased during the decade.[41]

The railroad clearly altered decisions that planters made about where to send their goods. Talladega County, Alabama, planter James Mallory sent his cotton to the new railroad "every three days" in the 1850s: "It often took seven days to make a trip to Wetumpka with four bales," but the railroad allowed him to send five or six bales per trip via railroad to Selma. Thus, Wetumpka was cut out as a market and Selma elevated because of the railroad connection. Planters, of course, shipped goods other than cotton. John Bickley wrote to the SCRR in 1838 wondering if it would be possible for him to ship wheel spokes made at his plantation. The board of directors granted the president the ability to negotiate with Mr. Bickley and determine if such an arrangement would be "practicable." Drawn to the new form of transit, planters enthusiastically used it as an outlet for their output.[42]

Even Frederick Law Olmsted found a positive benefit for railroads in the South. "A gentleman, near Raleigh, who had a quantity of wheat to dispose of, seeing it quoted at high prices, in a paper of Petersburg, Va., and seeing, at the same time, the advertisement of a commission-house there, wrote to the latter, making an offer of it. The next day he received a reply, by mail, and by the train a bundle of sacks, in which he immediately forwarded the wheat, and, by the following return mail, received his pay, at the rate of $1.20 a bushel, the top price of the winter. At the same time, only forty miles from where he lived, off the line of the railroad, wheat was selling at 60 cents a bushel." Here, the railroad was not only convenient but increased the speed of business and increased profit.[43]

People anticipating a large amount of business could request that the railroad build a turnout or depot at their place. A group of citizens petitioned the Spartanburg and Union Railroad to create a depot on one Mrs. Shelton's land. The petition was approved, so long as the land was granted to the company without charge. After a Mr. Martin lobbied the Virginia and Tennessee Railroad to build a turnout on his land, it

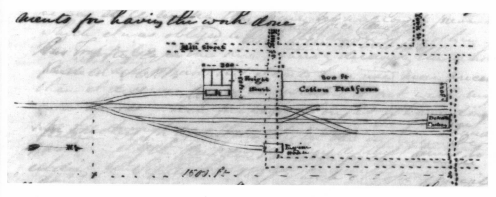

Figure 15. Sketch of the depot in Camden, South Carolina, drawn by John McRae. Letter from John McRae to James Gadsden, October 19, 1847, JML-2. Image number WHi-54987. Courtesy Wisconsin Historical Society, Madison.

was granted. Martin agreed to give up three acres of land in exchange for $300 in stock and also to relinquish any claims against the company for right of way.[44]

Depots were more public than turnouts.[45] Even for those not traveling on the railroad, the depot could become an important part of the community, although its design first had to meet the demands of freight and passengers. John McRae described in 1847 all the different needs the SCRR's depot in Camden (his sketch is figure 15) would serve:

> The freight House and passenger House to be under one roof. The passenger house being at the South end will have a separate & independent track for itself which will deposit the Passenger cars under the roof of the house. The Passenger House will be divided into four compartments. One contains the cars with a platform for Passengers and Baggage. One a Ladies room. Other a gentlemans Room and the fourth the Agents Office. Attached to the north end of the Freight house a cotton platform will extend if necessary up to the end of the Company lot 600 or 800 or even 1000 ft. Cars coming up loaded after depositing their freight in the house can be pushed up to the cotton platform or by a siding on to a separate track or the empty cars can be brought up to the cotton platform by a siding without interfering with the freight. A third track will hold empty cars & the De Kalb[?] Factory might put up a store house at the north end of the Lot. The dimensions of the freight and passenger house are large, but Boling and all at the Depot here say it ought to be at least 50 ft *wide.*[46]

The letter makes clear the complex needs the depot was meant to serve: space for passengers and baggage, a separate area for ladies, a substantial cotton platform, space for manipulating empty and loaded cars, and potential storage for a local company.

Some travelers felt that southern depots were not well kept. "One of the most striking things noticed by a northern traveller, upon all southern railroads, is the difference in the appearance of the depôts and more particularly the way stations," commented Solon Robinson. German immigrant Louis Heuser also found southern stations distasteful: "A person should not think of the American railway stations as being similar to those in Germany, but rather as simple shacks similar to our pigsties." And simply because companies decreed that passengers should have certain amenities did not mean that employees followed through. The Virginia and Tennessee Railroad set aside separate rooms for ladies in their depots. But in 1854 the board of directors chastised employees for not using them as such. "Should they be now used for sleeping apartments," the board declared, " . . . the beds [must] be removed therefrom without delay."[47]

Regardless of appearance, depots were the site of substantial activity. When passengers arrived at a depot, they found people there willing to meet their every need. One northerner related his treatment at a Louisiana depot: "Here we are at the depot and oh, what a collection of porters, cabmen & carmen, Irish, American & niggers. 'Take your baggage massa,' 'want a cab,' 'a nice conveyance I've got for your honor' were about the kinds of salutations I rec[ieve]d." Jeremiah Harris declared that Gordonsville, Virginia, had "more bustle and excitement than is to be witnessed almost any where else in the country, occasioned by the meeting of three mail trains, with their freight of live lumber." Mary Moragne compared a train's arrival to "the unloading of Noah's Ark, when the people began descending from it; for there seemed to be every variety of the human species. A great many men, like ourselves drawn by curiosity, or friendship, had been waiting their arrival; & two or three omnibusses were drawn up ready to transport them to town. In one second after the car had stopped, a man ran up to the window on the outside & enquired, 'Where do you stop, gentlemen'— 'at the United States.' 'Well, I am just in that route, & will carry you there in ten minutes—'" Changing trains could require a great deal of effort at a busy station. Thomas Hobbs wrote in 1859 that, when switching, "I had seven pieces of baggage to check; besides an armful of blankets, satchels, etc., etc., and what with getting seats for 3 ladies, a child & servant and checking baggage, I was kept pretty busy at each change of cars, for I . . . had to attend to it at every change."[48]

Even when the novelty of railroads had worn off, depots could still attract a crowd. Although surely familiar with railroads by 1855, James Frederick could still write of going to the train station as a "most magnificent affair." And why not? It gave him the opportunity to reconnect with those he loved, he could "rejoice once more to see Muts & the baby he doubled in size." Riding on the Charleston and Savannah Railroad in 1860, Samuel Burges noted that the train "stopped at quite a number of stations, where crowds had assembled to see Steam Horses, or to meet, or part with friends."[49]

Railroad stations became sites of community gatherings.[50] Peter Howell, an itinerant preacher, gave sermons from railroad cars in Virginia. When he was unable to find a place to preach in Suffolk, Virginia, he preached by candlelight while standing on a car. The impromptu accommodations worked well enough that he did it again the following night. That night the car had a table "with two candles and a Bible." He found that the audience "large and very attentive." When famed senator Daniel Webster stopped at a Georgia station in 1847, he was received by a "great crowd" with the "Mayor and Aldermen . . . there to give official welcome." A small boy dashed up to Webster "with the cry 'Howdy do-o-o Mr Webster' uttered in the shrill tones in which he was accustomed to talking to his deaf grandmother," much to the delight of the crowd. When Whig stump speaker John Bear arrived in Savannah, Georgia, his train was welcomed by "over three hundred men who met the cars some three miles out of the city, where myself and the committee were received with great pomp, and escorted into the city to the head quarters of the Young Men's Clay Club, where there were hundreds of men, women and children, white and black" gathered to hear his speech. In addition to private commerce and travel, depots also became centers of public activity.[51]

Cut Rope and Loose Stock

The relationship of the railroad to the immediate community was not always cordial. The SCRR suffered vandalism in 1835, when someone cut the rope at the inclined plane near Aiken. The president noted that a public reward would not be offered but that the agent of transportation would keep a "strict watch" to ensure that it would not happen again. The SCRR suffered from robberies on its trains in 1835 and 1836; the president reported in March 1836 that "no clue had as yet been afforded which might lead to a detection." Passing trains were also targets for vandals. The East Tennessee and Virginia Railroad noted in 1858 that one of its passenger trains had been shot at and another pelted with three stones. The board of directors authorized the superintendent to advertise that a $100 reward would be given for evidence leading to conviction. Companies also had to be wary of thievery. "One Who Knows" informed the Raleigh and Gaston Railroad that slaves made off with goods at night.[52]

While vandals and thieves were a nuisance, railroads also had to combat legal challenges from local landowners. Iverson Brooks complained to the SCRR in 1837 that the company had damaged his land by cutting down trees and removing dirt for an embankment.[53] That same year, some citizens of Charleston complained that the railroad's culverts were not draining properly, creating pools of stagnant water. The citizens also complained that between Line and Mary Streets there was a particular danger of fire from sparks; the company agreed to use a spark arrester.[54] Evidently the attempts to prevent sparks were not satisfactory to the Charleston community, because the railroad was fined $1,000 in January 1838 for being a public nuisance. The board responded by petitioning the governor for a remission of the fine, arguing that it had every right to run the trains and had placed "caps" on the smokestacks so as to prevent danger from fire. Moreover, the company noted that once people had complained that trains should not be allowed beyond Line Street, the company replaced steam engines with mules to haul the cars the remainder of the route.[55] Yet complaints continued to come in. Samuel Maverick, a businessman in Charleston, demanded $6,000 from the SCRR as the result of his buildings burning at the corner of King and Meeting Streets in Charleston.[56]

Railroad companies handled a diverse set of claims, as demonstrated by a September 1854 board of directors meeting on the RDRR. Jordon Cox sent his lawyer to claim $100 for "damages to his Buggie, run over by the coal train in Jan[uar]y last, in attempting to cross the main Street in Manchester," and this claim was denied. Joseph Gibbs, of Manchester, wanted $50 for a cow that had been killed; he was awarded $25. Martha Nobles wanted compensation "for the burning of a tobacco Barn," and upon conferring with legal counsel, the board decided to pay $25. In each case, the board had to decide what was prudent compensation.[57]

The most consistent claims that railroads faced were for dead livestock. Southern courts did not offer much relief to the corporations. Most southern states held that stock should be allowed to roam free, stockowners had no obligation to fence in the stock, and therefore railroads were responsible for animals injured on their tracks. In the 1850s the SCRR tried to argue that it should not be held responsible because damage to stock was inevitable, but state courts did not support their claim. Given the lack of sympathy they received from courts, railroads had to develop their own strategies to address this ongoing problem. In 1836 the SCRR paid $33 to W. D. Wade for several animals that were killed; the board of directors moved to instruct the superintendent to determine if the engineers should be held responsible and fined. Other railroads decided upon this solution. The East Tennessee and Virginia Railroad set up the following rules in 1856: first, if stock was killed it was the duty of conductors to report it immediately to the president or the "nearest Director" of the road. Second, engineers were required to take all care required when running on the road, and the superintendent was asked to post the state of Tennessee's regulations about stock killing in a "conspicuous place" on every passenger and freight train. Finally, the president or director to whom the report was made was authorized to negotiate settlement with the claimant or hire legal counsel as the case seemed to warrant. Eventually, the ETVRR shifted this responsibility to the engine drivers. In 1859 the company authorized engineers to be paid a $5 bonus for every month in which they killed no stock.[58]

But the company was continually bothered by stock wandering onto the track, and in its tenth annual report the company blasted the state legislature for allowing the problem to spin out of control. The stock problem was a grave one, not least of which because it caused a "loss of time, which the trains were compelled to regain by fast running, or lose the mails; thus wearing out machinery and road, and endangering the lives of passengers, and the property of both passengers and the road." The worry about the mail was serious—if it did not meet the Post Office's schedule, it would lose the contract. The company fumed that the railroad had a right to operate over the track that it had built, yet the courts routinely sided with animal owners in forcing the railroad to pay damages for stock killed. Should the legislature protect the "lives of the cattle or the travelers?" According to the company, the legislature had "protected the cattle and abandoned the travelers." It hoped that the legislature would take positive action in this matter and hold owners of stock liable for the damages that they caused to trains.[59]

The ETVRR's positive incentive of paying engineers when they did *not* kill stock contrasted with other railroads, who adopted the tactic of fining engineers for the value of stock killed by their engines. In 1857 the Memphis and Charleston Railroad decreed that engineers would be "charged with the value of said stock, when proven to have been killed or injured by the train run by said Engineer." In 1854 the Virginia and Tennessee Railroad required all conductors to immediately report stock killed; if they failed to do so, they would be held liable for the entire amount claimed by the farmer.[60]

Other railroads grudgingly agreed to pay, but set a high evidentiary bar in order to ensure that the claims were legitimate ones. In 1856 the Southside Railroad's board of directors authorized the treasurer "upon the warrant of the President" to pay one-half of any claim for stock killed. Claims had to be given by the claimant under oath, and the value of the stock had to be assessed "by two disinterested freeholders under oath taken before a justice of the peace." The following year the company revised its policy and would pay only for killed stock if the company "receives the stock killed through its agents, or half value when the owner retains the same." Two "disinterested freeholders" still had to give their assessment of the value under oath. Although the company seemed resigned to the existence of claims, it hoped to dissuade frivolous claims.[61]

Alternate Use

Railroad tracks were used by people who had no intent of riding trains. One such group included those who slept on the tracks. It is difficult to know why people would have selected such a dangerous spot, although because the roadbed was designed to drain water it may have been a relatively dry place to sleep. In 1855 a train on the Virginia and Tennessee Railroad struck a "negro man" asleep on the tracks. The board of directors believed that the man would recover, and in any event "no blame can be attached to the engineer, as it was after dark and he could not see the negro."[62] When slaves were the victims of such accidents and their owners demanded compensation, it entered a new area of law. Courts treated slaves differently from animals. Early court cases held that railroads were not liable for damage to slaves. In some cases, the engineer was not even required to blow the whistle. Thus, despite slaves' and stock's equivalent status as property, southern courts initially recognized that slaves should be able to hear a train, sense danger, and get out of the way just as any white person would. As time went on, "courts refined their rulings to require engineers to blow a whistle whenever they saw something on the tracks." Nevertheless, slaves were still assumed to have the power of reason, and slaves killed by railroads had to have acted "like a reasonable person" before their deaths if their owners were to claim any damages for the loss.[63]

Railroads may have also influenced how people walked. The railroad created a clear path for the train, but the majority of time, of course, the track was unoccupied by a train. Therefore, tracks constituted a level path for walkers where one may not have existed before. Moreover, the fact that roadbeds had to be constructed so that they would properly drain water meant that it could be a dry patch of land raised above rainfall in inclement weather. Nonriders walking on the track created an unintended use of the track. Figure 16 shows pedestrians using the tracks when not used by trains. Walkers demonstrated how the railroad represented the modern world but did not completely erase former ways of travel. The tracks across the South represented the most sophisticated engineering available at the time, yet these walkers were merely content to have a straight, level, and well-marked path. Railroad companies attempted to prevent such trespassing. "Please instruct contractors to put brush or some other obstructions on their sections before leaving them," civil engineer John McRae wrote, "to prevent people from travelling on the road bed and where cuttings are made across common road let them stop up the old roads so as to keep people from driving into the cuts."[64]

Such efforts notwithstanding, many took advantage of the path offered by the railroads. The Irish immigrant Andrew O'Brien had resolved in 1838 to travel from Charleston to Augusta in order to seek employment there. However, he arrived at

Figure 16. A pedestrian in the lower right makes use of the train tracks near Christiansburg, Virginia, for his own purposes, in this image from the Lewis Miller sketchbook. Courtesy Virginia Historical Society, Richmond, Virginia.

the train station just after the train had departed for the day. There he met another Irishman, who proposed that they travel on foot to see what work they could find. O'Brien agreed, and they set out, "peddler fashion" on the railroad, walking thirty miles to Summerville. They continued walking the next day to Midway, where they secured employment painting chimneys. James Frederick related a similar tale.

When he and his companion missed the train, the railroad still provided a clear trail of where to go. "Rise early, sunrise and no cars concluded to walk, Jim Fudge and myself started accordingly to walk — get to Marshallville at 9," Frederick wrote in 1855. Former slave Hanna Fambro recalled that on Sundays slaves would "walk togedder 'long the railroad track until we come to de church—'bout a half mile 'way—an' we look, an' dere, all roun' de church, was de hosses an' carriages of de white people. An' we slide down de railroad bank an' go in de side door" of the church. Former slave Anna Smith recalled that her father "was takin' a walk along de railroad tracks" when he unfortunately "dropped daid." Waddy Thompson walked on the train tracks during a snow storm in 1840. An observer wrote that the train knocked Thompson down and passed completely over him. Although Thompson was "somewhat bruised about the face," he did not sustain any larger injuries. Despite such risks, pedestrians were obviously not deterred from using railroad tracks, appropriating the company's investment to their own advantage.[65]

Local citizens also recognized that railroad bridges provided new, convenient access over waterways or swamps and demanded that they be allowed to make use of them. People simply asserted that right by walking over the bridges, which some companies did not appreciate. The ETVRR's board of directors authorized the employment of watchmen at several bridges in order to ward off trespassers. The railroad even built a house for watchmen to live in at one of the bridges. Citizens petitioned the ETVRR in 1859 to allow the construction of a footbridge on one of the railroad's bridges. The company allowed construction to go forward, under the conditions that the bridge was built under the railroad's supervision, it could determine the rules for using the bridge, and it could disallow common use of the bridge at any time.[66]

Other railroads took the view that bridges could be a potential moneymaker for the company. Indeed, the chief engineer of the Hiwassee Railroad (later the East Tennessee and Georgia Railroad) argued that bridges that the company would build across the Tennessee and Hiwassee Rivers should not "constitute an item in the expense of the road, as the toll derived from the common travelling alone, over them, will probably pay for their construction." Pedestrians would likely want to make use of any bridges, and so some companies felt that they may as well also be a source of income.[67]

Fares and Special Trains

For those who *did* pay to ride the railroad, however, the issue of how much they would have to pay was a critical one. Individuals and groups petitioned the railroad to accommodate their needs by scheduling special trains or reducing fares. In 1835 the SCRR received a petition from several citizens of Charleston, asking that a special engine be run to Summerville during the "sickly season," presumably to help residents escape the lowcountry. During the fever season in 1838, the president of the company noted that many people would like to leave Charleston yet were "destitute of pecuniary means." A unanimous resolution passed to allow such people to use the railroad to leave Charleston "at the discretion of the President."[68]

Some accommodations were made for much pleasanter circumstances. Companies could offer excursion tickets at reduced rates and also send extra trains for special passenger trips. In the spring of 1844, two Methodist camp meetings attended by between five thousand and six thousand people brought an enormous spike in passenger traffic to the SCRR. Numerous special trains were offered for the Methodist camp meetings in 1850 as well. People used the railroad to celebrate the Fourth of July in 1848, as noted by one South Carolina diarist: "It is said about 2,000 Persons came down. The fare was put down per $1 each way." The SCRR also offered a special deal for John C. Calhoun's funeral: "Persons from the interior of the State, who may be desirous of being present at the Funeral Ceremonies attendant on the arrival, in this city, on the 26th inst., of the remains of Mr. CALHOUN, will be enabled to do so, and return, by the payment of a single fare." Other railroads also granted dispensations to certain groups or locations. The RDRR ran special trains on Saturday nights to the Huguenot and Aurelia Springs during the summer of 1853. In 1856 the Southside Railroad extended a reduced rate to the Independent Order of Red Men for one of its excursions. The ETVRR issued special tickets in 1860 for people attending a Sons of Temperance meeting in Knoxville.[69]

The most problematic special ticket, however, was the "free pass," which plagued railroads for decades. Numerous individuals believed that their station in life warranted special treatment from the railroad. Newspaper editor R. H. Whitaker recalled that when he received his free pass "my importance had assumed the dimensions of the meadow frog, in Æsop's Fables, when he thought he was as large as the ox." The effect was similar on others, making them unwilling to part with such privilege. The determination of who was and was not eligible for a pass often sparked considerable controversy. Direct connection with the railroad company was not necessarily a prerequisite. One group that was often extended a pass was the clergy. The citizens of Aiken, South Carolina, asked the SCRR to grant a free pass to a Rev. D. People, who preached in that city; it was granted. The Tuscumbia, Courtland and Decatur Railroad in 1835 extended free passes to "ministers of the Gospel whilst going to or returning from preaching," but it also required conductors to carefully note who was receiving the free passes in each instance. Unusual circumstances could also gain a person a free pass. Former slave Hanna Fambro recalled that "Mauss John he had a life pass for hisself and family ter ride on dat railroad—

it was called Central of Georgia—'cause one time a train of cotton caught fire as it run thru de plantation, an' Mauss John he call out all his people an' day put out de fire, so Maus John he cud ride on de train whenever he feel like."[70]

The experience of the RDRR demonstrates how railroads attempted to bring the plethora of passes under control. The RDRR considered an extensive list of potential free passengers in 1851, including directors of the railroad and their families; former directors; presidents, directors, chief engineers, and superintendents of other railroads; the governor; members of Virginia's Board of Public Works; current and former civil engineers for the company; and contractors. The railroad moved to trim this list in 1855, when the board of directors moved to exclude former directors, subordinate engineers, newspaper editors, and the company's lawyers. The list was trimmed again in 1857, when the board of directors decided that "hereafter all persons attending Conventions, religious meetings, or associations as Delegates, all preachers, and ministers of the Gospel, and members of the Legislature as well during the session of the Legislature as at all other times, be charged the regular rate of fare on this road." Railroads wanted to maintain the goodwill of the communities they served, but not at the expense of the bottom line.[71]

Other Transportation Firms

In addition to the groups mentioned thus far, railroads also worked with other forms of transportation. Historians have been quick to point out that the South lacked the comprehensive railroad network of the North and that there were still substantial parts of the South that remained untouched by the iron horse. Yet taking a broader view creates a slightly different picture. We rarely expect one form of transportation to serve our entire needs without interruption, particularly for long-distance travel. Today, goods move smoothly on an array of options—ship, train, and truck—all linked by the common container. Although lacking the completely standardized container, transportation firms in the antebellum era worked in a similar fashion to supply a usable transportation network, even if one mode by itself could not satisfy all needs. Contemporaries recognized that the system that was being built was an improvement over alternatives. Railroad travel may have involved walking through Petersburg laden with baggage because there was no union station or catching a train in the wee hours of the morning. Yet few people clamored for a return to the stagecoach, and southern states pursued railroads voraciously in the 1850s. Indeed, the idea that southern railroads somehow failed because they did not constitute a total network that could accommodate all needs is particularly ironic given that the twentieth-century renaissance of freight railroading in America has been based in part on railroads' ability to cooperate with water and truck transportation.[72]

One of the first alternative forms of transport that railroads encountered was the public road. Just as railroad construction could interrupt public roads, these roads could be problematic for railroads. In 1835 the SCRR sent a complaint to the commissioner of roads "explaining to them the injury to the Rail Road from having the Public Road immediately by its side, and requesting them to remove such public Road, when it has been made, since the construction of the Rail Road." The railroad could also interrupt the land in ways that antagonized locals. In 1852 some citizens petitioned the RDRR to build a bridge over the railroad. Four years later, the same company allowed the Boston Toll Bridge Company to build a bridge over its track, with the understanding that "this company will not be held responsible for any injury resulting to persons or property on account of such crossing being over the road or near the depôt."[73]

Whatever problems may have come up because of the physical proximity of public roads and railroads, railroad companies soon found it to their advantage to work with those who used the roads. To boost their own passenger service, railroads wanted to offer up as many potential destinations as possible, which required maintaining close relationships with stagecoach companies. But a close relationship did not necessarily mean a coequal one. Railroads wanted to retain control over scheduling. The RFPRR thus informed one stagecoach line in 1836 that the stagecoach should plan on running "at the hours fixed by" the railroad.[74]

The inability of other companies to maintain schedules could frustrate the efforts of railroads to provide good service. When a Knoxville stage company was unable to deliver passengers on time, the ETVRR found that "great complaints are made against our company & agent." The agent was instructed to quit issuing through tickets on the stagecoach until the stage company was prepared to uphold its end of the bargain. Although fraught with such opportunities for disappointment, stages were still important for extending the railroad's reach. The CSCRR advertised that rail and stage connections were available to take passengers from Charlotte, North Carolina, to Danville, Virginia, where one could connect with the RDRR, the "Direct Route to Virginia Springs."[75]

Railroads also formed lasting connections with steamboat companies. Indeed, as Wayne Cline has argued, Alabama's railroads were "projected to haul farm and plantation products to river landings, where steamboats took over." Therefore, the two types of transportation worked in tandem. Likewise, Charles Johnson noted that the "chief reason for building a railroad in Mississippi was to make it possible to transport goods from different parts of the state to the Mississippi River." In 1835 the

Tuscumbia, Courtland and Decatur Railroad authorized a committee to either contract with a steamboat to carry passengers beyond the area served by the railroad or look into the railroad's acquisition of a boat. Railroads may have not been able to provide an entire transportation network alone, but their cooperation with steamboats allowed them to have broader impact.[76]

Railroad companies undertook improvements to stabilize water connections when necessary. The Greenville and Columbia Railroad's chief engineer urged the company to improve the navigability of the Broad River in South Carolina. Such improvements would "open to market a region of country which has no prospect of any other outlet; whose productions, lime, iron, meat and bread stuffs, require cheap transportation. . . . To the Greenville and Columbia Railroad, Broad River is destined to prove a most important feeder." In 1859 the CRRG responded quickly when it felt that steamboat service out of Savannah was going to be damaged. Because "the loss of this line of steamships would be very injurious, if not disastrous, to this Company, the Board of Directors did not hesitate to make the necessary arrangements for its continuance." In a bit of unintended irony, railroad promotional rhetoric had argued against depending on steamboats; once constructed, railroads wanted to ensure that steamboats were reliable.[77]

Steamboats even figured in railroad advertising, as demonstrated by the Petersburg Railroad. This company advertised its services in the *Charleston Courier*, attempting to attract passengers going north. It advertised a route that included both railroad and steamboat travel. The advertisements emphasized the quality of rolling stock: "The Engines new and of the most approved construction, the Cars are eight wheeled, with private apartments for Ladies, and there is a new and splendid steamboat on the Potomac." Transferring from one form of transportation to another was made easy once travelers reached Petersburg: "Here they pay through to Baltimore, and receive tickets for their baggage which relieve them of all trouble and expense on that score." As for accommodation, passengers would "dine in Petersburgh, sup in Fredericksburgh, sleep on board the Potomac steam boat, breakfast next morning in Baltimore, whence they can immediately proceed on to Philadelphia and New-York the same day." Charlestonians going west to Columbus, Georgia, were also tempted by advertisements promoting stagecoaches that were "daily in connection with the Georgia Rail Road, (84 miles of which is now finished,) and splendid Steam Packets on the Alabama River." The advertisements assured passengers that they "need feel under no apprehension of any attack being made upon them by the Seminole Indians."[78]

Not all relationships with other transportation firms were positive. The SCRR was rebuffed in its attempts to build a depot nearer to Charleston's wharf. In 1844 railroad president James Gadsden complained that the "agricultural interests of the interior," which were not exercising their necessary strength, were being "made subservient to a few wharfowners & drivers of drays." Two years later, civil engineer John McRae wrote to D. K. Minor, editor of the *American Railroad Journal:* "The wharf owners in this city have caused the company to abandon the water location for the depot in this city. The greater fools they." McRae noted that some local businessmen were disturbed by potential loss of business if travelers would not have to stop as long in the city: "The fraternity of *Black Barbers* . . . were alarmed by the idea that they were to be deprived, as well as other equally important interests, of an opportunity of *shaving* the travellers, as they passed from the cars to the steamboat."[79]

These complaints revealed a stark reality: whatever cooperation railroads and boats attempted, they could also be competitors for freight. The Southwestern Railroad reported that such competition could deny the railroad valuable cotton: "when the public roads are in bad order [planters] will make the shortest haul possible, whether to a river or the Railroad; and in consequence of the condition of things above stated, thousands of bales of cotton sought the Gulf markets by the Alabama, Chattahoochee and Flint Rivers, which in an ordinary season would have passed to Savannah over our line." The RFPRR dueled with Bay Steamship lines, and each company "slashed fares and distributed handbills condemning the rival company." While planters may have benefited from this competition, it revealed how friendly companies could turn foe.[80]

Finally, railroads cooperated with other railroads. Sometimes, they chose to assist in construction. In 1835 the SCRR received a request from the Georgia Railroad to transport "500 to 1000 Irish Labourers" to work on the Georgia Railroad. The SCRR transported the workers "in Freight cars, at the same rates as negroes, with the usual allowance of baggage." The SCRR also agreed to transport iron for the Georgia Railroad during the summer "at a lower rate, than at any other time of the year." As more railroads were built across the South, they began to coordinate their efforts in the sale of tickets. The SCRR declined the offer of the Georgia Railroad to carry passengers at the rate of five cents per mile in 1839, but the board of directors did express an interest in "the proposed arrangement for the sale of tickets."[81] The Virginia and Tennessee Railroad's general ticket agent reported in December 1859 that he kept accounts with fifteen other companies, demonstrating the level of cooperation and coordination that took place.[82] Increased railroads meant increased connections, and the Virginia and Tennessee Railroad boasted its potential connections with many other companies—the East Tennessee and Georgia, the East Tennessee and Virginia, the Western and Atlantic, the Memphis and Charleston, and the Alabama and Tennessee—as construction continued.[83]

Any cooperation among companies meant that they had to negotiate time. As with Sabbatarians, planters, and the Post Office, railroads were again not complete masters of their time, but were willing to negotiate away some control in return for larger benefits. The Wilmington and Raleigh Railroad informed the SCRR in 1837 that it "was about to put into operation a line of steam boats, railway and stages from Charleston via Wilmington to Halifax on the Roanoke river," and hoped that the two companies would be able to coordinate their schedules. The SCRR's board of directors promised "hearty cooperation" in this endeavor. In 1855 the RDRR found that its morning train that left Richmond was "attended with inconvenience both to passengers and the employees of the Railroad," and that the Virginia and Tennessee Railroad and the Southside Railroad also had problems with the connection. Therefore, the board of directors authorized the president of the railroad to consult with the presidents of the other two roads, to agree on a mutually acceptable time, and to "urge its adoption by the Post Master General." In April 1858 the CSCRR coordinated its schedules with the NCRR and the Raleigh and Gaston Railroad. The agreement, however, required the train to run at or above twenty miles per hour at points, which meant that it could pick up passengers only "at the more prominent points."[84]

Cooperation made some managers skittish because of the potential competition for traffic. When considering which route to construct, a minority of the CSCRR's committee on location urged cooperation with the Greenville and Columbia Railroad. "Companies as individuals ought to agree, when it is their mutual interest to do so," the committee noted. Yet the CSCRR felt the sting of competition less than a decade later. The company reported in 1857 that their connection to the North Carolina Central Railroad had "no doubt increased the through travel on our road." At the same time, though, the NCCRR "has opened a communication for trade and travel in another direction, which formerly found their outlet only over ours." Yet the committee hoped to persevere in cooperation, noting that passengers would benefit from coordinated scheduling by the companies. The lack of a national or sectional organization of railroads made setting up more permanent and uniform structures difficult. Agreements tended to be between individual roads and were not section- or nationwide. In 1853, for example, the SCRR announced that it had entered into an agreement with the Wilmington and Manchester Railroad to build a new bridge over the Wateree River and to build a depot for joint use where the Wilmington and Manchester intersected with the Camden branch of the SCRR. Even when direct cooperation was not possible, railroads still hoped for the benefits of cooperation. Although the CSCRR built its own depot near Columbia, the SCRR noted that it hoped to integrate its schedule with that railroad so that passengers would not suffer from delays.[85]

Linking up with other railroads could increase everybody's business, so railroads recognized when connecting with other railroads was to their advantage. When the North Carolina legislature approved a railroad to extend toward South Carolina, the Greenville and Columbia Railroad argued that it was of vital importance that it be selected as the connecting route. In 1856 the Southwestern Railroad reported that its "business" was "considerably increased by the Girard and Mobile Road," thanks to an interchange. There could be a delicate balance between cooperation and competition with any form of transit, but southern railroads hoped to realize the benefits of cooperation whenever they could.[86]

The railroads of the antebellum South were not simply faceless corporations, but were vital parts of the communities that they served. The railroad did not, however, hold the same meaning for all southerners. Rather, different groups invested railroads with different meanings, or made their own demands of the iron horse. Many of these demands revolved around time: towns, the federal government, other transportation firms, and religious leaders all wanted a say in when the railroad would stop. In some cases, the railroad was willing to negotiate with such groups, although occasionally it did so against its will. People also used railroads for purposes not intended when it was constructed: as a walking path or a place to sleep. Pedestrians took the most modern form of transportation available and used it in service of the oldest form of transportation, their feet. In sum, there was no one "southern" response to or use of the iron horse. Community groups and individuals made use of the technology in their own way, and companies had to address their demands.

Epilogue

Memory

The rapid construction of the 1850s was brought to a halt by the Civil War. Some corporations reacted to the changing political situation of the 1850s with political statements of their own. With disunion on the horizon, the RDRR's board of directors declared that they "will hereafter abstain from procuring supplies for the use of the Railroad Co from the Northern or Nonslaveholding states in all cases when they can be procured elsewhere of a suitable character and at reasonable rates of charge." They thus pledged their loyalty to the southern cause, although hedging a bit to ensure that rates were "reasonable." As war came closer, railroads brought news of rapidly changing events with them. John Hamilton Cornish reported in 1860 that at "the Cars" he found out "stirring news about the forts in Charleston Harbour." When secession was decided in Mississippi, the news was communicated at railroad depots. Reuben Davis rode on the Mobile and Ohio Railroad and remarked that he was "scarcely out of the sound of cannon all the way," as people celebrated the news with loud noise. "At those stations where cannon could not be procured, anvils were brought into requisition, and were managed with so much skill as to produce an equal uproar."[1]

Once the war began, railroads continued to bring news to anxious southerners. People longed to hear about their loved ones. "In passing up and down on the railroad," around August 1862, William Grayson wrote that "we saw at every stopping place a gathering of upturned, anxious faces seeking among the passengers someone who could give them intelligence of friends and relatives in the army of Virginia." Two former slaves recounted that whistles communicated information in advance of train's arrival. Ike Derricotte of Georgia recalled that "dere's one thing dat's still as fresh in my memory now as den, and dat's how people watched and waited to hear dat old Georgia train come in. . . . De way dat old train brought 'em de news was lak dis: if de southern troops was in de front, den dat old whistle jus' blowed continuously, but if it was bad news, den it was jus' one short, sharp blast. In dat way, from de time it got in hearin', evvybody could tell by de whistle if de news was good or bad and, believe me, evvybody sho' did listen to dat train." Jerry

Moore of Texas recounted a different system: "Soon as the train pulled into town it signaled. Three long, mournful whistles meant bad news. Three short, quick whistles meant good news." Southerners were long accustomed to the sounds of trains, but during war those sounds carried additional, weightier meaning.[2]

As Union troops advanced, the railroad could serve as an escape route for white southerners. Grayson wrote that as rumors spread that Federal troops were approaching Charleston, South Carolina, the railroads were filled with "flying multitudes" attempting to escape to the upcountry. Planters also used railroads to evacuate their slaves. Former slave Tim Thornton remembered, "When we was 'bout to hab freedom, dey thought the Yankees was a-goin' to take all de slaves, so dey put us all on trains and run us down South." Whereas masters had once feared that railroads would allow slaves to escape, now railroads were used to *help* slaves escape the approaching Union troops. Moreover, whites joined their slaves in using railroads to take flight. Sarah Morgan did not need to physically escape, but a train ride served as psychological escape for her: "I enjoyed that ride," she wrote of her trip. "It had but one fault; and that was, that it came to an end. I would have wished it to spin along until the war was over, or we in a settled home." Railroads could offer mental dislocation, but not enough to erase the reality of war.[3]

The success of railroads in the 1850s may have inflated hopes that they could sustain the fledgling Confederacy. Southern railroads did not provide adequate support for the Confederacy. But here they failed a task that they were never designed to do. Antebellum arguments about the military necessity of railroad routes were more geared toward securing government funding than actual military needs. The Confederate rail network's overall inability (despite some successes) to meet the challenges of war confirms this thesis. Problems such as the lack of union stations may have been a nuisance to travelers, but these problems became debilitating when the military needed to move large numbers of men and matériel quickly. In addition to the practical problems imposed by the state of the rail network, the Confederacy faced a significant problem of management and control. The new government did not impose any centralized control on the railroads. This could give companies veto power over whether troops were carried. The uniform rates offered by the government did not cover the costs of transportation, so railroads turned to more profitable traffic first. Thus, the lack of management of the complete rail system proved damaging to the war effort. Other factors also played a role in the Confederacy's difficulties. Skilled workers—slave and free—were drafted into the army or pulled away from the work by their owners. Confederate supplies of iron were never adequate to railroads' demand, and companies competed for supplies with the navy's ironclad ships. Finally, southern railroads had to overcome the tremendous disadvantage that the Civil War was fought largely on southern soil. Southern railroads not only had to work hard to satisfy the Confederacy's needs but also had to recover from damage by Union troops. These dual tasks were ultimately too much to bear. Although southern railroads had previously dealt with a wide range of users, during the Civil War railroads were suddenly confronted with a very new user—the military—with new demands, and they were unable to match the task.[4]

After the war, southern railroad men looked around and saw considerable destruction. Whether from the actions of the northern troops or the wearing out of southern equipment, many southern railroads saw their hard-won gains of the 1850s quickly erased. Yet that destruction did not last long. Railroads were soon rebuilt in the postbellum South. Construction took place "feverishly" and resulted in a more comprehensive network than what had come before; "by 1890," historian Edward Ayers observes, "nine out of every ten Southerners lived in a railroad county." As Ayers writes, railroads were "surrounded by an aura of glamour throughout the New South era." But this glamour was not a New South phenomenon. White southerners had made the decision to accept railroads long before the Reconstruction era. The Civil War denied them the opportunity to do it as they wished: with slavery intact.[5]

As the nineteenth century turned to the twentieth and the antebellum era faded into memory, railroads held a prominent position in southerners' reminiscences about the antebellum years. For some, the presence or absence of a railroad served as a marker in one's memory: a major event around which other events were timed. Tennessee Civil War veteran William Winston Eads, when asked about his ancestry, recalled that "my father moved to Sulivan county Tenn in 52 and to Louden in 53 that was the terminus of the East Tenn & Georgia railroad. the East Tenn & Va was built shortly after that time." The railroad itself could also provoke powerful memories. William Harden recalled that receiving a tour of the CRRG shops when he was a boy allowed him to "witnes[s] such machinery as I had never seen before" and "caused me to wonder." Charles Olmstead always knew he would remember his first train trip. "I was quite a small boy when my first railway journey was made," he noted. He remembered such details as the time and speed: "We started at 6 o'clock in the morning and did not reach Macon until 6 o'clock in the afternoon—good twelve hours. The speed of the train was never more than twenty miles an hour and there were long waits at every station. Yet it seemed to me that we were getting along at a terrific rate. . . . Every detail of that ride is impressed on my memory." David Gregg McIntosh recalled that on his first trip as a youth "it was my wonder and delight to watch and count the mile posts as we dashed by them." R. H. Whitaker recalled that the railroad made a strong impression on him as a boy, even before he saw one himself. His father traveled to Petersburg, Virginia, and came home to say that "he had not only seen a railroad, but had ridden on one; that the wheels ran on iron rails; the cars were propelled by a thing they called an engine; that the train, engine and two or three cars, ran at the rate of ten to fifteen miles an hour, and that said train could carry fifty passengers. I looked at him with amazement, and wondered how I would feel if I had seen as much as he had, and knew as much as he knew." N. J. Bell recalled that as a boy in Anderson District, South Carolina, he heard of the SCRR around 1831 and "wondered what kind of a road a railroad could be. I finally concluded it must be one built of fence rails, not being old enough to think of asking any one what a railroad was." A few years later, he traveled to Hamburg and finally saw the railroad for the first time. Although he admitted that he could "not remember of seeing any engine or coaches," he did recall "seeing some box-cars standing on a track on the outskirts of the town."[6]

Slaves also remembered railroads. When interviewed by the Works Progress Administration, former slave Al Rosboro recalled the train as his earliest memory: "Let me see. Fust thing I 'members well, was a big crowd wid picks and shovels, a buildin' de railroad track right out de other side of de big road in front of old marster's house. . . . A train of cars a movin' 'long is still de grandest sight to my eyes in de world. Excite me more now than greyhound busses, or airplanes in de sky ever do." When John Brown of South Carolina was asked by the WPA interviewer for the circumstances of his birth, Brown responded that he "found in a basket, dressed in nice baby clothes, on de railroad track at Dawkins, S. C. De engineer stop de train, got out, and found me sumpin' like de princess found Moses, but not in de bulrushes." Although interviewed by the WPA in Texas, Abraham Coker was enslaved in Georgia and remembered that his master allowed his father and siblings to take the train around 1856 to visit his sick mother, who worked on another plantation. His first ride stuck in his memory much as Charles Olmstead's did. "Dis was de first time dat I ever took a train ride," Coker reminisced. "De seats in de cars was padded lak the ones ob today. I thought dat ridin' on dat train was wunnerful. I was about twelb years old at de time, but I can remembah it jes' lak it was yesterday. I have never forgot dat train ride." For whites and blacks alike, the memory of a railroad journey was a powerful one, even decades removed from the event. By the time the men quoted here had been interviewed or set their thoughts to paper, they had doubtless ridden trains many times. As commonplace as the action may have become, the first encounter with the new technology was still memorable.[7]

Having experienced multiple decades of transportation improvement, some did not look back on their antebellum experiences in a positive light. "Going to Washington in those days was a very different affair from that of the present," Mrs. John A.

Logan informed the readers of her autobiography. "The crude railroading, the uncomfortable, barren, low-berthed sleeping-cars can never be forgotten. . . . All trains were late, overcrowded, and uncomfortable. We had to change frequently." Although Logan's trip from the West to Washington, D.C., represented the first time she crossed the Allegheny Mountains, she remembered that she was unable to enjoy the experience, the desire to take in the majestic scenery battling with her attempts to overcome "desperate car-sickness and fatigue." The places for eating were "few and far between," and the vocal complaints of the children "added additional annoyance to passengers." Even leaving the train was painful for Logan: upon arriving in Washington she was "almost deafened and completely terrified" when set upon by the "rush of burly hackmen" attempting to persuade people to go to their hotels. Others focused on the dangers they remembered. W. C. Curtis wrote that a traveler who arrived in Wilmington via the Wilmington and Weldon Railroad in the 1850s "could congratulat[e] himself on having escaped the dangers of a journey upon those rails which existed, not only on that railroad but nearly every other railroad in the country."[8]

Southerners were aware that railroads had definitely improved since the time they were youths. "As I look back to 1848 and behold the little thing called a train," recalled Whitaker, "poking along on the level, running like fury down grade, but puffing up grade, and compare it with the train I see sweeping along majestically, paying no heed to grades, curves nor bridges, I realize that a mighty change has taken place which only he who has seen the past as well as the present, can appreciate." But Whitaker recognized that future generations would regard his own time with equivalent bemusement: "I suppose that the man who writes reminiscences of these stirring days, when, in the time to come, balloons shall have taken the places of vestibuled trains and palatial steamers, will look down from the balloon window as his airship makes its thousand miles an hour, and sigh as he thinks of the good old times when, as a boy, he could take his time along through life at the snail's pace of a hundred miles an hour." Antebellum train travel may have sparked some bitter memories because of the improvements that had since taken place. But Whitaker and others realized that time had clouded their judgment, and antebellum innovations had struck them as remarkable at the time.[9]

Southerners embraced railroads from their inception in the United States. Moreover, southern railroad development had much in common with that in the North. Both regions of the country experimented with a new technology that they imported jointly from Great Britain. Railroad engineers, North and South, addressed issues of technology, finance, and labor control. Like northern railroads, southern railroads worked hard to introduce system, regularity, and bureaucracy into their works. Indeed, southern railroads were pioneering in this regard, as the experience of the SCRR demonstrates. The common thread of northern and southern development confirms the South's commitment to technology, but the South did not simply mimic the North. White southerners enthusiastically pursued modern transport, but wanted to do so without adopting northern labor relations. As a result, slaves performed the bulk of the labor involved in building and operating the South's railroads.

Examining the travel experience of southerners makes clear just how quickly they became enamored with travel. The experience of riding the train engaged all the senses. "Modernity" was not an abstract concept; it was something that travelers saw, heard, and felt. Historians often overemphasize problems faced by early transportation: the lack of network, the variability in gauge, the slow travel times compared to contemporary facilities. From the perspective of the early nineteenth century, however, travel did not always seem tedious or tiresome. Travelers learned to take the difficulties in stride, and appreciated the convenience railroads offered.

The passenger experience was but one among many that people could have with a train. There was no single "southern" railroad experience. Rather, it is possible to assess a range of different relationships that preachers, politicians, planters, pedestrians, and others had with the iron horse. Some of these relationships reflected unintended uses or even looked backward to older modes of transit—pedestrians who used tracks as a walking path, for example. But other relationships looked to the future as railroad companies negotiated with each other to build the South's railroad network. Various groups in the South were able to integrate the railroad into their own needs.[10]

Many of these relationships are best understood through the prism of time, and time was crucial for southern railroads. It was a guarantor of safety, a point of pride for railroad boosters, and a negotiating point with community groups. Railroad companies never fully controlled time, despite their best efforts. Southern railroads were fully aware of time's import and southern passengers expected and demanded timely service. The degree that multiple times penetrated contemporary discussions about the railroad demonstrates the Old South's modern mentality.

This fuller picture of southern railroads also gives us insight into antebellum political economy: we can see how free wage labor and slavery were integrated before the Civil War. Railroads expanded through antebellum America, built by both free whites and slaves. Railroads in the South illustrate perfectly how white southerners committed to racial slavery funded and developed a technological system associated with modern advancement. Their success in doing so forces us to reassess the meaning of modernity. Modernity, progress, and technology are so thoroughly intertwined in the American imagination—"Americans have seized upon the machine as their birthright," Leo Marx has written—that it may seem odd to see these very same qualities in the antebellum South, the region and time period that represent the path not taken by the American people.[11] But antebellum southerners seized their birthright as well. Moreover, white southerners sought to make it in their image by making modern technology compatible with slavery. It is a sobering reminder that technology and modernity do not automatically bring freedom with them. In the case of the South, slaves built railroads that, once in operation, facilitated their continued enslavement by helping sustain the cotton economy. The rapid embrace of the railroad by southern travelers and shippers demonstrates that railroads affected the Old South's development, and the ease with which slavery was integrated into railroading shows that the Old South also determined how railroads developed. We can no longer characterize the Old South as resolutely premodern. Rather, it was a society that pursued modernity in service of its own demands.

Abbreviations

ARJ	*American Railroad Journal*
BARC	Boston and Albany Railroad Collection, BHC
BARC-BW	Boston and Worcester Railroad, BARC
BARC-WR	Western Railroad, BARC
BHC	Baker Library Historical Collections, Harvard Business School, Boston, Massachusetts
ETGRR	East Tennessee and Georgia Railroad
ETGRR Minutebook	East Tennessee and Georgia Railroad, Minutebook of the Stockholders and Directors (1850–1869), Ms84-047, NSHC
ETVRR Minutebook-1	East Tennessee and Virginia Railroad, Minutebook of the Board of Directors (1853–1858), Ms84-048, NSHC
ETVRR Minutebook-2	East Tennessee and Virginia Railroad, Minutebook of the Stockholders and Directors (1858–1869), Ms84-048, NSHC
Glass Letterbook	Letterbook (1826–1873), Glass Family Papers, SCL
Hiwassee Minutebook	Hiwassee Railroad Company, Hiwassee Railroad Directors Minutebook (1836–1842, 1846–1848), East Tennessee & Georgia Railroad Directors Minutebook (1848–1850), Ms84-075, NSHC
JJP	*John B. Jervis Papers* (microfilm; Rome, N.Y.: Jervis Public Library, 1966)
JML	John McRae Letterbooks [nos. 1–8], WHS
HL	Courtesy of Hargrett Rare Book and Manuscript Library, University of Georgia Libraries, Athens
LCCRRJT	Louisville, Cincinnati and Charleston Rail Road Company Journal of Transactions (1837–1840), SCL
MCRR	Memphis and Charleston Railroad
MCRR Minutebook	Memphis and Charleston Railroad, Minutebook of the Board of Directors (1850–1870), Ms84-103, NSHC
Napier Letterbook	Letterbook (1846–1848), Thomas Napier Papers, SCL
Nott Letterbook	Letterbook (1841–1844), Samuel Nott Collection, BHC
NSHC	Norfolk Southern Historical Collection, Norfolk, Virginia
RASP	*Records of Ante-bellum Southern Plantations from the Revolution*

to the Civil War (microfilm; Frederick, Md., and Bethesda, Md.: University Publications of America, 1985–)

RDRR Minutebook   Richmond and Danville Railroad, Minutebook of the Directors (1847–1869), Ms83-133, NSHC

RFPRR-VHS   Richmond, Fredericksburg and Potomac Railroad Company Records (1833–1909), VHS

RWRRC   Rutland and Washington Railroad Collection, BHC

SASI   *Slavery in Antebellum Southern Industries* (microfilm; Bethesda, Md.: University Publications of America, 1991–)

SCHS   South Carolina Historical Society, Charleston

SCL   South Caroliniana Library, University of South Carolina, Columbia

SCRR Minutebook   South Carolina Canal and Rail Road Company, Minutebook of the Board of Directors (1835–1841), Ms84-149, NSHC

SHC   Southern Historical Collection, Wilson Library, University of North Carolina, Chapel Hill

Southside Minutebook-1   Southside Railroad, Minutebook of the Stockholders and Directors (August 4, 1849–October 19, 1853), Ms81-063, NSHC

Southside Minutebook-2   Southside Railroad, Minutebook of the Board of Directors (October 28, 1853–March 18, 1865), Ms81-063, NSHC

StHC   Stone and Harris Collection, BHC

SURR Minutebook   Spartanburg and Union Railroad, Minutebook of the Stockholders and Directors (1851–1873), Ms84-153, NSHC

SWRR   Southwestern Railroad

TCDRR Minutebook   Tuscumbia, Courtland and Decatur Railroad, Minutebook of the Stockholders and Directors (1832–1843), Ms84-159, NSHC

VHS   Virginia Historical Society, Richmond

VTRR Minutebook   Virginia and Tennessee Railroad, Minutebook of the Board of Directors (1853–1857), Ms81-073, NSHC

Wales Letterbook   Letterbook of T. B. Wales, vol. 145, BARC-WR

WHS   Wisconsin Historical Society, Madison

Introduction

1. These quotations are taken from Ebenezer Hiram Stedman, *Bluegrass Craftsman: Being the Reminiscences of Ebenezer Hiram Stedman, Papermaker, 1808–1885,* ed. Frances L. S. Dugan and Jacqueline P. Bull (Lexington: University of Kentucky Press, 1959), 105–6. The "Munchawsen" reference is to Baron Munchausen of *Baron Munchausen's Travels* (1785), a satirical collection of tall tales. See John Carswell, *The Romantic Rogue: Being the Singular Life and Adventures of Rudolph Eric Raspe, Creator of Baron Munchausen* (New York: Dutton, 1950), 184–89.

A few words about evidence: in quotations I have not chosen to litter the text with *sic* but have left spelling as it was in the original. I have silently inserted punctuation to ease reading. Of course, if the editor of a published primary source included the notation *sic* or other notations, I have included that when quoting from such sources. Annual reports from railroad companies formed a critical foundation of my research. These reports had no uniform titles when published: some companies numbered them, others issued them every six months instead of yearly, and one company once issued two "annual" reports in a single year. These and other problems led me to use the full title found on the cover of each report for the first reference in the notes. The subsequent references include the year (and month, when available) of each report in order to aid identification. Unless otherwise noted, I have left the emphasis as provided in the original source in all quotations.

2. Thomas W. Chadwick, ed., "The Diary of Samuel Edward Burges, 1860–1862," *South Carolina Historical and Genealogical Magazine* 48 (April 1947): 66, 67, 73. "T.O." is an abbreviation for "turnout," or siding. The diary continues as Thomas W. Chadwick, ed., "The Diary of Samuel Edward Burges, 1860–1862," *South Carolina Historical and Genealogical Magazine* 48 (July 1947): 141–63.

3. Eugene Genovese, *The Slaveholders' Dilemma: Freedom and Progress in Southern Conservative Thought, 1820–1860* (Columbia: University of South Carolina Press, 1992), 10 ("sealed the triumph"); see also Gavin Wright, *Slavery and American Economic Development* (Baton Rouge: Louisiana State University Press, 2006), 67; Elizabeth Fox-Genovese and Eugene Genovese, *Fruits of Merchant Capital* (New York: Oxford University Press, 1983), 59; and Eugene Genovese, *The Political Economy of Slavery,* 2d ed. (Middletown, Conn.: Wesleyan University Press, 1989).

4. Leo Marx, *The Machine in the Garden: Technology and the Pastoral Ideal in America,* rev. ed. (New York: Oxford University Press, 2000), 27, 219; Joyce Appleby, *Inheriting the Revolution: The First Generation of Americans* (Cambridge, Mass.: Belknap Press of Harvard University Press, 2000), 266; Richard D. Brown, *Modernization: The Transformation of American Life, 1600–1865* (New York: Hill and Wang, 1976), 181.

5. Albert Fishlow, "Internal Transportation in the Nineteenth and Early Twentieth Centuries," in *The Cambridge Economic History of the United States,* vol. 2, *The Long Nineteenth Century,* ed. Stanley L. Engerman and Robert E. Gallman (New York: Cambridge University Press, 2000), 575; Ralph Waldo Emerson, "Address Delivered in Concord on the Anniversary of the Emancipation of the Negroes in the British West Indies, August 1, 1844," in *Miscellanies,* vol. 11 (1878; reprint, Cambridge, Mass.: Riverside Press, 1904), 125; John F. Stover, *Iron Road to the West: American Railroads in the 1850s* (New York: Columbia University Press, 1978), 26, 61; Ulrich Bonnell Phillips, *A History of Transportation in the Eastern Cotton Belt to 1860* (1908; reprint, New York: Octagon Books, 1968), 1. For an example of how Phillips's arguments have remained with us, see Scott R. Nelson, *Iron Confederacies: Southern Railways, Klan Violence, and Reconstruction* (Chapel Hill: University of North Carolina Press, 1999), 13.

6. Stover, *Iron Road,* 26, 61; Fishlow, "Internal Transportation," 578.

7. Walter Johnson, "The Pedestal and the Veil: Rethinking the Capitalism/Slavery Question," *Journal of the Early Republic* 24 (Summer 2004): 305. For the capitalist vs. noncapitalist debate, see Mark M. Smith, *Debating Slavery: Economy and Society in the Antebellum American South* (Cambridge: Cambridge University Press, 1998), 92–94, and Tom Downey, *Planting a Capitalist South: Masters, Merchants, and Manufacturers in the Southern Interior, 1790–1860* (Baton Rouge: Louisiana State University Press, 2006), introduction. For the richness and complexity of southern intellectual life, see Michael O'Brien, *Conjectures of Order: Intellectual Life and the American South, 1810–1860* (Chapel Hill: University of North Carolina Press, 2004).

8. The railroad booster quotation is drawn from *Proceedings of the Citizens of Charleston, Embracing the Report of the Committee and the Address & Resolutions Adopted at a General Meeting in Reference to the Proposed Rail-Road from Cincinnati to Charleston* (Charleston, S.C.: A. E. Miller, 1835), 22. For the reforms described in this paragraph, see Joyce E. Chaplin, *An Anxious Pursuit: Agricultural Innovation and Modernity in the Lower South, 1730–1815* (Chapel Hill: University of North Carolina Press, 1993), 45, 259; Mark M. Smith, *Mastered by the Clock: Time, Slavery, and Freedom in the American South* (Chapel Hill: University of North Carolina Press, 1997), 95; Downey, *Planting a Capitalist South*, 226; Steven G. Collins, "System, Organization, and Agricultural Reform in the Antebellum South, 1840–1860," *Agricultural History* 75 (Winter 2001): 4; Richard Follett, *The Sugar Masters: Planters and Slaves in Louisiana's Cane World, 1820–1860* (Baton Rouge: Louisiana State University Press, 2005), chap. 3; Stephen B. Hodin, "The Mechanisms of Monticello: Saving Labor in Jefferson's America," *Journal of the Early Republic* 26 (Fall 2006): 415; John W. Quist, *Restless Visionaries: The Social Roots of Antebellum Reform in Alabama and Michigan* (Baton Rouge: Louisiana State University Press, 1998); Jeffrey Robert Young, *Domesticating Slavery: The Master Class in Georgia and South Carolina, 1670–1837* (Chapel Hill: University of North Carolina Press, 1999), 111, 124, 133; Genovese, *Slaveholders' Dilemma*, 27; and Stanley L. Engerman, "Southern Industrialization: Myths and Realities," in *Global Perspectives on Industrial Transformation in the American South*, ed. Susanna Delfino and Michele Gillespie (Columbia: University of Missouri Press, 2005), 14–25. That some of these men of capital were northerners did not prevent them from siding with their adoptive homes when the Civil War broke out. See Charles B. Dew, *Bond of Iron: Master and Slave at Buffalo Forge* (New York: Norton, 1994), 288, and Curtis J. Evans, *The Conquest of Labor: Daniel Pratt and Southern Industrialization* (Baton Rouge: Louisiana State University Press, 2001), 214.

9. Letter from John McRae to D. K. Minor, December 6, 1846, JML-1.

10. For outstanding examples of economic history, see George Rogers Taylor, *The Transportation Revolution, 1815–1860* (New York: Rinehart, 1951), and Albert Fishlow, *American Railroads and the Transformation of the Ante-Bellum Economy* (Cambridge, Mass.: Harvard University Press, 1965). Leading railroad historian Maury Klein has lamented the neglect of social history within American railroad history; see Maury Klein, *Unfinished Business: The Railroad in American Life* (Hanover, N.H.: University Press of New England, 1994), 168. Historians of Europe have been more willing to explore the social significance of transportation developments. See, for example, Michael Freeman, "The Railway as Cultural Metaphor: 'What Kind of Railway History?' Revisited," *Journal of Transport History*, 3d ser., 20 (September 1999): 160–66; Ralph Harrington, "The Railway Journey and the Neuroses of Modernity," in *Pathologies of Travel*, ed. Richard Wrigley and George Revill (Atlanta: Rodopi, 2000), 229–59; and Wolfgang Schivelbusch, *The Railway Journey* (New York: Urizen, 1979).

11. Peter Kolchin, *A Sphinx on the American Land: The Nineteenth-Century South in Comparative Perspective* (Baton Rouge: Louisiana State University Press, 2003), 15; Chaplin, *Anxious Pursuit*, 278.

12. Walter Licht, *Industrializing America: The Nineteenth Century* (Baltimore: Johns Hopkins University Press, 1995), 47–48. This theme of multiple understandings draws from the work of historians who argue that technology is best understood through the study of "relevant social groups." See, in particular, Trevor J. Pinch and Wiebe E. Bijker, "The Social Construction of Facts and Artifacts: Or How the Sociology of Science and the Sociology of Tech-

nology Might Benefit Each Other," in *The Social Construction of Technological Systems: New Directions in the Sociology and History of Technology*, ed. Wiebe E. Bijker, Thomas P. Hughes, and Trevor J. Pinch (Cambridge, Mass.: MIT Press, 1987), 17–50; Ronald Kline and Trevor J. Pinch, "Users as Agents of Technological Change: The Social Construction of the Automobile in the Rural United States," *Technology and Culture* 37 (October 1996): 763–95; and Wiebe E. Bijker, *Of Bicycles, Bakelites, and Bulbs: Toward a Theory of Sociotechnical Change* (Cambridge, Mass.: MIT Press, 1995).

13. James A. Ward, "On Time: Railroads and the Tempo of American Life," *Railroad History* 151 (1984): 87–95; L. S. Kemnitzer, "Another View of Time and the Railroader," *Anthropological Quarterly* 50 (January 1977): 25–29; Ian R. Bartky, *Selling the True Time: Nineteenth-Century Timekeeping in America* (Stanford: Stanford University Press, 2000); W. F. Cottrell, "Of Time and the Railroader," *American Sociological Review* 4 (April 1939): 190–98; and Frederick C. Gamst, *The Hoghead: An Industrial Ethnology of the Locomotive Engineer* (New York: Holt, Rinehart and Winston, 1980). For a general exploration of the multiplicity of times, see Barbara Adam, *Timewatch: The Social Analysis of Time* (Cambridge, Mass.: Polity, 1995), 12.

Chapter One • Dreams

1. *Report of a Special Committee Appointed by the Chamber of Commerce, to Inquire into the Cost, Revenue and Advantages of a Rail Road Communication between the City of Charleston and the Towns of Hamburg & Augusta* (Charleston, S.C.: A. E. Miller, 1828), 22; Peter A. Coclanis, *The Shadow of a Dream: Economic Life and Death in the South Carolina Low Country, 1670–1920* (New York: Oxford University Press, 1989), 112.

2. I thus dispute Richard Brown's claim that "initiatives for transportation . . . often came from outside" the South. Richard D. Brown, *Modernization: The Transformation of American Life, 1600–1865* (New York: Hill and Wang, 1976), 114. For a more recent restatement of a similar argument, see Jonathan Daniel Wells, *The Origins of the Southern Middle Class, 1800–1861* (Chapel Hill: University of North Carolina Press, 2004), 171. For the early projects cited in this paragraph, see James G. Holmes, "Account of Coolridge's Rail Road and Canal Earth Cart," *The Southern Agriculturalist, and Register of Rural Affairs; Adapted to the Southern Section of the United States* 3 (May 1830): 254; *Kentucky Reporter*, August 18, 1830, p. 1, col. 6; Hoyt Paul Canady, "Internal Improvements in Georgia in the Pre-Railroad Era, 1817–1833" (Master's thesis, Georgia Southern College, 1970), 51–52; Charles Ripley Johnson, "Railroad Legislation and Building in Mississippi, 1830–1840," *Journal of Mississippi History* 4 (October 1942): 196; Wayne Cline, *Alabama Railroads* (Tuscaloosa: University of Alabama Press, 1997), 10. For the *Advocate* and *Journal*, see David P. Forsyth, *The Business Press in America, 1750–1865* (Philadelphia: Chilton, 1964), 175–86. For names of southern subscribers to the *ARJ*, see the following issues of the *ARJ*, volume 3: (January 18, 1834): 32; (January 25, 1834): 48; (February 8, 1834): 80; (February 22, 1834): 112; (March 22, 1834): 176; (April 19, 1834): 240; (May 10, 1834): 288; and volume 6: (January 14, 1837): 17; (February 4, 1837): 65; (February 11, 1837): 81; (February 18, 1837): 97; (February 25, 1837): 113; (March 11, 1837): 160; (March 18, 1837): 161; (March 25, 1837): 191; (April 1, 1837): 207; (April 8, 1837): 223; (April 15, 1837): 239; (April 22, 1837): 255; (June 3, 1837): 351; (June 17, 1837): 383; (September 16, 1837): 567. Publisher D. K. Minor secured agents to sell subscriptions for the *ARJ*

and his other publications. In 1834 his agents were present in six cities in Virginia, three in Kentucky, two apiece in Mississippi and Louisiana, and one apiece in Tennessee and South Carolina. See *ARJ* 3 (April 12, 1834): 224. Information on the SCRR can be found in reports for the South Carolina Canal and Rail Road Company; South Carolina Railroad Company; and the Louisville, Cincinnati and Charleston Railroad Company. For simplicity's sake, in the text I will refer to the company as the South Carolina Railroad; the specific report will be clear from my notes. For the history of the railroad, see Samuel Melanchthon Derrick, *Centennial History of South Carolina Railroad* (Columbia, S.C.: State Company, 1930); Donald A. Grinde Jr., "Building the South Carolina Railroad," *South Carolina Historical Magazine* 77 (April 1976): 84–96; and Tom Downey, *Planting a Capitalist South: Masters, Merchants, and Manufacturers in the Southern Interior, 1790–1860* (Baton Rouge: Louisiana State University Press, 2006). For a general view of railroad development in the 1830s, see Frederick C. Gamst, "The Transfer of Pioneering British Railroad Technology to North America," in *Early Railways: A Selection of Papers from the First International Early Railways Conference*, ed. Andy Guy and Jim Rees (London: Newcomen Society, 2001), 256–61.

3. *Report of a Special Committee*, 32; *Charleston Courier*, March 14, 1850, 3; James A. Ward, "On Time: Railroads and the Tempo of American Life," *Railroad History* 151 (1984): 87–88.

4. George Rogers Taylor, *The Transportation Revolution, 1815–1860* (New York: Rinehart, 1951), 79; N. P. Renfro, *The Beginning of Railroads in Alabama* (Auburn, Ala.: n.p., 1910), 2n2; *Annual Report of the Board of Directors of the South Carolina Canal and Rail-Road Company to the Stockholders, with Accompanying Documents Submitted at Their Meeting, May 6, 1833* (Charleston, S.C.: A. E. Miller, 1833), 6.

5. *Report of a Special Committee*, 18; Elias Horry, *An Address Respecting the Charleston & Hamburgh Railroad, and on the Railroad System as Regards a Large Portion of the Southern and Western States of the North American Union* (Charleston, S.C.: A. E. Miller, 1833), 11.

6. Horry, *Address Respecting the Charleston & Hamburgh Railroad*, 5; Renfro, *Beginning*, 9; *Proceedings of the Knoxville Convention, in Relation to the Proposed Louisville, Cincinnati and Charleston Rail-Road, Assembled at Knoxville, Tennessee, July 4th, 1836* (Knoxville: Ramsey and Craighead, 1836), 10. Although the issue of trade balance is addressed in chapter 5, it is worth pointing out here that consumerism in the antebellum South has yet to receive sustained historiographic attention. For some recent scholarship, see Frank J. Byrne, *Becoming Bourgeois: Merchant Culture in the South, 1820–1865* (Lexington: University Press of Kentucky, 2006).

7. W. Kirk Wood, "U. B. Phillips and Antebellum Southern Rail Inferiority: The Origins of a Myth," *Southern Studies* 26 (Fall–Winter 1987): 181 ("limited, local, and conservative"); Douglas J. Puffert, "Path Dependence in Spatial Networks: The Standardization of Railway Track Gauge," *Explorations in Economic History* 39 (July 2002): 291 ("did not foresee the future value"—Puffert was not speaking of southern railroads in particular, but all railroads generally; therefore, while his point may be valid elsewhere, it does not appear to apply to the antebellum South); *Rail-Road Advocate* 1 (July 19, 1831): 11 ("emoluments"); *Charleston Courier*, June 20, 1831, 2 ("much cheaper rate"); *Proceedings of the Citizens of Charleston, Embracing the Report of the Committee and the Address & Resolutions Adopted at a General Meeting in Reference to the Proposed Rail-Road from Cincinnati to Charleston* (Charleston, S.C.: A. E. Miller, 1835), 13 ("cotton and rice").

For southern desire to connect to the Pacific, see R. S. Cotterill, "Southern Railroads and

Western Trade, 1840–1850," *Mississippi Valley Historical Review* 3 (March 1917): 427–42; Jere W. Roberson, "To Build a Pacific Railroad: Congress, Texas, and the Charleston Convention of 1854," *Southwestern Historical Quarterly* 78 (October 1974): 117–39; Jere W. Roberson, "The South and the Pacific Railroad, 1845–1855," *Western Historical Quarterly* 5 (April 1974): 161–86; and David E. Woodard, "Sectionalism, Politics, and Foreign Policy: Duff Green and Southern Economic and Political Expansion, 1825–1865" (Ph.D. diss., University of Minnesota, 1996), chap. 6.

For a summary of opinion on the "limited" character of southern railroads, see John E. Clark Jr., *Railroads in the Civil War: The Impact of Management on Victory and Defeat* (Baton Rouge: Louisiana State University Press, 2001), 42, and Barbara Young Welke, *Recasting American Liberty: Gender, Race, Law, and the Railroad Revolution, 1865–1920* (New York: Cambridge University Press, 2001), 250–51.

8. Richard V. Francaviglia and Jimmy L. Bryan Jr., "'Are We Chimerical in This Opinion?': Visions of a Pacific Railroad and Westward Expansion before 1845," *Pacific Historical Review* 71 (May 2002): 179–202; *ARJ* 4 (January 31, 1835): 49; *ARJ* 6 (February 4, 1837): 66–68, quotation on 67; *Second Annual Report of the Directors of the Pennsylvania Railroad Co. to the Stockholders, October 31, 1848* (Philadelphia: Crissy & Markley, 1848), 9.

9. *ARJ* 1 (September 1, 1832): 569; 3 (August 2, 1834): 466; 2 (May 25, 1833): 321; 4 (April 18, 1835): 225. See also James A. Ward, *Railroads and the Character of America, 1820–1887* (Knoxville: University of Tennessee Press, 1986), chap. 1, and John Lauritz Larson, "'Bind the Republic Together': The National Union and the Struggle for a System of Internal Improvements," *Journal of American History* 74 (September 1987): 363–87.

10. *ARJ* 3 (May 17, 1834): 289; 1 (March 3, 1832): 146; 3 (January 11, 1834): 5.

11. *ARJ* 3 (January 11, 1834): 1 (*"partizan* politics"). The announcement was made in *ARJ* 2 (December 28, 1833): 817. Literary material also left the magazine shortly thereafter, so this may be interpreted as a broader professionalization of the journal. Minor was clearly no abolitionist, and had he published anything too incendiary, his paper would have never found a southern audience. See William W. Freehling, *The Road to Disunion: Secessionists at Bay, 1776–1854* (New York: Oxford University Press, 1990), 290–95. Although Minor once praised the efforts of the American Colonization Society, he also published a brief news item complimentary of slave labor used in factories and another article that belittled African intelligence: *ARJ* 2 (August 10, 1833): 509 (colonization); 2 (November 23, 1833): 748 (factory labor); 3 (September 6, 1834): 557 (intelligence).

12. *Proposed Rail-Road from Cincinnati to Charleston,* 10, 13, 16, 17; A. Blanding, *Address of Col. A. Blanding, to the Citizens of Charleston Convened in Town Meeting, on the Louisville, Cincinnati, and Charleston Rail Road* (Charleston, S.C.: A. S. Johnston, 1836), 30 ("death blow"). Southern sentiments about links to other sections are even more remarkable coming immediately after South Carolina's failed attempt at nullification. See William W. Freehling, *Prelude to Civil War: The Nullification Controversy in South Carolina, 1816–1836* (New York: Harper and Row, 1966).

13. Entry for June 3, 1834, in James Z. Rabun, ed., "Alexander H. Stephens's Diary, 1834–1837, Part I," *Georgia Historical Quarterly* 36 (March 1952): 89; letter from G. Stabler[?] to Hugh McLean, November 23, 1841, Hugh McLean Papers, SCL; letter from "Julia" to Catharine Barnwell, July 10, 1831, Barnwell Family Papers, SCHS.

14. Downey, *Planting a Capitalist South,* 106–9; Horry, *Address Respecting the Charleston*

& *Hamburgh Railroad*, 27; *Reports of the Presidents, Engineers-in-Chief and Superintendents, of the Central Rail-Road and Banking Company of Georgia, from No. 1 to 19 Inclusive, with Report of Survey by Alfred Cruger, and the Charter of the Company* (Savannah, Ga.: John M. Cooper, 1854), 4, 74; *Proceedings of the Greenville and Columbia Rail Road Company, at the Extra Meeting, 1st December, 1848, and Annual Meeting, 7th and 8th May, 1849* (Columbia, S.C.: John G. Bowman, 1849), 11. The ease with which the railroads obtained their right of way makes it difficult to accept uncritically John Stover's assertion: "Men of commerce in the city were generally not too hard to convince [to support railroads], but often the backcountry folk would not buy the high-flown schemes of the railroad promoter." John F. Stover, *Iron Road to the West: American Railroads in the 1850s* (New York: Columbia University Press, 1978), 60.

15. Allen W. Trelease, *The North Carolina Railroad, 1849–1871, and the Modernization of North Carolina* (Chapel Hill: University of North Carolina Press, 1991), 31; David E. Paterson, *Frontier Link with the World: Upson County's Railroad* (Macon, Ga.: Mercer University Press, 1998), 41; entry for May 13, 1852, SURR Minutebook; *Annual Report of the President and Chief Engineer to the Stockholders of the Blue Ridge Rail Road Company, at the Annual Meeting Held in Charleston, Nov. 6, 1855* (Charleston, S.C.: Walker & Evans, 1855), 7; *Report of the President and Directors to the Annual Meeting of the Stockholders of the Blue Ridge Rail Road Company, in South Carolina, Held in Charleston, the 17th November, 1857* (Charleston, S.C.: Walker, Evans and Company, 1857), 5.

16. William Kauffman Scarborough, *Masters of the Big House: Elite Slaveholders of the Mid-Nineteenth-Century South* (Baton Rouge: Louisiana State University Press, 2003), 228–30; Lacy K. Ford Jr., *Origins of Southern Radicalism: The South Carolina Upcountry, 1800–1860* (New York: Oxford University Press, 1988), 223; John Majewski, *A House Dividing: Economic Development in Pennsylvania and Virginia before the Civil War* (New York: Cambridge University Press, 2000), 64; letter from J. Newton Dexter to John Jervis, November 27, 1830, reel 11, *JJP*; letter from William Woods Holden to David Lowry Swain, August 1, 1849, in William Woods Holden, *The Papers of William Woods Holden*, ed. Horace W. Raper and Thornton W. Mitchell, vol. 1, *1841–1868* (Raleigh: Division of Archives and History, North Carolina Department of Cultural Resources, 2000), 30.

17. James W. Ely Jr., *Railroads and American Law* (Lawrence: University Press of Kansas, 2001), 30–31. For opposition, see Harry L. Watson, "Squire Oldway and His Friends: Opposition to Internal Improvements in Antebellum North Carolina," *North Carolina Historical Review* 54 (April 1977): 109. For a history of governmental involvement in public works projects, see John Lauritz Larson, *Internal Improvement: National Public Works and the Promise of Popular Government in the Early United States* (Chapel Hill: University of North Carolina Press, 2001). John Majewski characterizes state spending in Virginia as inefficient, with "political infighting" that led to "wasteful duplication of effort." Majewski, *A House Dividing*, 113. For foreign investment in southern railroads, see Scott R. Nelson, *Iron Confederacies: Southern Railways, Klan Violence, and Reconstruction* (Chapel Hill: University of North Carolina Press, 1999), 22.

18. Entry for the Blue Ridge Railroad, South Carolina, vol. 7, p. 405, R. G. Dun & Co. Collection, BHC.

19. Rachel N. Klein, *Unification of a Slave State: The Rise of the Planter Class in the South Carolina Backcountry, 1760–1808* (Chapel Hill: University of North Carolina Press, 1990); Larson, *Internal Improvement*, 222–23.

20. Letter from Horatio Allen to John Jervis, December 25, 1829, reel 7, *JJP*; *Proceedings of the Fourth Annual Meeting of the Richmond & Danville Railroad Company, Held in the Town of Danville; with the President's, Chief Engineer's and Treasurer's Reports* (Richmond, Va.: H. K. Ellyson, Printer, 1851), 11.

21. James Henry Hammond, *Secret and Sacred: The Diaries of James Henry Hammond, a Southern Slaveholder*, ed. Carol Bleser (New York: Oxford University Press, 1988), 97, 186–87, 189; Anti-Debt, *The Railroad Mania: A Series of Essays, by Anti-Debt, Published in the "Charleston Mercury"* (Charleston, S.C.: Burges, James, and Paxton, 1847), 13–14, 9. Ironically, given the emphasis that later historians would place on the importance of cotton to southern railroads, Hammond himself did not discount the nonagricultural traffic: "The amount of up freights on these roads will be something, for as the goods, &c. will usually depart from Charleston, they will follow the Railroad to the depots most convenient to the purchasers. But the down [agricultural] freight cannot for a long period amount to much. Farmers in the immediate neighborhood of the road will, no doubt, send their produce on it." Rather, Hammond believed (although he could not offer any evidence) that producers would *still* rather cart their goods in a wagon than use the railroad. Anti-Debt, *Railroad Mania*, 7–8. The principal response to Hammond's work argued that Hammond's true target was the Bank of the State of South Carolina, and so the bulk of the response focused on the bank, not on railroads. Fair Play, *Reply to Anti-Debt, on the Bank of the State of South Carolina, in a Series of Articles Originally Published in the Charleston Mercury and Columbia South Carolinian, by Fair Play* (Columbia, S.C.: South Carolinian Office, 1848), 3. See also Eugene Genovese, *The Slaveholders' Dilemma: Freedom and Progress in Southern Conservative Thought, 1820–1860* (Columbia: University of South Carolina Press, 1992), 12.

22. William Gregg, *Speech of William Gregg, of Edgefield, on a Bill to Amend an Act, Entitled "An Act to Authorize Aid to the Blue Ridge Railroad Company in South Carolina." In the House of Representatives, December 8, 1856* (Columbia, S.C.: R. W. Gibbes, 1857), 28–29; *Reports of the Presidents and Superintendents of the Central Railroad and Banking Co. of Georgia, from No. 20 to 32 Inclusive, and the Amended Charter of the Company* (Savannah, Ga.: G. N. Nichols, 1868), 105–6.

23. Letter from [S. A. White?] to John M. Allen, March 3, 1859, John Mebane Allen Papers, #4118-z, SHC.

24. Larson, *Internal Improvement*, 226; Michael F. Holt, *The Rise and Fall of the American Whig Party: Jacksonian Politics and the Onset of the Civil War* (New York: Oxford University Press, 1999), 81–82; Arthur Charles Cole, *The Whig Party in the South* (1914; reprint, Gloucester, Mass.: Peter Smith, 1962), 77–78; Ford, *Origins of Southern Radicalism*, 173, 220; Frederick Beck Gates, "Building the 'Empire State of the South': Political Economy in Georgia, 1800–1860" (Ph.D. diss., University of Georgia, 2001), 105; Carter Goodrich, *Government Promotion of American Canals and Railroads, 1800–1890* (New York: Columbia University Press, 1960), 120; Albert Fishlow, "Internal Transportation in the Nineteenth and Early Twentieth Centuries," in *The Cambridge Economic History of the United States*, vol. 2, *The Long Nineteenth Century*, ed. Stanley L. Engerman and Robert E. Gallman (New York: Cambridge University Press, 2000), 579.

25. *Proceedings of the Charlotte and South Carolina Rail Road Company, at Their First Annual Meeting, at Chesterville, 11th October, 1848. Also the Annual Reports of the President, Chief Engineer and Treasurer, &c. &c* (Columbia, S.C.: A. S. Johnston, 1848), 7; *Tenth Annual Report*

of the President and Other Officers of the East Ten. and Va. Railroad Co. to the Stockholders, at Their Annual Meeting at Greeneville, November 24, 1859 (Jonesborough, Tenn.: W. A. Sparks & Co., 1859), 15; Derrick, *Centennial History*, 34; Steven Hahn, *The Roots of Rural Populism* (New York: Oxford University Press, 1983), 36–37; letter from John McRae to DeSaussure, November 10, 1846, JML-1.

26. Entries for October 1, 1838, and May 16, 1840, Hiwassee Minutebook; entry for April 9, 1832, TCDRR Minutebook.

27. Letter from John McRae to D. K. Minor, January 8[?], 1849, JML-3.

28. *ARJ* 2 (August 17, 1833): 517; 3 (February 22, 1834): 97; 6 (September 30, 1837): 581; 3 (May 17, 1834): 290.

29. A. R. Newsome, ed., "Simeon Colton's Railroad Report, 1840," *North Carolina Historical Review* 11 (July 1934): 213–14.

30. Letter from Thomas Napier to J. P. DeVeaux, March 15, 1847, Napier Letterbook; John McRae, *Report of the Chief Engineer, on the Preliminary Survey of the Charleston and Savannah Rail Road* (Charleston, S.C.: Walker & Evans, 1854), 19. For the effects of the rival roads on Charleston, see Frederick Burtrumn Collins Jr., "Charleston and the Railroads: A Geographic Study of a South Atlantic Port and Its Strategies for Developing a Railroad System, 1820–1860" (Master's thesis, University of South Carolina, 1977), 69–70, 99. For another example of changing fortunes on the SCRR, see Downey, *Planting a Capitalist South*, 195–203.

31. Thomas Field Armstrong, "Urban Vision in Virginia: A Comparative Study of Ante-Bellum Fredericksburg, Lynchburg, and Staunton" (Ph.D. diss., University of Virginia, 1974), 89–94.

32. *Fourth Annual Report of the President and Directors to the Stockholders of the Louisville, Cincinnati & Charleston Rail-Road Company. September, 1840* (Charleston, S.C.: A. E. Miller, 1840), 33; *Proceedings of the Stockholders of the South-Carolina Rail-Road Company, and the South-Western Rail-Road Bank, at Their Annual Meeting, in the Hall of the Bank, on the 10th, 11th, and 12th February, 1846* (Charleston, S.C.: Miller and Browne, 1846), 19.

33. Letter from John McRae to D. K. Minor, December 1, 1846, JML-1. The SCRR did not receive permission from the South Carolina General Assembly to bridge the Savannah River until 1845. Despite opposition from Hamburg, the right was reaffirmed in 1848; permission from Augusta came in 1852, and the bridge was finished the following year. Downey, *Planting a Capitalist South*, 199–201.

34. Lynn Willoughby, *Flowing through Time: A History of the Lower Chattahoochee River* (Tuscaloosa: University of Alabama Press, 1999), 78–79; Anti-Debt, *Railroad Mania*, 11. For a rivalry in Tennessee that led one side to lay trees across the railroad track, see Forrest Laws, "The Railroad Comes to Tennessee: The Building of the LaGrange and Memphis," *West Tennessee Historical Society Papers* 30 (1976): 39–40.

35. Letter from Myron Safford Webb to William Webb, May 28, 1841, in Myron Safford Webb, *Myron Safford Webb, 1810–1871: His Log Book, 1840* (Sarasota, Fla.: Aceto Bookmen, 1985), 25; Charles B. George, *Forty Years on the Rail*, 3d ed. (Chicago: R. R. Donnelley and Sons, 1887), 44.

36. Stephen J. Buck, "A Vanishing Frontier: The Development of a Market Economy in DuPage County," *Journal of the Illinois State Historical Society* 93, no. 4 (2000–2001): 370; Colleen A. Dunlavy, *Politics and Industrialization: Early Railroads in the United States and*

Prussia (Princeton: Princeton University Press, 1994), 121; Michael J. Connolly, *Capitalism, Politics and Railroads in Jacksonian New England* (Columbia: University of Missouri Press, 2003), 1–72, quotation on 4.

37. Philip S. Klein and Ari Hoogenboom, *A History of Pennsylvania* (New York: McGraw-Hill, 1973), 188; Ely, *Railroads and American Law*, 14–15; Jeffrey P. Roberts, "Railroads and the Downtown: Philadelphia, 1830–1900," in *The Divided Metropolis: Social and Spatial Dimensions of Philadelphia, 1800–1975*, ed. William W. Cutler III and Howard Gillette Jr. (Westport, Conn.: Greenwood, 1980), 33, 48; Michael Feldberg, "Urbanization as a Cause of Violence: Philadelphia as a Test Case," in *The Peoples of Philadelphia: A History of Ethnic Groups and Lower-Class Life, 1790–1940*, ed. Allen F. Davis and Mark H. Haller (Philadelphia: Temple University Press, 1973), 58–61; Michael Bruce Kahan, "Pedestrian Matters: The Contested Meanings and Uses of Philadelphia's Streets, 1850s–1920s" (Ph.D. diss., University of Pennsylvania, 2002), chap. 1. Ely goes on to note troubles in another Pennsylvania city, Pittsburgh. The city labored to keep a "foreign" company—Maryland's Baltimore and Ohio Railroad—from entering the city until 1871. Ely, *Railroads and American Law*, 15–16.

38. *Report of the Directors of the Boston and Worcester Rail-Road, Presented at a Special Meeting of the Stockholders, on the Eighteenth of January, 1833* (Boston: Stimpson and Clapp, 1833), 7; letters from George Bliss to Thomas B. Wales, July 4 and 7, 1837, Wales Letterbook.

39. Untitled and undated report [ca. 1851] to the board of directors, folder 2, vol. 39b, New York, New Haven and Hartford Railroad Collection, BHC.

40. Letter from John McRae to James Gadsden, June 5, 1847, JML-2; letter from John McRae to D. K. Minor, July 29, 1847, JML-3; letter from Lawrence O'B. Branch to David Settle Reid, January 27, 1855, in David Settle Reid, *The Papers of David Settle Reid*, ed. Lindley S. Butler, vol. 2, *1853–1913* (Raleigh: Department of Cultural Resources, Division of Archives and History, 1997), 112. For railroad enthusiasm in upcountry South Carolina, see Ford, *Origins of Southern Radicalism*, 220–21.

41. *Edgefield Advertiser*, February 24, 1847, 3; letter from Thomas Napier to J. P. DeVeaux, March 15, 1847, Napier Letterbook; *Self-Instructor* 1 (October 1853): cover page; letter from Jennie Speer to A. B. Patterson, October 20, 1851, in Allen Paul Speer and Janet Barton Speer, eds., *Sisters of Providence: The Search for God in the Frontier South (1843–1858)* (Johnson City, Tenn.: Overmountain Press, 2000), 87–88.

42. Elijah Millington Walker, *A Bachelor's Life in Antebellum Mississippi: The Diary of Dr. Elijah Millington Walker, 1849–1852*, ed. Lynette Boney Wrenn (Knoxville: University of Tennessee Press, 2004), 139, 227. For further preparations, see pp. 228–29.

43. Walker, *Bachelor's Life*, 230.

44. Walker, *Bachelor's Life*, 231.

45. George Little, *Memoirs of George Little* (Tuscaloosa, Ala.: Weatherford Printing Company, 1924), 24; *Seventh Annual Report of the Directors to the Stockholders in the Memphis and Charleston Railroad Company, July 1, 1857* (Memphis: Bulletin Company's Cheap Steam Presses, 1857), 12–13; *The Blue Ridge Railroad: Letters of a Constituent to L. W. Spratt, Esq* (Charleston, S.C.: Evans & Cogswell, 1860), 10, 9.

46. Thomas P. Hughes, "Technological Momentum," in *Does Technology Drive History? The Dilemma of Technological Determinism*, ed. Merritt Roe Smith and Leo Marx (Cambridge, Mass.: MIT Press, 1994), 101–13; *Speeches of Maj. B. F. Perry & Hon. J. D. Allen on the Blue Ridge Railroad* (n.p.: n.d.), 5, 17–18. Although this pamphlet is undated, Perry makes refer-

ence to the failure of Anson Bangs and Company, so the speeches were probably given after 1856. Evidently the speech was given at or near Stumphouse Mountain Tunnel.

Chapter Two • Knowledge

1. Joseph B. Cumming, *A History of Georgia Railroad and Banking Company and Its Corporate Affiliates, 1833–1958* (n.p.: [1959?]), 3–4.

2. Letter from John McRae to B. H. Latrobe, March 28, 1849, JML-4; Robert G. Angevine, *The Railroad and the State: War, Politics, and Technology in Nineteenth-Century America* (Stanford: Stanford University Press, 2004), 81. For the Baltimore and Ohio's debate over steam, see John F. Stover, *History of the Baltimore and Ohio Railroad* (West Lafayette, Ind.: Purdue University Press, 1987), 34–36.

3. Elting E. Morison, *From Know-How to Nowhere: The Development of American Technology* (New York: Basic, 1974), 21–23; Angevine, *Railroad and the State*, 64; Mark Aldrich, "Earnings of American Civil Engineers, 1820–1859," *Journal of Economic History* 31 (June 1971): 413. See also Daniel Hovey Calhoun, *The American Civil Engineer: Origins and Conflict* (Cambridge, Mass.: Technology Press, Massachusetts Institute of Technology, 1960), 182; and for developments in Europe, see Colleen A. Dunlavy, *Politics and Industrialization: Early Railroads in the United States and Prussia* (Princeton: Princeton University Press, 1994), 63.

4. Letter from E. L. Miller to John Jervis, June 2, 1829, reel 7, *JJP*; letter from Henry Bird to Mary Fox, September 9, 1832, folder 7, section 1, Bird Family Papers, VHS; letter from John McRae to E. Boykin, November 9, 1849, JML-4.

5. Letter from John McRae to William Hamilton, March 2, 1849, JML-4. McRae later referred to the "uncertain and wandering life of an Engineer in this country." Letter from John McRae to "My Dear Duncan," June 14, 1850, JML-5.

6. Raymond H. Merritt, *Engineering in American Society, 1850–1875* (Lexington: University Press of Kentucky, 1969), 91. Given the mobility detailed later in this chapter, Merritt's statement on page 99 that it is "obvious . . . at that early stage" that engineers "clung to an obsolescent agrarian individualism and identified with one section or another of the country" seems unpersuasive.

7. *First Annual Report of the Directors of the Pennsylvania Railroad Co. to the Stockholders, October 30, 1847* (Philadelphia: Crissy & Markley, 1847), 5. As Steven Collins has pointed out, it was in Georgia that Thomson "first developed the incipient managerial hierarchy that historian Alfred Chandler has considered the critical component in the development of a corporate bureaucracy." Steven G. Collins, "Progress and Slavery on the South's Railroads," *Railroad History* 181 (Autumn 1999): 10.

8. Robert E. Carlson, "British Railroads and Engineers and the Beginnings of American Railroad Development," *Business History Review* 34 (Summer 1960): 147; entry for December 14, 1839, LCCRRJT; *Biographical Sketches of the Leading Men of Chicago* (Chicago: Wilson, Pierce, 1876), 67; Stephen Ray Henson, "Industrial Workers in the Mid Nineteenth-Century South: Atlanta Railwaymen, 1840–1870" (Ph.D. diss., Emory University, 1982), 23; letter from John McRae to Alfred Sears, December 27, 1854, JML-7; Merritt, *Engineering in American Society*, 79; Charles B. Stuart, *Lives and Works of Civil and Military Engineers of America* (New York: D. Van Nostrand, 1871), 182–87. For more examples, see Darwin H. Stapleton, "Moncure Robinson: Railroad Engineer, 1828–1840," in *Benjamin Henry Latrobe & Moncure*

Robinson: The Engineer as Agent of Technological Transfer, ed. Barbara Benson (Greenville, Del.: Eleutherian Mills Historical Library, 1975), 50; and James A. Ward, "J. Edgar Thomson and the Georgia Railroad, 1834–1847," *Railroad History* 134 (Spring 1976): 4–33.

9. *Self-Instructor* 1 (October 1853): 39–40; J. Patrick McCarthy Jr., "Commercial Development and University Reform in Antebellum Athens: William Mitchell as Entrepreneur, Engineer, and Educator," *Georgia Historical Quarterly* 83 (Spring 1999): 19, 23, 26; Terry S. Reynolds, "The Engineer in 19th-Century America," in *The Engineer in America: A Historical Anthology from Technology and Culture,* ed. Terry S. Reynolds (Chicago: University of Chicago Press, 1991), 20.

10. Merritt, *Engineering in American Society,* 3; entry for May 20, 1836, SCRR Minute-book; *Fourth Annual Report of the Directors of the Pennsylvania Railroad Co. to the Stockholders, December 31, 1850* (Philadelphia: Crissy & Markley, 1851), 41–42 (although Haupt had not worked for the Georgia Railroad, his supervisor, J. Edgar Thomson, did); letter from Horatio Allen to John Jervis, February 24, 1828, reel 7, *JJP.* Allen's diary from that trip has been published as Horatio Allen, "Diary of Horatio Allen," *Bulletin* [of the Railway and Locomotive Historical Society] 89 (November 1953): 97–138. See also Carlson, "British Railroads and Engineers," 138; Dunlavy, *Politics and Industrialization,* 203–4; and Frederick C. Gamst, "The Transfer of Pioneering British Railroad Technology to North America," in *Early Railways: A Selection of Papers from the First International Early Railways Conference,* ed. Andy Guy and Jim Rees (London: Newcomen Society, 2001), 251–65. Professional workers have not received a great deal of historiographic attention in the antebellum South, but for another professional network in the antebellum South, see Jennifer R. Green, "Networks of Military Educators: Middle-Class Stability and Professionalization in the Late Antebellum South," *Journal of Southern History* 73 (February 2007): 39–74.

11. Letter from John McRae to J. D. Fleming, November 26, 1848, JML-3; letters from John McRae to Raymond Lee, to B. H. Latrobe, to L. O. Reynolds, and to J. R. Trimble, all dated February 19, 1849, JML-4; letters from Benjamin Henry Latrobe to John McRae, July 12, 1845, and June 22, 1846, Benjamin Henry Latrobe Papers, SCL; letter from E. Chesbrough to John McRae, April 28, 1849, letter-size folder 21, Cantey Family Papers, SCL.

12. Engineering literature available in Charleston included the works of James Renwick and Dionysius Lardner (*Charleston Courier,* January 7, 1831, 1); George Stephenson (*Charleston Courier,* April 14, 1831, 1); and M. J. Saganzin, James Walker, and Henry Booth (*Charleston Courier,* May 18, 1831, 4). Engineering literature continued to be sold in the city after the railroad was complete to Hamburg. See *Charleston Courier,* January 15, 1840, 4; February 18, 1840, 1; and February 26, 1840, 1. For McRae's complaints, see letter from John McRae to Messers. Wiley and Putnam, October 16, 1845, JML-1; letter from John McRae to E. Baldwin, October 2, 1846, JML-1; letter from John McRae to D. K. Minor, October 3, 1847, JML-3.

13. *Semi-Annual Report of the Board of Directors of the South-Carolina Canal and Rail Road Company* (Charleston, S.C.: A. E. Miller, 1829), 3; N. J. Bell, *Southern Railroad Man: Conductor N. J. Bell's Recollections of the Civil War Era,* ed. James A. Ward (DeKalb: Northern Illinois University Press, 1993), 6–7; Balthasar Henry Meyer, ed., *History of Transportation in the United States before 1860* (Washington, D.C.: Carnegie Institute of Washington, 1917), 310.

14. Horatio Allen, *The Railroad Era: The First Five Years of Its Development* (New York: n.p., 1884), 25–26. Another description is available in Franz Anton Ritter von Gerstner, *Early American Railroads,* ed. Frederick C. Gamst (Stanford: Stanford University Press, 1997),

711–12. Michel Callon correctly argues that multiple considerations—"technical, scientific, social, economic, or political"—are always "bound" into an "organic whole" for engineers; engineers never do work that is purely (or solely) technical. Michel Callon, "Society in the Making: The Study of Technology as a Tool for Sociological Analysis," in *The Social Construction of Technological Systems: New Directions in the Sociology and History of Technology,* ed. Wiebe E. Bijker, Thomas P. Hughes, and Trevor J. Pinch (Cambridge, Mass.: MIT Press, 1987), 83–84.

15. Horatio Allen, *Reports to the Board of Directors of the South-Carolina Canal and Rail Road Company* (Charleston, S.C.: J. S. Burges, 1831), 3–4, 12; entry for December 8, 1836, Herbert A. Kellar, ed., "A Journey through the South in 1836: Diary of James D. Davidson," *Journal of Southern History* 1 (August 1935): 373; *Annual Report of the Board of Directors of the South Carolina Canal and Rail-Road Company to the Stockholders, with Accompanying Documents Submitted at Their Meeting, May 6, 1833* (Charleston, S.C.: A. E. Miller, 1833), 3. For a description of the construction process, see the letter from Horatio Allen to John Jervis, February 5, 1830, reel 7, *JJP.*

16. *Annual Report of the Direction of the South Carolina Canal and Rail Road Company, to the Stockholders, May 6th, 1834, with Accompanying Documents* (Charleston, S.C.: W. S. Blain, 1834), 10–11; *Annual Report of the Board of Direction of the So. Canal and R. R. Company, with the Semi-Annual Statement of Accounts, to the 31st December, 1835* (Charleston, S.C.: A. E. Miller, 1836), 4. It does not appear that the entire road was embanked; higher portions may have simply been redone with woodwork framing. The board of directors noted that "piles have been cut down and the Road rebuilt upon framed work. The most extensive piece so rebuilt is between the Edisto Bridge and Branchville and is about Fifteen hundred Feet long." Entry for July 1, 1835, SCRR Minutebook.

17. Frederic Trautmann, ed., "South Carolina through a German's Eyes: The Travels of Clara von Gerstner, 1839," *South Carolina Historical Magazine* 85 (July 1984): 232; *Report of the Committee Appointed by the Inhabitants of the Town of Columbia, and Its Vicinity, for Causing a Survey of the Route of Rail Road to Branchville. Together with the Reports of the Engineers* (Columbia, S.C.: Printed at the Telescope Office, 1834), 22, 18.

18. Letters from Horatio Allen to Messers J. and J. Townsend, November 4, 1832, and April 2, 1834, Horatio Allen Papers, SCL.

19. Letter from Henry Bird to Mary Fox, October 22, 1832, folder 7, section 1, Bird Family Papers, VHS. See also Wiebe E. Bijker and John Law, "General Introduction," in *Shaping Technology / Building Society: Studies in Sociotechnical Change,* ed. Wiebe E. Bijker and John Law (Cambridge, Mass.: MIT Press, 1992), 4; John Law, "Technology and Heterogeneous Engineering: The Case of Portuguese Expansion," in *The Social Construction of Technological Systems: New Directions in the Sociology and History of Technology,* ed. Wiebe E. Bijker, Thomas P. Hughes, and Trevor J. Pinch (Cambridge, Mass.: MIT Press, 1987), 111–34; and McCarthy, "Commercial Development," 15.

20. *Annual Report of the South-Carolina Canal and Rail Road Company, by the Direction. Submitted and Adopted May 2, 1831* (Charleston, S.C.: Office of the Irishman & Democrat, 1831), 14–15; letter from Andrew Talcott to "Messrs Stone, Harris, and Birnie," August 18, 1849, folder 10-2, vol. 10, StHC.

21. Letter from John McRae to "My dear Craven," August 6, 1847, JML-3; *Proceedings of the Stockholders of the Louisville, Cincinnati & Charleston Rail-Road Company, and the South-*

Western R. Road Bank. At Their Annual Meeting, in the Hall of the Bank, on the 16th, 17th, and 18th of Nov. 1841 (Charleston, S.C.: A. E. Miller, 1841), 6–7. There was a silver lining: after the rains subsided and the sickness cleared from the air, the work proceeded quickly, thanks in part to "a more efficient organization of the Rail-layers themselves." *Proceedings of the Stockholders of the Louisville, Cincinnati & Charleston Rail-Road Company, and the South-Western R. Road Bank, at Their Annual Meeting, in the Hall of the Bank, on the 22d, 23d, 24th, and 25th of Nov. 1842* (Charleston, S.C.: A. E. Miller, 1842), 20.

22. Letter from John McRae to Walter Gwynn, January 16, 1853, JML-6.

23. Manuscript copy of the "Report of the Directors of the Boston and Worcester Rail Road Corporation to the Stockholders, at Their Annual Meeting, June 1, 1835," vol. 11, BARC-BW; letter from Samuel Nott to D. A. Neale, February 5, 1842, Nott Letterbook; letter from George Bliss to Thomas Wales, September 16, 1837, Wales Letterbook.

24. Letter from A. L. Maxwell to Harris, September 12, 1854, folder 13-4, vol. 13, StHC; advertisement quoted in N. P. Renfro, *The Beginning of Railroads in Alabama* (Auburn, Ala.: n.p., 1910), 7.

25. Letter from J. R. Anderson to "Messrs Stone & Harris," June 9, 1849, folder 10-1, vol. 10, StHC; letter from J. R. Anderson to "Messrs Stone, Harris & Birnie," June 30, 1849, folder 10-1, vol. 10, StHC; entry for October 11, 1849, RDRR Minutebook.

26. Letter from John McRae to "My dear Craven," September 27, 1846, JML-1; letter from John McRae to "My dear Craven," August 15, 1847, JML-3. For disease on the Illinois Central, see David Lightner, *Labor on the Illinois Central Railroad, 1852–1900* (New York: Arno, 1977), 31.

27. Letter from John McRae to "My Dear Craven," December 22, 1845, JML-1; entry for May 22, 1840, LCCRRJT.

28. Entry for February 2, 1854, SURR Minutebook.

29. "Articles of Agreement," 1850, Hawkins Family Papers, SHC, *SASI*, series B, reel 13, frames 897–900.

30. Letter from John McRae to "My Dear Craven," September 15, 1845, JML-1; letter from John McRae to James Herron, February 16, 1847, JML-2; letter from James Herron to John McRae, March 6, 1847, letter-size folder 19, Cantey Family Papers, SCL; letter from John McRae to James Herron, March 30, 1847, JML-2.

31. Letter from J. R. Anderson to "Messrs Stone and Harris," April 30, 1849, folder 10-1, vol. 10, StHC; letter from A. L. Maxwell to D. L. Harris, August 10, 1854, folder 13-3, vol. 13, StHC; letter from William Birnie to "Messrs Stone & Harris," June 16, 1849, folder 10-1, vol. 10, StHC.

32. *Reports of the Presidents, Engineers-in-Chief and Superintendents, of the Central Rail-Road and Banking Company of Georgia, from No. 1 to 19 Inclusive, with Report of Survey by Alfred Cruger, and the Charter of the Company* (Savannah, Ga.: John M. Cooper, 1854), 34.

33. *LCCRR Report, November 1842,* 22.

34. Entry for May 12, 1852, SURR Minutebook; *Annual Report of the President and Chief Engineer to the Stockholders of the Blue Ridge Rail Road Company, at the Annual Meeting, Held in Charleston, July 11, 1854; Together with the Report of B. H. Latrobe, Esq. Chief Engineer Baltimore and Ohio Rail Road, of His Examination of the Route of the Blue Ridge Rail Road* (Charleston, S.C.: Walker & Evans, 1854), 11; Eugene Genovese, *The Political Economy of Slavery,* 2d ed. (Middletown, Conn.: Wesleyan University Press, 1989), 223.

35. Letter from John McRae to T. Stark, July 20, 1846, JML-1.

36. Letter from John McRae to T. T. Stark, August 16, 1846, JML-1; letter from John McRae to T. T. Stark, September 6, 1846, JML-1; letter from John McRae to John Canty, September 18, 1846, JML-1; letter from John McRae to Thos. T. Stark, October 3, 1846, JML-1.

37. Letter from John McRae to Dr. Stark, April 13, 1847, JML-2.

38. Entry for July 29, 1857, Thomas J. Shaw Diaries, VHS, *SASI*, series C, part 2, reel 11, frame 699; letter from John McRae to Robert M. Cochran, November 28, 1852, JML-6 (McRae hints in this letter that Cochran may have already been paid more than he should have been); letter from John McRae to Alfred Hargrave, December 14, 1852, JML-6; letter from David S. Taylor to Joseph Slocum, September 18, 1854, David S. Taylor Papers, SCL.

39. Letter from John McRae to Walter Gwynn, May 8, 1853, JML-6. McRae himself felt that the problem lay with the masons who were employed by Shaver: "I think the masons worked very leisurely there was much waste in the manner the stone were got out at the quarry." Letter from John McRae to Walter Gwynn, May 29, 1853, JML-6.

40. Letter from John McRae to J. M. Morehead, January 4, 1853, JML-6; letter from John McRae to James Gadsden, February 7, 1847, JML-2.

41. Sections 5, 9, and 10 had no comments on the quality of management. See "Notes of cost of several sections of Grading on First Division of Schen & Troy Rail Road," n.d., in folder 8-3 (Memos), vol. 8, StHC.

42. Letter from Samuel Nott to Mr. F. P. Becknap, April 2, 1842, Nott Letterbook; letter from J. Chamberlain to George Strong, September 8, 1848, folder labeled "Letters, 1845–1853," box 4, RWRRC; entry for May 15–20, 1836, Wilson Miles Cary Fairfax Diary, VHS.

43. Bent Flyvbjerg, Nils Brazelius, and Werner Rothengatter, *Megaprojects and Risk: An Anatomy of Ambition* (New York: Cambridge University Press, 2003), 11–12, 16, 19. See also Randall A. Dodgen, *Controlling the Dragon: Confucian Engineers and the Yellow River in Late Imperial China* (Honolulu: University of Hawai'i Press, 2001), 31; Bent Flyvbjerg, Mette K. Skamris Holm, and Søren L. Buhl, "What Causes Cost Overrun in Transport Infrastructure Projects?" *Transport Reviews* 24 (January 2004): 3–18; and Peter W. G. Morris and George H. Hough, *The Anatomy of Major Projects: A Study of the Reality of Project Management* (New York: John Wiley, 1987), 7, 12.

44. *Report of the Chief Engineers, Presidents and Superintendents of the South-Western R. R. Co., of Georgia, from No. 1 to 22, Inclusive, with the Charter and Amendments Thereto* (Macon, Ga.: J. W. Burke, 1869), 126; letter from John McRae to "My dear Manigault," August 9, 1854, JML-7; entry for May 14, 1850, RDRR Minutebook.

45. Entry for May 14, 1850, RDRR Minutebook. Harvey's case continued to be discussed on January 20 and February 13, 1851, and September 14, 1854. For other problematic contractors, see the entry for January 19, 1853, RDRR Minutebook, and entry for October 20, 1852, Southside Minutebook-1.

46. Voucher entitled "Abstract of Engineering Exp," folder 6, box 1, RFPRR-VHS; "Memorandum of Agreement," legal-size folder 7, Cantey Family Papers, SCL; entries for January 9 and July 30, 1838, October 25 and 28–30, 1839, LCCRRJT.

47. Letter from John McRae to "My dear Craven," September 27, 1846, JML-1; letter from Edward St. George Cooke to John Esten Cooke, January 11, 1855, VHS.

48. Entries for December 23–27, 1839, LCCRRJT.

49. Letter from William Birnie to "Stone and Harris," November 4, 1849, and letter from G. H. Burt to "Stone and Harris," November 11, 1849, both in folder 11-1, vol. 11, StHC.

50. John Smedberg to Hugh G. Young, November 19, 1837, John Smedberg Papers, SCL; entry for December 3, 1839, LCCRRJT; letter from John McRae to A. H. Hopkins, January 3, 1846, JML-1; *Fifth Annual Report of the President and Directors to the Stockholders of the Northwestern Virginia Railroad Company, Submitted at Their Annual Meeting, June 18, 1856* (Baltimore: John Murphy, 1856), 16.

51. Letter from John McRae to "My dear Sir," February 15, 1846, JML-1; letter from John McRae to "My dear Manigault," January 12, 1847, JML-1; letter from John McRae to "My dear Craven," February 15, 1850, JML-3; letter from John McRae to Thomas Lay, March 27, 1850, JML-3; letter from John McRae to L. D. Fleming, May 3, 1852, JML-6.

52. Letters from Jos. M. Sheppard to John H. Hopkins, September 12 and 20, 1836, both in folder 7, box 2, RFPRR-VHS; letter from John McRae to James Gadsden, June 8, 1846, JML-1.

53. Letter from John McRae to R. P. Harris, February 23, 1851, JML-5.

54. Entry for January 2, 1840, LCCRRJT; letter from John McRae to [J. R. Spann?], October 2, 1846, JML-1.

55. Aldrich, "Earnings of American Civil Engineers," 408; Merritt, *Engineering in American Society,* 72; entry for February 15, 1840, LCCRRJT; letter from John McRae to Walter Gwynn, January 14, 1855, JML-7; letter from John McRae to Thos. F. Drayton, February 10, 1855, JML-7. This could also be a problem in the North: in 1832 the Boston and Worcester Railroad had to determine what precisely were the duties (and who held ultimate authority) of the superintendent and the engineer. Entries for December 5 and 15, 1832, Minutebook of the Directors and Stockholders (1831–1835), vol. 1, BARC-BW.

56. Letter from J. T. Craig to "Cousin Lizzie," April 17, 1854, N. E. Craig Papers, SCL; entry for July 26, 1854, R. Conover Bartram, ed., "The Diary of John Hamilton Cornish, 1846–1860," *South Carolina Historical Magazine* 64 (April 1963): 83; Christopher Silver, "Immigration and the Antebellum Southern City: Irish Working Class Mobility in Charleston, South Carolina, 1840–1860" (Master's thesis, University of North Carolina at Chapel Hill, 1975), 32.

57. Bangs appears to have been operating in the South as early as 1848; there is an entry for his company in the R. G. Dun record books for that year that claims it has the "Reputation of a strong Company." The next entry is for 1854, in which it is identified as "B.R.R. contractors Northern men. Don't know them." Entry for Anson Bangs and Company, South Carolina, vol. 2, p. 59, R. G. Dun & Co. Collection, BHC. See also George Dewitt Brown, "A History of the Blue Ridge Railroad, 1852–1874" (Master's thesis, University of South Carolina, 1967), 50–51.

58. *Annual Report of the President and Chief Engineer to the Stockholders of the Blue Ridge Rail Road Company, at the Annual Meeting Held in Charleston, Nov. 6, 1855* (Charleston, S.C.: Walker & Evans, 1855), 6; Blue Ridge Railroad Company, *The Blue Ridge Rail Road Company, ads. Anson Bangs & Co. In Equity, In the Circuit Court of the U.S. for the Sixth Circuit Northern District of Georgia. A Statement of the Defendant's Case* (Charleston, S.C.: Walker, Evans and Company, 1857), 3.

59. *Report of the President and Directors to the Annual Meeting of the Stockholders of the Blue Ridge Rail Road Company, in South Carolina, Held in Charleston, the 17th November, 1857*

(Charleston, S.C.: Walker, Evans and Company, 1857), 3. Although this report gives the April date, another publication states that the contract was dissolved on March 25, 1856. *BRRR ads. Anson Bangs,* 10.

60. Entry for A. Birdsall and Company, South Carolina, vol. 2, p. 68, R. G. Dun & Co. Collection, BHC.

61. Bangs and Company, *Bangs, vs. the Blue Ridge Rail Road Company* (Charleston, S.C.: Walker, Evans & Co., 1857), 12; *BRRR ads. Anson Bangs,* 4, 10–11, 15.

62. *BRRR Report, November 1857,* 7–8. The Blue Ridge also had a smaller tunnel, Dick's Creek Tunnel, that was difficult enough that two contractors abandoned it. Unlike the Stumphouse Mountain tunnel, this tunnel could be dug only from the two ends, which made it slow going. *BRRR Report, November 1857,* 9.

63. "Articles of Agreement," August 1, 1857, Hawkins Family Papers, SHC, *SASI,* series B, reel 14, frame 259. Work evidently was done at night with the aid of primitive headlights: a twentieth-century local history of the area around the tunnel mentions a man who had once been given "a miner's lamp . . . in the form of a miniature watercan as the wick was in the spout and the tallow was in the reservoir. Instead of a handle there was a unique catch which was used to hook on the bill of the miner's cap." Charles Sloan Reid, Marguerite Brennecke, and R. C. Carter II, *Persons, Places and Happenings in Old Walhalla* (Walhalla, S.C.: Walhalla Historical Society, 1960).

64. *BRRR Report, November 1857,* 6.

65. *Report of the President and Directors and of the Chief Engineer to the Annual Meeting of the Stockholders of the Blue Ridge Rail Road Company, in South Carolina, Held in Charleston, the 10th November, 1858* (Charleston, S.C.: Walker Evans & Company, 1858), 8, 36; entry for Humbird Hickcock and Company, South Carolina, vol. 2, p. 65, R. G. Dun & Co. Collection, BHC; *Report of the President and Directors and of the Chief Engineer to the Annual Meeting of the Stockholders of the Blue Ridge Rail Road Company, in South Carolina, Held in Charleston, the 22d November, 1859* (Charleston, S.C.: Walker, Evans & Company, 1859), 3; Brown, "History of the Blue Ridge Railroad," 55. See also Jim Haughey, "Tunnel Hill: An Irish Mining Community in the Western Carolinas," *Proceedings of the South Carolina Historical Association* (2004): 51–62. The incomplete tunnel is still accessible to the public and was still a part of local lore in the twentieth century. See Reid, Brennecke, and Carter, *Persons, Places and Happenings,* 193.

66. Letter from Horatio Allen to John Jervis, February 5, 1830, reel 7, *JJP;* letter from Edward St. George Cooke to John Esten Cooke, January 8, 1856, VHS.

Chapter Three • Sweat

1. *Annual Report of the South-Carolina Canal and Rail Road Company, by the Direction. Submitted and Adopted May 2, 1831* (Charleston, S.C.: Office of the Irishman & Democrat, 1831), 7; Ulrich Bonnell Phillips, *A History of Transportation in the Eastern Cotton Belt to 1860* (1908; reprint, New York: Octagon Books, 1968), 150; Charles W. Turner, "The Louisa Railroad, 1836–1850," *North Carolina Historical Review* 24 (January 1947): 39; *Reports of the Presidents, Engineers-in-Chief and Superintendents, of the Central Rail-Road and Banking Company of Georgia, from No. 1 to 19 Inclusive, with Report of Survey by Alfred Cruger, and the Charter of the Company* (Savannah, Ga.: John M. Cooper, 1854), 44; entry for January 11, 1849, RDRR

Minutebook; *Report of the Chief Engineers, Presidents and Superintendents of the South-Western R. R. Co., of Georgia, from No. 1 to 22, Inclusive, with the Charter and Amendments Thereto* (Macon, Ga.: J. W. Burke, 1869), 131; Kenneth W. Noe, *Southwest Virginia's Railroad: Modernization and the Sectional Crisis* (Urbana: University of Illinois Press, 1994), 82; Charles W. Turner, "Early Virginia Railroad Entrepreneurs and Personnel," *Virginia Magazine of History and Biography* 58 (July 1950): 334. Historian Mark Aldrich, without reference to any particular section of the United States, notes that the difficulty in securing workers may have impaired the safety of early railroads. Mark Aldrich, *Death Rode the Rails: American Railroad Accidents and Safety, 1828–1965* (Baltimore: Johns Hopkins University Press, 2006), 16, 21.

2. William Kauffman Scarborough, *Masters of the Big House: Elite Slaveholders of the Mid-Nineteenth-Century South* (Baton Rouge: Louisiana State University Press, 2003), 440, 454, 482–84.

3. Letter from Horatio Allen to John Jervis, July 15, 1829, reel 7, *JJP*; *SCRR Report, May 1831*, 3–4; letter from John McRae to James Gadsden, February 7, 1847, JML-2.

4. Walter Licht, *Working for the Railroad: The Organization of Work in the Nineteenth Century* (Princeton: Princeton University Press, 1983), 65–68. See also Stephen Ray Henson, "Industrial Workers in the Mid Nineteenth-Century South: Atlanta Railwaymen, 1840–1870" (Ph.D. diss., Emory University, 1982), 24–25; and Phillips, *History of Transportation*, 149–50. Planters pulled hired slaves back onto plantations in contexts besides railroad work. See *Third Annual Report of the President and Directors of the Wilmington and Manchester Rail Road Company. With the Report of the Resident Engineer, and Proceedings of the Meeting, January 29, 1851* (Marion, S.C.: Office of the Star, 1851), 47; *SWRR Reports, 1 to 22 Inclusive*, 133; Jonathan D. Martin, *Divided Mastery: Slave Hiring in the American South* (Cambridge, Mass.: Harvard University Press, 2004), 30; David Williams, "Georgia's Forgotten Miners: African Americans and the Georgia Gold Rush of 1829," in *Appalachians and Race: The Mountain South from Slavery to Segregation*, ed. John C. Inscoe (Lexington: University Press of Kentucky, 2001), 42; and Wayne Cline, *Alabama Railroads* (Tuscaloosa: University of Alabama Press, 1997), 39.

5. Entry for August 20, 1836, Wilson Miles Cary Fairfax Diary, VHS; *Second Annual Report of the Directors of the Pennsylvania Railroad Co. to the Stockholders, October 31, 1848* (Philadelphia: Crissy & Markley, 1848), 25; *Third Annual Report of the Directors of the Pennsylvania Rail-Road Company to the Stockholders, October 31, 1849* (Philadelphia: Crissy & Markley, 1850), 54; *Seventh Annual Report of the Directors of the Pennsylvania Railroad Co. to the Stockholders, February 6, 1854* (Philadelphia: Crissy & Markley, 1854), 9, 28–29; letter from N. Chapin to Harris, July 4, 1854, folder 13-2, vol. 13, StHC; Paul Wallace Gates, *The Illinois Central Railroad and Its Colonization Work* (Cambridge, Mass.: Harvard University Press, 1934), 95. See also David Lightner, *Labor on the Illinois Central Railroad, 1852–1900* (New York: Arno, 1977), 16–18, and letters from N. Chapin to Harris, June 10, 12, 15, and 26, 1854 (folder 13-1); July 22 and 25 (two letters), 1854 (folder 13-2); and August 1 and 2, 1854 (folder 13-3), all in vol. 13, StHC. Sickness also affected canal construction, see Peter Way, *Common Labour: Workers and the Digging of North American Canals, 1780–1860* (New York: Cambridge University Press, 1993), 152–60.

6. Letter from William Birnie to "Stone & Harris," October 22, 1849, folder 10-3, vol. 10, StHC.

7. Theodore Kornweibel Jr., "Railroads and Slavery," *Railroad History* 189 (2003): 35;

Robert S. Starobin, *Industrial Slavery in the Old South* (New York: Oxford University Press, 1970), 28. For the overall history of African Americans and railroad work, see Eric Arnesen, *Brotherhoods of Color: Black Railroad Workers and the Struggle for Equality* (Cambridge, Mass.: Harvard University Press, 2001). For nonagricultural slavery in general, see Starobin, *Industrial Slavery*; and Charles B. Dew, *Bond of Iron: Master and Slave at Buffalo Forge* (New York: Norton, 1994). For railroads, see Kornweibel, "Railroads and Slavery," 34–59; Steven G. Collins, "Progress and Slavery on the South's Railroads," *Railroad History* 181 (Autumn 1999): 6–25; Noe, *Southwest Virginia's Railroad*, chap. 4; and Allen W. Trelease, *The North Carolina Railroad, 1849–1871, and the Modernization of North Carolina* (Chapel Hill: University of North Carolina Press, 1991), 35, 62–64. Licht argues that southern railroads were "forced" into purchasing slaves. To the contrary: as Kornweibel has shown, railroads regarded purchasing as a sound investment, even if they could not afford as many as they wanted. Licht, *Working for the Railroad*, 67.

8. *CRRG Reports, 1 to 19 Inclusive*, 34; *Report on the Preliminary Surveys for the Savannah and Albany Rail-Road* (Savannah, Ga.: Evening Journal Print, 1853), 20.

9. Kornweibel, "Railroads and Slavery," 39; letter from William Lewis to John McRae, March 23, 1838, letter-size folder 4, Cantey Family Papers, SCL; entry for May 10, 1841, in Susan E. O'Donovan and Lee W. Formwalt, eds., "The Journal of Nelson Tift, Part III: January–October 1841," *Journal of Southwest Georgia History* 5 (Fall 1987): 70; Cline, *Alabama Railroads*, 23; George P. Rawick, ed., *The American Slave: A Composite Autobiography*, supplement, series 1, vol. 5, *Indiana and Ohio Narratives* (Westport, Conn.: Greenwood, 1977), 334; interviewed in Florida, Young Winston Davis was enslaved in Alabama: George P. Rawick, ed., *The American Slave: A Composite Autobiography*, vol. 17, *Florida Narratives* (Westport, Conn.: Greenwood, 1972), 86–87.

10. Ezra Michener, *Autobiographical Notes from the Life of Ezra Michener, M.D.* (Philadelphia: Friends' Book Association, 1893), 57.

11. Entry for September 20, 1836, SCRR Minutebook. In Tennessee, free blacks were legally prohibited from working as railroad engineers. Jonathan M. Atkins, "Party Politics and the Debate over the Tennessee Free Negro Bill, 1859–1860," *Journal of Southern History* 71 (May 2005): 249. One former slave used his expertise gained while in slavery to operate a train in Canada in 1855. See John Hebron Moore, ed., "A Letter from a Fugitive Slave," *Journal of Mississippi History* 24 (1962): 101. One letter uncovered by historian Ellen Eslinger recounts a slave, John Smith, who was described as having "had entire charge of the Locomotive on the Winchester and Harper's Ferry Railroad for *five years.*" It is unclear if "entire charge" meant that the slave was responsible for repair or actually drove the engine. Ellen Eslinger, "The Brief Career of Rufus W. Bailey, American Colonization Society Agent in Virginia," *Journal of Southern History* 71 (February 2005): 69.

12. Kornweibel, "Railroads and Slavery," 43; Samuel Melanchthon Derrick, *Centennial History of South Carolina Railroad* (Columbia, S.C.: State Company, 1930), 84; *Proceedings of the Stockholders of the Charlotte and South Carolina Rail Road Company, at Their Sixth Annual Meeting at Columbia, S. C., on the Seventh and Eighth of February, 1854. Also, the Annual Reports of the President, Chief Engineer and Treasurer* (Columbia, S.C.: R. W. Gibbes, 1854), 40; Trelease, *North Carolina Railroad*, 62; entry for October 14, 1852, SURR Minutebook; bond for Philip dated January 18, 1836, and bond for William dated January 21, 1836, both in folder 8, box 2, RFPRR-VHS; letter from John McRae to Thomas Stark, June 3, 1847, JML-2; letter

from Joseph Wharton to Anna Lovering, January 27, 1853, in H. Larry Ingle, ed., "Joseph Wharton Goes South, 1853," *South Carolina Historical Magazine* 96 (October 1995): 310; George P. Rawick, ed., *The American Slave: A Composite Autobiography*, supplement, series 1, vol. 8, *Mississippi Narratives, Part 3* (Westport, Conn.: Greenwood, 1977), 1015. For other examples of slaves as firemen, see James M. Bisbee, "The History of the South Side Rail Road, 1846–1870" (Master's thesis, University of Richmond, 1994), 52.

13. *Annual Report of the Board of Directors of the South Carolina Canal & Rail Road Company to the Stockholders, with Accompanying Documents, Submitted at Their Meeting, May 7, 1832* (Charleston, S.C.: William S. Blain, 1832), 3.

14. Entry for August 13, 1836, SCRR Minutebook; *Semi-Annual Report to the Stockholders of the South-Carolina Canal and Rail-Road Company, Made on the Third Monday of July, 20th, 1840, in Conformity with a By-Law of the Company* (Charleston, S.C.: A. E. Miller, 1840), 10–11.

15. *Proceedings of the Stockholders of the South-Carolina Rail-Road Company, and of the South-Western Rail-Road Bank, at Their Annual Meeting, in the Hall of the Bank, on the 9th, 10th, and 11th February, 1847* (Charleston, S.C.: Miller and Brown, 1847), 9–10, 16–17.

16. *Proceedings of the Stockholders of the South-Carolina Rail-Road Company, and of the South-Western Rail-Road Bank, at Their Annual Meeting, in the Hall of the Bank, on the 13th and 14th February 1849* (Charleston, S.C.: Miller and Browne, 1849), 15; *Proceedings of the Stockholders of the South-Carolina Rail-Road Company, and of the South-Western Rail-Road Bank, at Their Annual Meeting, in the Hall of the Bank, on the 11th, and 12th of February, 1851* (Charleston, S.C.: A. E. Miller, 1851), table B; *Proceedings of the Stockholders of the South-Carolina Rail-Road Company, and of the South-Western Rail-Road Bank, at Their Annual Meeting, in the Hall of the Bank, on the 10th, 11th, and 12th February 1852* (Charleston, S.C.: A. E. Miller, 1852), table B.

17. *Proceedings of the Stockholders of the South-Carolina Railroad Company, and of the South-Western Railroad Bank, at at* [sic] *Their Annual Meeting in the Hall of the Bank, on the on the* [sic] *8th and 9th of February, 1853* (Charleston, S.C.: A. E. Miller, 1853), 7, tables B and E, 32.

18. *Annual Reports of the President and Directors and the General Superintendent of the South Carolina Railroad Company, for the Year Ending 31st December, 1856* (Charleston, S.C.: Walker, Evans & Co., 1857), 17, 20; *Annual Reports of the President and Directors and the General Superintendent of the South Carolina Railroad Company, for the Year Ending December 31st, 1857* (Charleston, S.C.: Walker, Evans & Co., 1858), 15; chart reproduced in Derrick, *Centennial History*, 312–13; Michael Tadman, *Speculators and Slaves: Masters, Traders, and Slaves in the Old South* (Madison: University of Wisconsin Press, 1989), 290. Prices for such men ranged from $1,400 to $1,575.

19. James A. Ward, "J. Edgar Thomson and the Georgia Railroad, 1834–1847," *Railroad History* 134 (Spring 1976): 15; *Fourth Annual Report of the President and Directors to the Stockholders of the Louisville, Cincinnati & Charleston Rail-Road Company. September, 1840* (Charleston, S.C.: A. E. Miller, 1840), 19; *Proceedings of the Stockholders of the South-Western R. Road Bank, and of the Stockholders of the Louisville, Cincin. and Charleston Rail-Road Company. At Their Annual Meetings, in Their Banking Institution, on the 17th, 19th, and 20th November, 1840* (Charleston, S.C.: A. E. Miller, 1840), 13, 15; and *Proceedings of the Stockholders of the Louisville, Cincinnati & Charleston Rail-Road Company, and the South-Western R. Road Bank. At Their Annual Meeting, in the Hall of the Bank, on the 16th, 17th, and 18th of Nov. 1841* (Charleston, S.C.: A. E. Miller, 1841), 5–6. The LCCRR's argument here adopts the paternalist stance that historian Jeffrey Young has described, that "masters should maintain highly personalized,

affectionate relations with their slaves." Jeffrey Robert Young, *Domesticating Slavery: The Master Class in Georgia and South Carolina, 1670–1837* (Chapel Hill: University of North Carolina Press, 1999), 124. Nothing came of the proposal, but the LCCRR was soon folded into the SCRR.

20. *Proceedings of the Charlotte and South Carolina Rail Road Company, at Their Second Annual Meeting, at Winnsboro, 10th October, 1849. Also the Annual Reports of the President, Chief Engineer and Treasurer, &c. &c* (Columbia, S.C.: A. S. Johnston, 1849), 4; Jefferson Max Dixon, "An Abstract of the Central Railroad of Georgia, 1833–1892" (Ph.D. diss., Georgia Peabody College for Teachers, 1953), 63–66; Thomas D. Clark, *A Pioneer Southern Railroad from New Orleans to Cairo* (Chapel Hill: University of North Carolina Press, 1936), 91; Arnesen, *Brotherhoods of Color*, 8; *Proceedings of the Fourth Annual Meeting of the Stockholders of the Western North-Carolina Railroad Company; with the Reports of the Officers* (Raleigh, N.C.: Holden & Wilson, "Standard" Office, 1858), 17–18.

21. Account book (1839–53), Joseph Franklin White Papers, SCL; *Fifth Report of the President of the Charleston and Savannah Railroad, to the Stockholders, January 19, 1859* (Charleston, S.C.: Walker, Evans & Co., 1859), 7. Thus, slave hiring could be more flexible than Henson allows. Henson, "Industrial Workers," 25.

22. Contract in the Thomas Larkin Turner Papers, HL.

23. Bond for Randle dated January 11, 1836; bond for Armistead dated January 6, 1836; bond for Jesse and Ben dated January 27, 1836; bond for Peter and Overton dated January 22, 1836; and bond for Henry dated January 22, 1836, all in folder 8, box 2, RFPRR-VHS; David T. Gleeson, *The Irish in the South, 1815–1877* (Chapel Hill: University of North Carolina Press, 2001), 49–50.

24. Bond for Nat and Sam dated January 2, 1836; bond for Bob and Henry dated January 15, 1836; bond for Ben, Lewis, and Braxton, dated January 6, 1836; bond for John, George, and Tarleton, dated January 15, 1836, all in folder 8, box 2, RFPRR-VHS; Martin, *Divided Mastery*, 98–99.

25. Contract of William Boyd, March 8, 1849, folder 5, Isaac Nelson Papers, SCL; letter from John McRae to Edward Lucas, July 28, 1847, JML-2.

26. Letter from John McRae to J. Campbell, July 1, 1852, JML-6.

27. Letter from John McRae to M. W. Haley, October 21, 1847, JML-2.

28. Letter from John McRae to Thos Bracy, September 30, 1847, JML-2; Sharon Ann Murphy, "Securing Human Property: Slavery, Life Insurance, and Industrialization in the Upper South," *Journal of the Early Republic* 25 (Winter 2005): 645; insurance claim in the John Buford Papers, RASP, series F, part 3, reel 34, frame 380; letter from the Richmond Fire Association to Buford, February 17, 1855, ibid., frame 453; Martin, *Divided Mastery*, 79.

29. Letter from John McRae to Thos. Lang, October 29, 1846, JML-1 (McRae also took the time to drop a subtle hint: "I suppose the crop season is now nearly over & that you will soon be at work on your contract"); letter from John McRae to Thos. Lang, December 28, 1846, JML-1; letter from John McRae to Thos. Lang, January 15, 1847, JML-1.

30. Letter from John McRae to Thomas Lang, June 3, 1847; letter from John McRae to Thomas Lang, June 29, 1847; letter from John McRae to Thomas Lang, July 27, 1847, all in JML-2. Richmond had been injured by a man hunting runaways. See John McRae to Thos. Lang, May 24, 1847, JML-2.

31. Peter A. Coclanis, "Off Track: The Railroading of Antebellum Southern Economic

History," *Social Science Quarterly* 84 (September 2003): 738–43; letter from John McRae to Dr. Deas, February 14, 1847, JML-2; letter from John McRae to Dr. Deas, February 17, 1847, JLM-2. As the material later in the chapter demonstrates, $50 per month was a considerable sum for slave rental. Purchased skilled slaves also had a high value. In 1835 the SCRR paid $1,000 for a slave carpenter. Entry for September 10, 1835, SCRR Minutebook.

32. Entry for November 16, 1835, SCRR Minutebook; Martin, *Divided Mastery*, 65; Cline, *Alabama Railroads*, 23; letter from John McRae to E. Lucas, September 30, 1847, JML-2 (two more of Lucas's slaves ran away in October, see letter from John McRae to E. Lucas, October 9, 1847, JML-2); undated voucher paid to E. Meeds, folder 43, box 8, RFPRR-VHS; entry for November 13, 1855, RDRR Minutebook.

33. *Semi-Annual Report of the Direction of the South-Carolina Canal and Rail-Road Company to the Stockholders, November 4th, 1833, with Accompanying Documents* (Charleston, S.C.: A. E. Miller, 1833), 4; *Semi-Annual Report of the South-Carolina Canal and Rail-Road Company, to December 31, 1837* (Charleston, S.C.: J. S. Burges, 1837), 6; letter from George Mills to "Sir," April 19, 1837, and "Memorandum" dated April 29, 1837, both in folder 10, box 2, RFPRR-VHS.

34. *Semi-Annual Report of the Direction of the South-Carolina Canal and Rail-Road Company, to the Stockholders, October 31, 1834. With Accompanying Documents* (Charleston, S.C.: J. S. Burges, 1834), 5.

35. Entry for October 22, 1835, SCRR Minutebook; *Semi-Annual Report of the South-Carolina Canal and Rail Road Company. Accepted Jan. 18, 1840* (Charleston, S.C.: A. E. Miller, 1840), 12, 6–7; *Thirtieth Semi-Annual Report of the South-Carolina Canal and Rail-Road Company, to January 1st, 1843, Accepted at the Adjourned Meeting, 26th February, 1843* (Charleston, S.C.: Miller and Browne, 1843), 25–26. Not everyone was pleased with the system. In May 1835 Benjamin F. Wish made an application to be released from service to the company. The board of directors appointed two directors to examine the situation and "dispose of it as they may think fit." See entry for May 22, 1835, SCRR Minutebook. Other companies created apprenticeships as well, although the details are not as well known as in the case of the SCRR. When the Blue Ridge Railroad was being surveyed, the president reported that the railroad had in its employ twenty-nine engineers, of whom twenty-one were paid and eight were "learners as yet" and worked without pay. *Annual Report of the President and Chief Engineer to the Stockholders of the Blue Ridge Rail Road Company, at the Annual Meeting, Held in Charleston, July 11, 1854; Together with the Report of B. H. Latrobe, Esq. Chief Engineer Baltimore and Ohio Rail Road, of His Examination of the Route of the Blue Ridge Rail Road* (Charleston, S.C.: Walker & Evans, 1854), 11–12.

36. Lisa Denmark, "At the Midnight Hour: Optimism and Disillusionment in Savannah, 1865–1880" (Ph.D. diss., University of South Carolina, 2004), 23; Cecil Kenneth Brown, *A State Movement in Railroad Development: The Story of North Carolina's First Effort to Establish an East and West Trunk Line Railroad* (Chapel Hill: University of North Carolina Press, 1928), 117; *SCRR Report, January 1840*, 12.

37. Franz Anton Ritter von Gerstner, *Early American Railroads*, ed. Frederick C. Gamst (Stanford: Stanford University Press, 1997), 696; *Seventeenth Annual Report of the President and Directors of the Virginia Central Railroad Company to the Stockholders, at Their Annual Meeting, on the 29th October 1852* (Richmond, Va.: Colin and Nowlan, 1852), 30–31; *Seventh Annual Report of the President and Directors to the Stockholders of the Virginia and Tennessee Railroad Company, October 25th and 26th, 1854* (Lynchburg: "Virginian" Job Printing Estab-

lishment, 1854), 19–20; *Tenth Annual Report of the President and Directors to the Stockholders of the Virginia and Tenn. Railroad Co., September 16–17, 1857* (Lynchburg: Virginian Steam Power Press Print, 1857), 213.

38. Letter from John Glass to Peter Glass, May 22, 1851, Glass Letterbook; letter from John Glass to Peter Glass, November 6, 1851, Glass Letterbook. Ethnic battles among railroad construction crews can be found in other parts of the world. See David Brooke, *The Railway Navvy: "That Despicable Race of Men"* (North Pomfret, Vt.: David and Charles, 1983), 118, and Matthew J. Payne, *Stalin's Railroad: Turksib and the Building of Socialism* (Pittsburgh: University of Pittsburgh Press, 2001), 95, 118–19.

39. Joseph A. Waddell, *Home Scenes and Family Sketches* (Staunton, Va.: Stoneburner & Prufer, 1900), 152–53.

40. N. J. Bell, *Southern Railroad Man: Conductor N. J. Bell's Recollections of the Civil War Era*, ed. James A. Ward (DeKalb: Northern Illinois University Press, 1993), 6.

41. Letter from John McRae to James Gadsden, October 11, 1846, JML-1; letters from M. B. Pritchard to "Messrs Stone and Harris," May 2 and 21, 1849, both in folder 10-1, vol. 10, StHC.

42. "Articles of Agreement," August 1, 1857, Hawkins Family Papers, SHC, *SASI*, series B, reel 14, frame 259; letter from John McRae to Mr. Kennedy, July 27, 1852, JML-6 (McRae emphasized the need for horses again in a letter written on the same day to Walter Gwynn); *Proceedings of the Annual Meeting of Stockholders of the Western North Carolina Rail Road Company, Held in Statesville, August 30, 1860* (Salisbury, N.C.: Printed at the Carolina Watchman Office, 1860), 27. On horses generally, see Jason Hribal, " 'Animals Are Part of the Working Class': A Challenge to Labor History," *Labor History* 44 (November 2003): 435–53. On the railroad's role of reinforcing the role of the urban horse in nineteenth-century America, see Clay McShane and Joel A. Tarr, "The Centrality of the Horse in the Nineteenth-Century American City," in *The Making of Urban America*, 2d ed., ed. Raymond A. Mohl (Wilmington, Del.: Scholarly Resources, 1997), 105–30. See also Colin Divall, "Beyond the History of Early Railways," in *Early Railways 2: Papers from the Second International Railways Conference*, ed. M. J. T. Lewis (London: Newcomen Society, 2003), 1–9, and Ralph Turvey, "Horse Traction in Victorian London," *Journal of Transport History*, 3d ser., 26 (September 2005): 43.

43. Entry for April 1, 1850, RDRR Minutebook; entries for November 12, 1851, and October 20, 1852, Southside Minutebook-1; Phillips, *History of Transportation*, 295.

44. *SCRR Report, December 1837*, 4; von Gerstner, *Early American Railroads*, 715; *SWRR Reports, 1 to 22 Inclusive*, 275; *Reports of the Presidents and Superintendents of the Central Railroad and Banking Co. of Georgia, from No. 20 to 32 Inclusive, and the Amended Charter of the Company* (Savannah, Ga.: G. N. Nichols, 1868), 85–86.

45. Kornweibel, "Railroads and Slavery," 38; Starobin, *Industrial Slavery*, 60–62; James Redpath, *The Roving Editor, or, Talks with Slaves in the Southern States*, ed. John R. McKivigan (University Park: Pennsylvania State University Press, 1996), 125–26; Frederick Law Olmsted, *A Journey in the Seaboard Slave States in the Years, 1853–1854, with Remarks on Their Economy*, vol. 2 (1856; reprint, New York: Putnam's Sons, 1904), 1.

46. *Twelfth Annual Report of the President and Directors to the Stockholders of the Virginia and Tennessee R. R. Co., September 14, 1859* (Lynchburg: Printed at the Virginian Power Press Job Office, 1859), 146–49; voucher dated February 24, 1837, folder 9, box 2, RFPRR-VHS; letter from William Birnie to "Messrs Stone & Harris," June 16, 1849, folder 10-1, vol. 10,

StHC; entries for June 13, 1855, and July 27, 1857, RDRR Minutebook; entry for January 11, 1859, ETVRR Minutebook-2.

47. Charles H. Hildreth, "Railroads out of Pensacola, 1833–1883," *Florida Historical Quarterly* 37 (January–April 1959): 403; Way, *Common Labour*, 113; letter from John McRae to James Chesnut, May 28, 1847, JML-2; "Articles of Agreement," August 1, 1857, Hawkins Family Papers, SHC, *SASI*, series B, reel 14, frame 263; Trelease, *North Carolina Railroad*, 35; letter from John Glass to Peter Glass, February 5, 1851, Glass Letterbook; William Johnson, *William Johnson's Natchez: The Ante-Bellum Diary of a Free Negro*, ed. William Ransom Hogan and Edwin Adams Davis (Baton Rouge: Louisiana State University Press, 1951), 156. Northern railroads also worried about alcohol consumption. Samuel Nott, an engineer on the Portland, Saco and Portsmouth Railroad, expressed "great astonishment" that a store owner named Carr was "dealing out liquor on a large scale." Nott considered it "so very important for our business interests, as well as our private character" that the company should "persuade Carr to relinquish the objectionable part of his business." Letter from Samuel Nott to A. F. Edwards, August 20, 1841, Nott Letterbook. See also Henson, "Industrial Workers," 9.

48. This attitude changed after the Civil War: "By the 1890s, most of the major carriers had or were developing some form of medical organization." Mark Aldrich, "Train Wrecks to Typhoid Fever: The Development of Railroad Medicine Organizations, 1850 to World War I," *Bulletin of the History of Medicine* 75 (Summer 2001): 264. There were some exceptions: the Southside Railroad hired one nurse on a yearly basis. Entry for October 20, 1852, Southside Minutebook-1.

49. James W. Ely Jr., *Railroads and American Law* (Lawrence: University Press of Kansas, 2001), 214; see also John Fabian Witt, *The Accidental Republic: Crippled Workingmen, Destitute Widows, and the Remaking of American Law* (Cambridge, Mass.: Harvard University Press, 2004), 13–14.

50. Entry for May 4, 1858, ETVRR Minutebook-1.

51. The board of directors also resolved that it would be appropriate for the workers themselves to "create a sinking fund by voluntary subscription to meet such contingencies as happened to Mr Gros." Entry for April 20, 1840, SCRR Minutebook.

52. Entry for March 4, 1852, RDRR Minutebook.

53. Entry for June 28, 1852, RDRR Minutebook. They were still willing to look at some cases individually, however. In 1854 William Gilman received an unspecified amount to cover "injuries he received while in discharge of his duties as Conductor of the Freight Train." Entry for September 14, 1854, RDRR Minutebook.

54. Entry for June 10, 1857, RDRR Minutebook.

55. Entries for February 7 and 8, 1855, VTRR Minutebook. Northern railroads do not appear to have treated their employees much better. In 1848 the Nashua and Lowell Railroad agreed to pay a sum to John Jackman, an injured brakeman. The committee made it clear that it did "not acknowledge a legal liability" on the company's part but was merely acting because Jackman had "been rendered permanently lame in consequence of the injury" and intended to "avoid a Lawsuit." The Illinois Central Railroad offered "no sick pay or other compensation for men incapacitated by accident or disease; once dropped from the payroll they might as well have ceased to exist so far as the railroad company and its contractors were concerned." Committee report dated June 1848, folder 1, vol. 130, Nashua and Lowell Railroad Co. Collection, BHC; Lightner, *Labor on the Illinois Central*, 36.

56. Entry for September 14, 1859, RDRR Minutebook; entry for July 10, 1856, Southside Minutebook-2.

57. Entries for May 11 and 12, 1855, VTRR Minutebook.

58. Bond for Robert dated January 5, 1836, and bond for Major dated January 3, 1836, both in folder 8, box 2, RFPRR-VHS; entry for August 20, 1835, SCRR Minutebook; *Fifth Annual Report of the Board of Directors to the Stockholders in the Memphis and Charleston Railroad Company, for the Year Ending March 1st, 1855* (Memphis: Eagle and Enquirer Steam Printing House, 1855), 44. The necessity for such a precaution did not deter the MCRR from using slaves, because the "economy of slave labor upon Southern Roads has been frequently demonstrated."

59. Ely, *Railroads and American Law,* 136–37; Jenny Bourne Wahl, "The Bondsman's Burden: An Economic Analysis of the Jurisprudence of Slaves and Common Carriers," *Journal of Economic History* 53 (September 1993): 503.

60. Entries for November 20 and December 20, 1837, April 20 and June 7, 1839, and January 20, 1840, SCRR Minutebook.

61. Voucher for W. Malloy, January 22, 1837, voucher for James Cavanagh, January 22, 1837, voucher dated February 1, 1837, all in folder 9, box 2, RFPRR-VHS.

62. *Proceedings of an Adjourned Meeting of the Stockholders of the Louisville, Cincinnati and Charleston Rail-Road Company, Held in Columbia, on the 4th, 6th and 7th December, 1839* (Charleston, S.C.: A. E. Miller, 1840), 18; *Action of the Stockholders of the South-Carolina Railroad Company, on the Reports and Resolutions of the Committee of Inspection, Held at the Hall of the South-Western Railroad Bank, on Tuesday, 30th of May 1848. To which Is Annexed, in an Appendix, the Report of the President to Board of Directors, Referred to in the Proceedings; and the Report of S. M. Fox, (Civil Engineer) on the Inclined Plane* (Charleston, S.C.: Miller and Browne, 1848), 11–12; entry for July 1, 1859, ETVRR Minutebook-2.

63. Bond for Randle dated January 11, 1836, bond for Daniel dated January 18, 1836, bond for Robert dated January 4, 1836, bond for Henry dated January 6, 1836, and bond for Smith, not dated, all in folder 8, box 2, RFPRR-VHS; R. S. Cotterill, "Southern Railroads, 1850–1860," *Mississippi Valley Historical Review* 10 (March 1924): 404; letter from John McRae to Thos. F. Drayton, December 2, 1854, JML-7; letter from John McRae to C. S. Gadsden, December 12, 1854, JML-7; *Annual Report of the Board of Directors of the South Carolina Canal and Rail-Road Company to the Stockholders, with Accompanying Documents Submitted at Their Meeting, May 6, 1833* (Charleston, S.C.: A. E. Miller, 1833), 3.

64. Payroll figures have been taken from the payroll sheets and receipts for salaried workers located in folder 43 (January–February), folder 44 (March–May), folder 45 (June–July), all in box 8; folder 46 (August), folder 47 (September–October), folder 48 (November–December), all in box 9, RFPRR-VHS. Bonds for slaves hired to work in 1845 are in the folders for 1846: folder 49 (January) and folder 50 (February–March), both in box 9, RFPRR-VHS.

65. Weekly payroll sheets are available for the "Train," "Depot," and "Repairs of Engines, Cars, etc." departments. While nearly complete (for the week ending January 3 through the week ending December 26), three January payrolls are missing for the depot and four for repairs. Thus, the total wages paid—$10,458.76—represents a minimum, not the total paid for the year.

66. Three other men worked for the department. Elisha Fox, of indeterminate race,

worked for two weeks in July and August. Tom Scott, paid the same wages as Mr. Evans, suggesting he was white, worked from September through December. And Freeborn Woodson, whose first name and wages suggest he was not a slave, joined the division in November.

67. The highest amount that Waller accumulated was actually $3.60 for the week ending January 3; however, because nearly all the slaves were compensated an unusually high amount, this appears to be an anomaly—perhaps reflecting generosity at the end of the year or a period of unusually bad weather. Waller's take for that week put him above the $3.00 earned by Mr. Evans, as did the $4.20 earned by seven slaves. While Evans's feelings about the discrepancy have not been recorded, he did not quit his post as a result.

68. Their owners were also comparatively well compensated: Warner Burress earned his master $75, Johnson cost $70, and Jefferson cost $75. There are two Mats with bonds that have survived, but both are for similarly high amounts: one for $70 and the other for $85.

69. Voucher entitled "List of hands which are to have the amount of their winter clothing in money, by order of their masters," folder 43, box 8, RFPRR-VHS.

70. Entry for July 10, 1854, VTRR Minutebook; Noe, *Southwest Virginia's Railroad*, 83; *Annual Report to the Stockholders of the Petersburg R. Road Company, 1857* (Petersburg, Va.: O. Ellyson, 1857), 18.

71. Entries for March 20, 1850, and December 14, 1855, RDRR Minutebook.

72. Robert F. Hunter and Edwin L. Dooley Jr., *Claudius Crozet: French Engineer in America, 1790–1864* (Charlottesville: University Press of Virginia, 1989), 157; Christopher Silver, "Immigration and the Antebellum Southern City: Irish Working Class Mobility in Charleston, South Carolina, 1840–1860" (Master's thesis, University of North Carolina at Chapel Hill, 1975), 33; Martha Gallaudet Waring, ed., "Charles Seton Henry Hardee's Recollections of Old Savannah (Part I)," *Georgia Historical Quarterly* 12 (December 1928): 380–81.

73. Matthew E. Mason, " 'The Hands Here are Disposed to be Turbulent': Unrest among the Irish Trackmen of the Baltimore and Ohio Railroad, 1829–1851," *Labor History* 39 (August 1998): 253–72; *Fourth Annual Report of the Directors of the Pennsylvania Railroad Co. to the Stockholders, December 31, 1850* (Philadelphia: Crissy & Markley, 1851), 8; *Fifth Annual Report of the Directors of the Pennsylvania Railroad Co. to the Stockholders, February 2, 1852* (Philadelphia: Crissy & Markley, 1852), 41; entry for July 7, 1841, Engineering Department Diary, vol. 159, BARC-WR.

74. Bell, *Southern Railroad Man*, 7; Gustavus W. Dyer, John Trotwood Moore, Colleen Mores Elliott, and Louise Armstrong Moxley, eds., *The Tennessee Civil War Veterans Questionnaires*, vol. 4, *Confederate Soldiers (Lackey–Quarles)* (Easley, S.C.: Southern Historical Press, 1985), 1665; letter from John Glass to Peter Glass, January 6, 1852, Glass Letterbook.

Chapter Four • Structure

1. *Charleston Courier*, January 6, 1831, 1; Alfred D. Chandler Jr., *The Visible Hand: The Managerial Revolution in American Business* (Cambridge, Mass.: Belknap Press of Harvard University Press, 1977), 87, 120; Steven G. Collins, "Progress and Slavery on the South's Railroads," *Railroad History* 181 (Autumn 1999): 10; see also Steven G. Collins, "System, Organization, and Agricultural Reform in the Antebellum South, 1840–1860," *Agricultural History* 75 (Winter 2001): 4–5. Companies routinely began operating before the road was fully completed so that they could start to earn money on the investment they had made; see

Daniel Hovey Calhoun, *The American Civil Engineer: Origins and Conflict* (Cambridge, Mass.: Technology Press, Massachusetts Institute of Technology, 1960), 74; William Penn Dickinson Jr., "The Social Saving Generated by the Richmond & Danville Railroad: 1859" (Master's thesis, University of Virginia, 1975), 19; Allen W. Trelease, *The North Carolina Railroad, 1849–1871, and the Modernization of North Carolina* (Chapel Hill: University of North Carolina Press, 1991), 36; and Charles W. Turner, "The Early Railroad Movement in Virginia," *Virginia Magazine of History and Biography* 55 (October 1947): 356. For recent explorations of shareholder activity, see Charles W. Wootton and Christie L. Roszkowski, "Legal Aspects of Corporate Governance in Early American Railroads," *Business and Economic History*, 2d ser., 28 (Winter 1999): 335; and Colleen A. Dunlavy, "From Citizens to Plutocrats: Nineteenth-Century Shareholder Voting Rights and Theories of the Corporation," in *Constructing Corporate America: History, Politics, Culture*, ed. Kenneth Lipartito and David B. Sicilia (New York: Oxford University Press, 2004), 66–93.

2. Chandler, *Visible Hand*, 96–97.

3. *Annual Report of the Direction of the South Carolina Canal and Rail Road Company, to the Stockholders, May 6th, 1834, with Accompanying Documents* (Charleston, S.C.: W. S. Blain, 1834), 8–9, quotation on 9. While the reason for the choice of the title "overseer" was not explained, it does lend to at least an implicit plantation analogy. For more duties of the overseer, see the broadside, c. 1840, in the South Carolina Railroad Papers, SCL. Von Gerster observed the system; see Franz Anton Ritter von Gerstner, *Early American Railroads*, ed. Frederick C. Gamst (Stanford: Stanford University Press, 1997), 715.

4. Entry for June 25, 1835, SCRR Minutebook; entry for January 2, 1856, SURR Minutebook.

5. Letter from John McRae to Mr. Allison, August 8, 1849, JML-4. The letter also indicates that the railroad numbered cars for tracking.

6. A similar letter was sent to the superintendent of the inclined plane at Aiken. Letter from John McRae to N. D. Boxly, September 8, 1849, JML-4.

7. Letter from John McRae to William Magrath, September 9, 1849, JML-4.

8. Mark M. Smith, *Mastered by the Clock: Time, Slavery, and Freedom in the American South* (Chapel Hill: University of North Carolina Press, 1997), 79–92.

9. *SCRR Report, May 1834*, 12. The report does not specify which member of the train crew would have to pay the fine. In 1839 the rule was modified: "Upon satisfactory information that either passengers or freight have been delayed by contention between the Engineers for the right to go on or for any other cause, the Engineers shall be fined Five Dollars each, and in addition the value of the wages, fuel &c lost thereby." Entry for June 7, 1839, SCRR Minutebook. Walter Licht argues that, compared to Great Britain, "the relative absence of fining practices [in America] is striking." By contrast, the SCRR is one of several companies I studied that made use of fining, so the process may have been more widespread than Licht allows. It is worth noting that most of his examples of fining and opposition to fining come from the postbellum era. Walter Licht, *Working for the Railroad: The Organization of Work in the Nineteenth Century* (Princeton: Princeton University Press, 1983), 118–19.

10. Entry for May 28, 1835, SCRR Minutebook; *Charleston Courier*, January 6, 1831, 1. For more on southern punctuality, see Smith, *Mastered by the Clock*, 87.

11. *SCRR Report, May 1834*, 12; entry for June 7, 1839, SCRR Minutebook. For an alternative view, see Carlene Stephens, " 'The Most Reliable Time': William Bond, the New En-

gland Railroads, and Time Awareness in 19th-Century America," *Technology and Culture* 30 (January 1989): 2.

12. Samuel Melanchthon Derrick, *Centennial History of South Carolina Railroad* (Columbia, S.C.: State Company, 1930), 309, 311.

13. Letter from John McRae to George Lythgoe, January 9, 1850, JML-4; entry for January 2, 1856, SURR Minutebook; entry for February 7, 1855, VTRR Minutebook.

14. *Semi-Annual Report of the South-Carolina Canal and Rail-Road Company, to December 31, 1837* (Charleston, S.C.: J. S. Burges, 1837), 5.

15. Richard B. Du Boff, "Business Demand and the Development of the Telegraph in the United States, 1844–1860," *Business History Review* 54 (Winter 1980): 465; *Proceedings of the Stockholders of the South-Carolina Rail-Road Company, and of the South-Western Rail-Road Bank, at Their Annual Meeting, in the Hall of the Bank, on the 12th, 13th and 14th of February 1850* (Charleston, S.C.: Miller and Browne, 1850), n.p.; *Annual Reports of the President and Directors and the General Superintendent of the South Carolina Railroad Company, for the Year Ending 31st December, 1856* (Charleston, S.C.: Walker, Evans & Co., 1857), 9; entries for April 9 and 10, 1857, in the Minutebook of the Stockholders (April 6, 1853–November 21, 1861), Norfolk and Petersburg Railroad, NSHC; transcript of contract in folder marked "East Tennessee and Virginia Railroad, 1858," NSHC; contract between the North Carolina Railroad and the Magnetic Telegraph Company, July 24, 1858, box labeled "Contracts, Correspondence, etc. (1849–91)," North Carolina Railroad, NSHC; entry for May 19, 1858, ETGRR Minutebook; contract between the Richmond and Danville Railroad and the Lynchburg and Abingdon Telegraph Company, June 14, 1859, in folder marked "Miscellaneous contracts and correspondence, 1850–69," box marked "Contracts, Correspondence, etc. (1850–92)," Richmond and Danville Railroad, NSHC; entry for February 10, 1859, RDRR Minutebook.

16. *Fifth Annual Report of the Directors of the Pennsylvania Railroad Co. to the Stockholders, February 2, 1852* (Philadelphia: Crissy & Markley, 1852), 15–16, 52.

17. Manuscript copy of "At the Annual Meeting of the Stockholders of the Boston & Worcester Rail road, on the 5th of June 1837," vol. 11, BARC-BW; letter from G. Twichell to D. N. Pickering, January 7, 1850, G. Twichell Letterbook, vol. 119, BARC-BW.

18. Two of the accidents were in 1853 (which killed four and fourteen people) and one was in 1855 (which killed three people). Ian R. Bartky, *Selling the True Time: Nineteenth-Century Timekeeping in America* (Stanford: Stanford University Press, 2000), 24.

19. Horatio Allen, *Reports to the Board of Directors of the South-Carolina Canal and Rail Road Company* (Charleston, S.C.: J. S. Burges, 1831), 6; *Charleston Courier,* May 16, 1831, 2. Although it can be difficult to remember in our fast-paced age, "speed is not a value in itself." Barbara Adam, *Timewatch: The Social Analysis of Time* (Cambridge, Mass.: Polity, 1995), 100.

20. *Charleston Courier,* March 7, 1831, 2. The author took his trip two months before the SCRR announced its rate of travel described above.

21. See Smith, *Mastered by the Clock,* chap. 1, for an analysis of prerailroad time consciousness.

22. Entry for June 7, 1839, SCRR Minutebook. Von Gerstner reported similar speeds on other southern railroads. The Winchester and Potomac Railroad's passenger trains "as prescribed by regulation may never exceed 15 miles per hour. For freight trains, 10 miles an hour is the established maximum rate." The Richmond and Petersburg took the following speeds: "On passenger runs, then, the speed is 15 to 20 miles an hour including stops. For

freight trains it is 12 miles per hour. Speeds are always lower on the bridge at Richmond; the prescribed time to pass over it is 4 minutes for passenger trains, and that corresponds to 8 miles an hour." Von Gerstner, *Early American Railroads*, 677, 689.

23. Letter from John McRae to Thomas Lang, November 3, 1850, JML-5. In a letter to McRae, recently ousted company president James Gadsden blamed "mismanagement & neglect" for the collapse. Letter from James Gadsden to John McRae, November 9, 1850, folder titled "December 1849–November 1851," James Gadsden Papers, SCL.

24. Von Gerstner, *Early American Railroads*, 679, 681; letter from John McRae to N. W. Daniell, January 17, 1850, JML-4.

25. *Tenth Annual Report of the President and Other Officers of the East Ten. and Va. Railroad Co. to the Stockholders, at Their Annual Meeting at Greeneville, November 24, 1859* (Jonesborough, Tenn.: W. A. Sparks & Co., 1859), 12–13.

26. *Fourth Annual Report of the Directors of the Pennsylvania Railroad Co. to the Stockholders, December 31, 1850* (Philadelphia: Crissy & Markley, 1851), 52; letter from George Stark to William Parker, December 11, 1854, vol. 112, Nashua and Lowell Railroad Co. Collection, BHC; *Rules of the Boston & Lowell and Nashua & Lowell Railroad Companies, for the Government of the Executive Service, Adopted by the Board of Directors, February, 1857* (Nashua, N.H.: Albin Beard, 1857), 13.

27. *Report on the Preliminary Surveys for the Savannah and Albany Rail-Road* (Savannah, Ga.: Evening Journal Print, 1853), 10–11.

28. W. F. Cottrell, "Of Time and the Railroader," *American Sociological Review* 4 (April 1939): 191; *Charleston Courier*, January 6, 1831, 1. Another schedule was announced the next day, *Charleston Courier*, January 7, 1831, 2. The train reverted to the previous schedule twelve days later; *Charleston Courier*, January 19, 1831, 1.

29. Entry for November 12, 1833, TCDRR Minutebook; von Gerstner, *Early American Railroads*, 689, 749.

30. Entries for November 3 and 5, 1836, Blair Bolling Diary, Bolling Family Papers, VHS, *RASP*, series M, part 5, reel 9, frame 795.

31. *Report of the Committee of Inspection on the Condition of the South-Carolina Rail-Road. Made at an Extra Meeting of the Stockholders, Held on the 2d of May 1848* (Charleston, S.C.: Miller and Browne, 1848), 6; entry for September 11, 1854, MCRR Minutebook.

32. Entry for June 25, 1835, SCRR Minutebook; Mary G. Cumming, *Georgia Railroad & Banking Company, 1833–1945: An Historic Narrative* (Augusta, Ga.: Walton Printing Company, 1945), 57.

33. Entry for June 7, 1839, SCRR Minutebook.

34. Letter from John McRae to C. J. Bollin, January 16, 1850, JML-4; letter from John McRae to James Gadsden, February 7, 1847, JML-2. L. S. Kemnitzer has pointed out that railroaders need "multiple time senses." L. S. Kemnitzer, "Another View of Time and the Railroader," *Anthropological Quarterly* 50 (January 1977): 25. See also Adam, *Timewatch*, 12.

35. Entry for February 2, 1855, Southside Minutebook-2.

36. Letter from John Glass to Peter Glass, February 5, 1851, Glass Letterbook; entry for April 20, 1838, SCRR Minutebook.

37. Horatio Allen, *The Railroad Era: The First Five Years of Its Development* (New York: n.p., 1884), 28–29. John White comments on Allen's tale that "it is not known if night operations were continued at this time, but if they were, surely a more convenient plan of illumi-

nation was adopted." John H. White Jr., *American Locomotives: An Engineering History, 1830–1880*, 2d ed. (Baltimore: Johns Hopkins University Press, 1997), 215. Although Barbara Welke refers to the post–Civil War era as the "onset" of night travel, it is clear that night operations predate the Civil War. Barbara Young Welke, *Recasting American Liberty: Gender, Race, Law, and the Railroad Revolution, 1865–1920* (New York: Cambridge University Press, 2001), 140.

38. Voucher dated March 31, 1837, folder 9, box 2, RFPRR-VHS.

39. Von Gerstner, *Early American Railroads*, 681, 724; *Reports of the Presidents, Engineers-in-Chief and Superintendents, of the Central Rail-Road and Banking Company of Georgia, from No. 1 to 19 Inclusive, with Report of Survey by Alfred Cruger, and the Charter of the Company* (Savannah, Ga.: John M. Cooper, 1854), 126; *Proceedings of the Stockholders of the Louisville, Cincinnati & Charleston Rail-Road Company, and the South-Western R. Road Bank, at Their Annual Meeting, in the Hall of the Bank, on the 22d, 23d, 24th, and 25th of Nov. 1842* (Charleston, S.C.: A. E. Miller, 1842), 27.

40. *SCRR Report, February 1850*, 45; *Proceedings of the Stockholders of the South-Carolina Rail-Road Company, and of the South-Western Rail-Road Bank, at Their Annual Meeting, in the Hall of the Bank, on the 10th, 11th, and 12th February 1852* (Charleston, S.C.: A. E. Miller, 1852), 29, 30.

41. *SCRR Report, December 1856*, 8; *Annual Reports of the President and Directors and the General Superintendent of the South Carolina Railroad Company, for the Year Ending December 31st, 1857* (Charleston, S.C.: Walker, Evans & Co., 1858), 8.

42. *ETVRR Report, November 1859*, 13; entry for January 2, 1856, SURR Minutebook; *Report of the President and Directors of the East Tennessee and Georgia Rail Road Company, to the Stockholders in the Same, at Their Annual Convention, Held at Athens, Sept. 3, 1856* (Athens, Tenn.: Printed by Sam. P. Ivins, Office of the "Post," 1857), 18–19; letter from Henry Bird to Mary Fox, November 24, 1832, folder 8, section 1, Bird Family Papers, VHS.

43. Entry for April 12, 1834, Minutebook of the Directors and Stockholders (1831–1835), vol. 1, BARC-BW; entry for August 6, 1839, Minutebook of the Directors (January 8, 1836–October 16, 1841), vol. 2, BARC-BW; *Rules of the Boston & Lowell*, 20; Paul Johnson, *Shopkeeper's Millennium: Society and Revivals in Rochester, New York, 1815–1837* (New York: Hill and Wang, 1978), 81.

44. John H. White Jr., "Everyday Life of the Train Captain," *Railroad History* 187 (Fall–Winter 2002): 14–39; entry for January 2, 1856, SURR Minutebook; Solon Robinson, *Solon Robinson: Pioneer and Agriculturalist*, ed. Herbert Anthony Kellar, vol. 2, *1846–1851* (Indianapolis: Indiana Historical Bureau, 1936), 387; entry for January 2, 1856, SURR Minutebook. See also Amy G. Richter, *Home on the Rails: Women, the Railroad, and the Rise of Public Domesticity* (Chapel Hill: University of North Carolina Press, 2005), chap. 5.

45. Collins, "Progress and Slavery," 22; *Rules of the Boston & Lowell*, 9, 10.

46. Entries for October 22, 1835, and December 20, 1837, SCRR Minutebook; entry for February 2, 1852, RDRR Minutebook. See also entries for May 11, 1852, February 10 and October 29, 1853, and July 27 and October 15, 1857, all in RDRR Minutebook; and entry for July 10, 1856, Southside Minutebook-2.

47. Mary Elizabeth Pearson Boyce, *And God Still Spares Us: The Diary of Mary Elizabeth Pearson Boyce, September 28, 1854–November 2, 1855* (Hickory Grove, S.C.: Broad River Basin Historical Society, 1993), 39; Julia Wright Sublette, "The Letters of Anna Calhoun Clemson, 1833–1873" (Ph.D. diss., Florida State University, 1993), 556–57.

48. *SCRR Report,* May 1834, 11. This aura of personal responsibility extended to other railroads. On the Western and Atlantic, engines were "assigned to an engineer who cared for it as though it were his personal property." Raymond B. Carneal and James G. Bogle, "Locomotives of the Western & Atlantic Railroad," *Atlanta Historical Bulletin* 15, no. 1 (1970): 8.

49. *Semi-Annual Report of the Direction of the South-Carolina Canal and Rail-Road Company, to the Stockholders, October 31, 1834. With Accompanying Documents* (Charleston, S.C.: J. S. Burges, 1834), 5; *Proceedings of the Stockholders of the South-Carolina Rail-Road Company, and of the South-Western Rail-Road Bank, at Their Annual Meeting, in the Hall of the Bank, on the 11th, and 12th of February, 1851* (Charleston, S.C.: A. E. Miller, 1851), table 1. For locomotives in America, see White, *American Locomotives.*

50. John H. White Jr., *The American Railroad Freight Car: From the Wood-Car Era to the Coming of Steel* (Baltimore: Johns Hopkins University Press, 1993), 164–65, 176; letters from John McRae to F. C. Arms, April 13 and July 11, 1849, both in JML-4.

51. John H. White Jr., *The American Railroad Passenger Car* (Baltimore: Johns Hopkins University Press, 1978), 20; Burford quoted in N. P. Renfro, *The Beginning of Railroads in Alabama* (Auburn, Ala.: n.p., 1910), 4; entry for March 8, 1838, in Mary E. Moragne, *The Neglected Thread: A Journal from the Calhoun Community, 1836–1842,* ed. Delle Mullen Craven (Columbia: University of South Carolina Press, 1951), 80.

52. Entry for December 23, 1841, in Charles Lyell, *Travels in North America in the Years 1841–2, with Geological Observations on the United States, Canada, and Nova Scotia* (1845; reprint, New York: Arno, 1977), 113. See also entry for December 10, 1849, in Charles W. Nicholson, ed., "The Journal of Frederick William Muller," *South Carolina Historical Magazine* 86 (October 1985): 277. For the connection of parlors and antebellum transportation, see Katherine C. Grier, *Culture and Comfort: Parlor Making and Middle-Class Identity, 1850–1930* (Washington, D.C.: Smithsonian Institution Press, 1997), 47–48. There were some exceptions. Charles Hardee rode another type of passenger car in 1844 in Georgia. "There was a long bench running the whole length of the car, on each side, on which you had to sit bolt-upright all night. There was no such thing as getting a nap. If you wanted to get even 'forty winks' of sleep you would have to take them a wink at a time." Martha Gallaudet Waring, ed., "Reminiscences of Charles Seton Henry Hardee," *Georgia Historical Quarterly* 12 (June 1928): 174.

53. Louis Heuser Diary Transcription, p. 28, WHS.

54. For a general description of interiors, see White, *American Railroad Passenger Car,* 434–36.

55. William Harden, *Recollections of a Long and Satisfactory Life* (1934; reprint, New York: Negro Universities Press, 1968), 79; letters from John McRae to J. R. Trimble, May 3, May 22, and December 22, 1849, all in JML-4. Trimble worked for a railroad; McRae asked him to order the cars from a local company on his behalf.

56. *SCRR Report,* February 1850, 15; John R. deTreville, "The Little New South: Origins of Industry in Georgia's Fall-Line Cities, 1840–1865" (Ph.D. diss., University of North Carolina at Chapel Hill, 1985), 103.

57. Von Gerstner, *Early American Railroads,* 680; entry for November 18, 1847, in Thomas Hubbard Hobbs, *The Journals of Thomas Hubbard Hobbs,* ed. Faye Acton Axford (University: University of Alabama Press, 1976), 84; *Reports of the Presidents and Superintendents of the Central Railroad and Banking Co. of Georgia, from No. 20 to 32 Inclusive, and the*

Amended Charter of the Company (Savannah, Ga.: G. N. Nichols, 1868), 145; entry for October 19, 1853, Southside Minutebook-1.

58. *Thirtieth Semi-Annual Report of the South-Carolina Canal and Rail-Road Company, to January 1st, 1843, Accepted at the Adjourned Meeting, 26th February, 1843* (Charleston, S.C.: Miller and Browne, 1843), 27; letter from John McRae to J. R. Scott, January 25, 1850, JML-4. See also White, *American Railroad Passenger Car,* 14–15.

59. Frederick Law Olmsted, *A Journey in the Seaboard Slave States in the Years 1853–1854, with Remarks on Their Economy,* vol. 2 (1856; reprint, New York: Putnam's Sons, 1904), 22. For the skill of antebellum shop craftsmen, see John K. Brown, "Design Plans, Working Drawings, National Styles: Engineering Practice in Great Britain and the United States, 1775–1945," *Technology and Culture* 41 (April 2000): 213. John White explicitly excluded surviving locomotives as a source for his study of American locomotives, noting that "surviving nineteenth-century locomotives in this country are not authentic representatives of locomotives of their respective period. Most have been rebuilt to the point that little remains of the original mechanism." White, *American Locomotives,* xxvi.

60. *SCRR Report, May 1834,* 12; *Semi-Annual Report to the Stockholders of the South-Carolina Canal and Rail-Road Company. Made on the Third Monday in January, (the 17th) in Conformity with a By-Law of the Company* (Charleston, S.C.: A. E. Miller, 1842), 6; von Gerstner, *Early American Railroads,* 728; *CRRG Reports, 20 to 32 Inclusive,* 145.

61. Entry for April 11, 1851, ETGRR Minutebook; *Report of the Chief Engineers, Presidents and Superintendents of the South-Western R. R. Co., of Georgia, from No. 1 to 22, Inclusive, with the Charter and Amendments Thereto* (Macon, Ga.: J. W. Burke, 1869), 168, 169; L. Diane Barnes, *Artisan Workers in the Upper South: Petersburg, Virginia, 1820–1865* (Baton Rouge: Louisiana State University Press, 2008), 32. Indeed, in the case of Petersburg, Virginia, the railroad industry seems to have stimulated considerable demand for and concomitant increase in white labor in the city. Barnes, *Artisan Workers,* 72. The CRRG also built engines. Jefferson Max Dixon, "An Abstract of the Central Railroad of Georgia, 1833–1892" (Ph.D. diss., Georgia Peabody College for Teachers, 1953), 159–61.

Chapter Five • Motion

1. *Annual Report of the Direction of the South Carolina Canal and Rail Road Company, to the Stockholders, May 6th, 1834, with Accompanying Documents* (Charleston, S.C.: W. S. Blain, 1834), 4. Southern railroads had a grip on freight service for some time; independent express service did not arrive in the South in earnest until about 1850. According to one contemporary account of express service, "In 1850, up to which time the Express business had not been generally extended to the Southern States, a line to Mobile, New Orleans, and the far Southern and South-western States was commenced by John K. and A. L. Stimson. In 1852, this line was amalgamated with Adams & Co's. . . . In 1855, 'The Adams Express Company' (which was the designation now assumed by this association), commenced running from Charleston to Columbia, S. C., Montgomery, Ala., Atlanta and Augusta, Geo., and Nashville, Tenn.; and, at the time of the outbreak of the rebellion, had entire control of the Express business in the Southern and South-western States." H. Wells, *Sketch of the Rise, Progress, and Present Condition of the Express System* (Albany, N.Y.: Van Benthuysen's Steam Printing House, 1864), 11.

2. Ulrich Bonnell Phillips, *A History of Transportation in the Eastern Cotton Belt to 1860* (1908; reprint, New York: Octagon Books, 1968), 166; entry for March 10, 1851, RDRR Minutebook; entry for December 21, 1855, RDRR Minutebook.

3. Phillips, *History of Transportation*, 166. For a discussion of the law as applicable to warehousing by railroads, see James W. Ely Jr., *Railroads and American Law* (Lawrence: University Press of Kansas, 2001), 182–83.

4. Entry for June 13, 1855, RDRR Minutebook. This was rescinded a few months later. See entry for January 16, 1856, RDRR Minutebook.

5. Franz Anton Ritter von Gerstner, *Early American Railroads*, ed. Frederick C. Gamst (Stanford: Stanford University Press, 1997), 690, 675; *Annual Report to the Stockholders of the Petersburg R. Road Company, 1857* (Petersburg, Va.: O. Ellyson, 1857), 22.

6. Entry for January 2, 1856, SURR Minutebook; entry for March 10, 1851, RDRR Minutebook; Tim Ingold, "Work, Time and Industry," *Time and Society* 4 (February 1995): 5–28. Emma Bell and Alan Tuckman suggest that the division of "work" and "non-work" when it comes to "time ownership" is a "problematic distinction." Surely workers living at the loading site and perpetually on call would agree. Emma Bell and Alan Tuckman, "Hanging on the Telephone: Temporal Flexibility and the Accessible Worker," in *Making Time: Time and Management in Modern Organizations*, ed. Richard Whipp, Barbara Adam, and Ida Sabelis (New York: Oxford University Press, 2002), 125.

7. Frederick Law Olmsted, *A Journey in the Seaboard Slave States in the Years 1853–1854, with Remarks on Their Economy*, 2 vols. (1856; reprint, New York: Putnam's Sons, 1904), 2:19–20.

8. Lawrence W. Levine, *Black Culture and Black Consciousness* (New York: Oxford University Press, 1977), 208–15.

9. *Report of the Committee of Inspection on the Condition of the South-Carolina Rail-Road. Made at an Extra Meeting of the Stockholders, Held on the 2d of May 1848* (Charleston, S.C.: Miller and Browne, 1848), 21–22; *Reports by the President of So. Ca. Rail Road Company, and of a Committee of Seven Appointed by the Stockholders, on the Change of Location of the Present Depot, and Work Shops on Mary and Line Streets, with the Advantages, Objections, and Probable Cost of Each New Location Suggested. Also, the Engineer's Report upon Inclined Plane, &c* (Charleston, S.C.: Walker & Burke, 1846), 90–91; *Action of the Stockholders of the South-Carolina Railroad Company, on the Reports and Resolutions of the Committee of Inspection, Held at the Hall of the South-Western Railroad Bank, on Tuesday, 30th of May 1848. To Which is Annexed, in an Appendix, the Report of the President to Board of Directors, Referred to in the Proceedings; and the Report of S. M. Fox, (Civil Engineer) on the Inclined Plane* (Charleston, S.C.: Miller and Browne, 1848), 4; *SCRR Committee of Inspection, May 1848*, 26.

10. Entries for May 28 and July 23, 1835, SCRR Minutebook; entry for January 5, 1854, VTRR Minutebook.

11. George P. Rawick, ed., *The American Slave: A Composite Autobiography*, vol. 14, *North Carolina Narratives, Part 1* (Westport, Conn.: Greenwood, 1972), 332; letter from Joseph Wharton to Anna Lovering, January 27, 1853, in H. Larry Ingle, ed., "Joseph Wharton Goes South, 1853," *South Carolina Historical Magazine* 96 (October 1995): 310–11; letter from Joseph Wharton to William Wharton, January 28, 1853, in Ingle, "Joseph Wharton," 311. See also Steven Deyle, *Carry Me Back: The Domestic Slave Trade in American Life* (New York: Oxford University Press, 2005), 73, 111–12, 148.

12. Jacob Stroyer, "My Life in the South," in *Five Slave Narratives*, ed. William Loren Katz (New York: Arno, 1968), 40–42.

13. Olmsted, *A Journey in the Seaboard Slave States*, 1:33, 59–60; 2:1–2. The slave quoted Acts 16:28.

14. Wilma A. Dunaway, "Put in Master's Pocket: Cotton Expansion and Interstate Slave Trading in the Mountain South," in *Appalachians and Race: The Mountain South from Slavery to Segregation*, ed. John C. Inscoe (Lexington: University Press of Kentucky, 2001), 125. See also Deyle, *Carry Me Back*, 150.

15. Arthur Cunynghame, *A Glimpse of the Great Western Republic* (London: Richard Bentley, 1851), 270–71.

16. Olmsted, *Journey in the Seaboard States*, 2:22, 23.

17. *SCRR Committee of Inspection, May 1848*, 6–7, 11.

18. Phillips, *History of Transportation*, 1, 8; Richard D. Brown, *Modernization: The Transformation of American Life, 1600–1865* (New York: Hill and Wang, 1976), 143; Scott R. Nelson, *Iron Confederacies: Southern Railways, Klan Violence, and Reconstruction* (Chapel Hill: University of North Carolina Press, 1999), 13. See also Elizabeth Fox-Genovese and Eugene Genovese, *Fruits of Merchant Capital* (New York: Oxford University Press, 1983), 50. Peter Coclanis has argued that the lack of internal markets was a problem of long standing. Peter A. Coclanis, *The Shadow of a Dream: Economic Life and Death in the South Carolina Low Country, 1670–1920* (New York: Oxford University Press, 1989), 146–47 and 190n91.

19. Lacy K. Ford Jr., *Origins of Southern Radicalism: The South Carolina Upcountry, 1800–1860* (New York: Oxford University Press, 1988), 235–37.

20. *Tenth Annual Report of the President and Directors to the Stockholders of the Virginia and Tenn. Railroad Co., September 16–17, 1857* (Lynchburg: Virginian Steam Power Press Print, 1857), 106–11; *Report of the Chief Engineers, Presidents and Superintendents of the South-Western R. R. Co., of Georgia, from No. 1 to 22, Inclusive, with the Charter and Amendments Thereto* (Macon, Ga.: J. W. Burke, 1869), 159; letter from C. C. Jones to "My dear Son," December 18, 1856, folder 8, box 3, Charles Colcock Jones Collection, HL.

21. Measured in pounds of cotton per farm; see Ford, *Origins of Southern Radicalism*, tables 7.1 and 7.2.

22. *Speeches of Maj. B. F. Perry & Hon. J. D. Allen on the Blue Ridge Railroad* (n.p.: n.d.), 8–9.

23. Donald W. Buckwalter, "Effects of Early Nineteenth Century Transportation Disadvantage on the Agriculture of Eastern Tennessee," *Southeastern Geographer* 27 (May 1987): 33–34.

24. *Semi-Annual Report of the South-Carolina Canal and Rail-Road Company, to December 31, 1837* (Charleston, S.C.: J. S. Burges, 1837), 5; *Proceedings of an Adjourned Meeting of the Stockholders of the Louisville, Cincinnati and Charleston Rail-Road Company, Held in Columbia, on the 4th, 6th and 7th December, 1839* (Charleston, S.C.: A. E. Miller, 1840), 18. Other railroads, such as the CRRG, had to wrestle with the problem of empty cars, as they noted in 1845: "We had made provision in motive power, cars, and outfit generally, for a full business both ways, and as our upward freight has been equal to that of the last year, we have been obliged to run trains of empty cars, on our down trips, for a great portion of the last quarter of the year." *Reports of the Presidents, Engineers-in-Chief and Superintendents, of the Central Rail-Road and Banking Company of Georgia, from No. 1 to 19 Inclusive, with Report of Survey by Alfred Cruger, and the Charter of the Company* (Savannah, Ga.: John M. Cooper, 1854), 125–26.

25. Likewise, Stephen Henson found that the CRRG carried about 8% of its cotton from May to September when the railroad began operations; this increased to 17% in 1858. Stephen Ray Henson, "Industrial Workers in the Mid Nineteenth-Century South: Atlanta Railwaymen, 1840–1870" (Ph.D. diss., Emory University, 1982), 15–16.

26. Entry for July 9, 1835, SCRR Minutebook; *Annual Report of the Board of Direction of the So. Canal and R. R. Company, with the Semi-Annual Statement of Accounts, to the 31st December, 1835* (Charleston, S.C.: A. E. Miller, 1836), 4; *LCCRR Report, December 1839*, 18; *Semi-Annual Report of the South-Carolina Canal and Rail-Road Company, Accepted 20th of July, 1843* (Charleston, S.C.: Miller and Browne, 1843), 8–9. For another railroad's effort to handle the business cycle, see von Gerstner, *Early American Railroads*, 677.

27. Entries for December 7 and 8, 1836, in Herbert A. Kellar, ed., "A Journey through the South in 1836: Diary of James D. Davidson," *Journal of Southern History* 1 (August 1935): 373; *Proceedings of the Stockholders of the South-Carolina Rail-Road Company, and of the South-Western Rail-Road Bank, at Their Annual Meeting, in the Hall of the Bank, on the 10th, 11th, and 12th February 1852* (Charleston, S.C.: A. E. Miller, 1852), 30; Joseph A. Waddell, *Home Scenes and Family Sketches* (Staunton, Va.: Stoneburner & Prufer, 1900), 177–79.

28. *Charleston Courier*, May 29, 1840, 2; May 30, 1840, 2. The repairs came a few days later: "We are gratified to learn that the damage done to the Rail Road, by the recent freshet, has been so far repaired by the energetic exertions of those concerned, that freight is landed and delivered as usual, at the upper Depository, at Hamburg." *Charleston Courier*, June 3, 1840, 2.

29. *Proceedings of the Stockholders of the South-Carolina Railroad Company, and of the South-Western Railroad Bank, at at* [sic] *Their Annual Meeting in the Hall of the Bank, on the on the* [sic] *8th and 9th of February, 1853* (Charleston, S.C.: A. E. Miller, 1853), 8–9; *Proceedings of the Stockholders of the South-Carolina Rail Road Company, and of the South-Western Railroad Bank at Their Annual Meeting in the Hall of the Bank, on the 14th and 15th of February 1854* (Charleston, S.C.: A. E. Miller, 1854), 6; entry for the Greenville and Columbia Railroad, South Carolina, vol. 12, p. 57, R. G. Dun & Co. Collection, BHC.

30. Letter from Robert Habersham to Barnard Elliot Habersham, December 5, 1840, folder 11, Barnard Elliot Habersham Papers, HL; Alexander Mackay, *The Western World; or, Travels in the United States in 1846–47: Exhibiting Them in Their Latest Development, Social, Political, and Industrial; Including a Chapter on California. With a New Map of the United States, Showing Their Recent Territorial Acquisitions, and a Map of California*, vol. 2 (1849; reprint, New York: Negro Universities Press, 1968), 157–58. Of course, freshets also damaged railroads in the North. The Nashua and Lowell Railroad reported to its stockholders in 1852 that, as the result of water flooding "several feet" over the rails, passengers had to be briefly transported in canoes. Manuscript copy of the "Seventeenth Annual Report of the Directors of the Nashua and Lowell Rail Road to the Stockholders May 26, 1852," folder 3, vol. 129, Nashua and Lowell Railroad Co. Collection, BHC.

31. *LCCRR Report, December 1839*, 17–18.

32. *Proceedings of the Stockholders of the South-Carolina Rail Road Company and of the South-Western Rail Road Bank, at Their Annual Meeting, in the Hall of the Bank, on the 13th and 14th of February, 1855* (Charleston, S.C.: A. E. Miller, 1855), 7, n.p. The company had found itself in similar circumstances in another occasion; see *Proceedings of the Stockholders of the South-Carolina Rail-Road Company, and of the South-Western Rail-Road Bank, at Their Annual*

Meeting, in the Hall of the Bank, on the 12th, 13th and 14th of February 1850 (Charleston, S.C.: Miller and Browne, 1850), 5–6.

33. *CRRG Reports, 1 to 19 Inclusive,* 73; *Reports of the Presidents and Superintendents of the Central Railroad and Banking Co. of Georgia, from No. 20 to 32 Inclusive, and the Amended Charter of the Company* (Savannah, Ga.: G. N. Nichols, 1868), 100.

34. Mark Aldrich, *Death Rode the Rails: American Railroad Accidents and Safety, 1828–1965* (Baltimore: Johns Hopkins University Press, 2006), 14–15; Robert B. Shaw, *A History of Railroad Accidents, Safety Precautions, and Operating Practices,* 2d ed. (n.p.: Vail-Ballou, 1978), 3; Wolfgang Schivelbusch, *The Railway Journey* (New York: Urizen, 1979), 131–33; N. J. Bell, *Southern Railroad Man: Conductor N. J. Bell's Recollections of the Civil War Era,* ed. James A. Ward (DeKalb: Northern Illinois University Press, 1993), 11.

35. Samuel Melanchthon Derrick, *Centennial History of South Carolina Railroad* (Columbia, S.C.: State Company, 1930), 310. Northern railroads employed a similar system; see *Rules of the Boston & Lowell and Nashua & Lowell Railroad Companies, for the Government of the Executive Service, Adopted by the Board of Directors, February, 1857* (Nashua, N.H.: Albin Beard, 1857), 19.

36. Henry Benjamin Whipple, *Bishop Whipple's Southern Diary, 1843–1844,* ed. Lester B. Shippee (1937; reprint, New York: Da Capo, 1968), 76; Shaw, *History of Railroad Accidents,* 20. I was unable to uncover any verifiable snakeheads in the South. One possible exception is in von Gerstner, *Early American Railroads,* 700.

37. *Semi-Annual Report of the Direction of the South-Carolina Canal and Rail-Road Company, to the Stockholders, October 31, 1834. With Accompanying Documents* (Charleston, S.C.: J. S. Burges, 1834), 8; *SCRR Report, December 1837,* 4; entry for March 20, 1840, SCRR Minutebook; letter from John McRae to Thomas Lang, July 25, 1847, JML-2; Daniel W. Cobb, *Cobb's Ordeal: The Diaries of a Virginia Farmer, 1842–1872,* ed. Daniel W. Crofts (Athens: University of Georgia Press, 1997), 107–8.

38. Entry for October 18, 1854, VTRR Minutebook.

39. Jones's attempt to be reinstated was rebuffed by the board of directors. Entries for July 25 and August 23, 1855, VTRR Minutebook.

40. Entry for September 24, 1858, RDRR Minutebook.

41. Shaw, *History of Railroad Accidents,* 41–42; *Proceedings of the Stockholders of the Louisville, Cincinnati & Charleston Rail-Road Company, and the South-Western R. Road Bank. At Their Annual Meeting, in the Hall of the Bank, on the 16th, 17th, and 18th of Nov. 1841* (Charleston, S.C.: A. E. Miller, 1841), 20; *SCRR Report, February 1853,* 34 [misnumbered 43].

42. *SWRR Reports, 1 to 22 Inclusive,* 317 (some nonpassengers were killed; see 197, 303); *CRRG Reports, 1 to 19 Inclusive,* 200; *Proceedings of the Annual Convention of Stockholders in the East Tenn. & Georgia Railroad Co., Held at Knoxville, Tenn. Wednesday, the 5th Day of September, 1860* (Athens, Tenn.: Sam. P. Ivins, Office of the "Post," 1860), 8. In the case of the East Tennessee and Georgia, the injured passenger was a female slave; the owner's claim for damage was settled when the company purchased the woman with the intent of allowing her to recuperate and then selling her.

43. Letter from Joseph Wharton to Anna Lovering, February 12, 1853, in Ingle, "Joseph Wharton," 325. While Wharton comforted himself with the fact that cars were of northern construction, they probably received substantial care in the South; see my discussion in chapter 4.

44. Entry for January 17, 1838, Blair Bolling Diary, Bolling Family Papers, VHS, *RASP*, series M, part 5, reel 9, frame 861.

45. Letter from G. J. Kollock to his wife, January 31, 1851, in Susan M. Kollock, ed., "Letters of the Kollock and Allied Families, 1826–1884, Part II," *Georgia Historical Quarterly* 34 (March 1950): 36.

46. Letter from James C. Dobbin to David Settle Reid, March 7, 1847, in David Settle Reid, *The Papers of David Settle Reid*, ed. Lindley S. Butler, vol. 1, *1829–1852* (Raleigh: Department of Cultural Resources, Division of Archives and History, 1993), 187; entry for October 4, 1852, Jeremiah C. Harris Diary, vol. 3, VHS; letter from Stephen A. Douglas to David Settle Reid, March 25, 1847, in Reid, *Papers of David Settle Reid*, 1:188.

47. Letter from Benjamin Babb to Samuel Babb, April 27, 1844, VHS; Carolina Seabury, *The Diary of Carolina Seabury, 1854–1863*, ed. Suzanne L. Bunkers (Madison: University of Wisconsin Press, 1991), 29.

48. Entry for October 5, 1853, RDRR Minutebook.

49. Letter from Jennie Speer to Elizabeth Speer, May 25, 1853, in Allen Paul Speer and Janet Barton Speer, eds., *Sisters of Providence: The Search for God in the Frontier South (1843–1858)* (Johnson City, Tenn.: Overmountain Press, 2000), 189; T. C. Girardeau to "My dear Mother," April 2, 1859, T. C. Girardeau Papers, SCL.

50. Thomas W. Chadwick, ed., "The Diary of Samuel Edward Burges, 1860–1862," *South Carolina Historical and Genealogical Magazine* 48 (April 1947): 72; Anna Marie Resseguie, *A View from the Inn: The Journal of Anna Marie Resseguie, 1851–1867* (Ridgefield, Conn.: Keeler Tavern Preservation Society, 1993), 151–52.

Chapter Six • Passages

1. Mark M. Smith, *Mastered by the Clock: Time, Slavery, and Freedom in the American South* (Chapel Hill: University of North Carolina Press, 1997), 79–80. As Smith notes, even modern passengers "moan about mechanical delays on runways and fret about making their next appointment." Indeed, before judging antebellum transportation too harshly, we should question to what degree timeliness is a universal marker of transportation success. Economist Christopher J. Mayer has argued that airline passengers' insistence on cheap fares means that "airlines do not think they can successfully market timeliness" in the twenty-first century. This quotation is drawn from Micheline Maynard, "Will the Flight Be on Time? It's Anybody's Guess," *New York Times*, August 8, 2005, A12. "Airlines' on-time performance" has been on the decline in the twenty-first century. See Del Quentin Wilber, "Flying Late, Arriving Late: Air Carriers Are Delaying More Flights and Losing Your Shirts," *Washington Post*, February 8, 2007, D1, D6. For the story of a fifty-three-minute flight with an average delay of seventy-nine minutes, see Del Quentin Wilber, "Like Clockwork: Hour of Delay, Hour of Flight," *Washington Post*, November 13, 2006, A1, A15.

2. Thomas W. Parsons, *Incidents and Experiences in the Life of Thomas W. Parsons from 1826 to 1900*, ed. Frank Furlong Mathias (Lexington: University Press of Kentucky, 1975), 11; letter from M. F. Kollock to Dr. P. M. Kollock, September 11, 1826, in Edith Duncan Johnston, ed., "The Kollock Letters, 1799–1850 (Part II)," *Georgia Historical Quarterly* 30 (December 1946): 313; letter from M. F. Kollock to Dr. P. M. Kollock, September 18, 1829, in Edith

Duncan Johnston, ed., "The Kollock Letters, 1799–1850 (Part III)," *Georgia Historical Quarterly* 31 (March 1947): 35.

3. Parmenas Taylor Turnley, *Reminiscences of Parmenas Taylor Turnley, from the Cradle to Three-score and Ten, by Himself, from Diaries Kept from Early Boyhood, with a Brief Glance Backward Three Hundred and Fifty Years at Progenitors and Ancestral Lineage* (Chicago: Donohue and Henneberry, 1892), 32, 36. Turnley rode 55 miles in a stagecoach.

4. Letter from Anna Calhoun Clemson to Patrick Calhoun, March 4, 1838, in Julia Wright Sublette, "The Letters of Anna Calhoun Clemson, 1833–1873" (Ph.D. diss., Florida State University, 1993), 149; entry for December 6, 1836, in Herbert A. Kellar, ed., "A Journey through the South in 1836: Diary of James D. Davidson," *Journal of Southern History* 1 (August 1935): 372–73.

5. *Report of the President and Directors of the Cheraw & Darlington Railroad Co.* (Columbia, S.C.: R. W. Gibbes, 1856), 6.

6. *Charleston Courier,* January 5, 1831, 2.

7. Entry for December 8, 1832, Samuel Cram Jackson Diary, #4048-z, SHC; entry for April 12, 1836, Anne Carolina Lesesne Diary, SCHS; entry for December 8, 1836, in Kellar, "Diary of James D. Davidson," 373–74.

8. Historians have only begun to explore the social history of travel in the United States. See, for example, Amy G. Richter, *Home on the Rails: Women, the Railroad, and the Rise of Public Domesticity* (Chapel Hill: University of North Carolina Press, 2005). For the importance of sensory history, see Mark M. Smith, *Sensing the Past: Seeing, Hearing, Smelling, Tasting, and Touching in History* (Berkeley: University of California Press, 2007).

9. Leo Marx, *The Machine in the Garden: Technology and the Pastoral Ideal in America,* rev. ed. (New York: Oxford University Press, 2000), 191, 207–8.

10. Letter from Maria Glass to Martha Fenton Hunter, June 10, 1854, Hunter Family Papers, VHS; George P. Rawick, ed., *The American Slave: A Composite Autobiography,* supplement, series 1, vol. 9, *Mississippi Narratives, Part 4* (Westport, Conn.: Greenwood, 1977), 1462; George P. Rawick, ed., *The American Slave: A Composite Autobiography,* vol. 3, *South Carolina Narratives, Part 4* (Westport, Conn.: Greenwood, 1972), 38–39; George P. Rawick, ed., *The American Slave: A Composite Autobiography,* vol. 3, *South Carolina Narratives, Part 3* (Westport, Conn.: Greenwood, 1972), 77. Not all slaves were frightened by the prospect of riding the train: one ex-slave told a Fisk University interviewer that "when the [Civil] War broke out, Missy [her mistress] asked me if I wanted to go to Mississippi with her. I was so glad I just hopped and skipped, 'cause I was going to git to ride the train', I was foolish you know." George P. Rawick, ed., *The American Slave: A Composite Autobiography,* vol. 18, *Unwritten History of Slavery (Fisk University)* (Westport, Conn.: Greenwood, 1972), 202.

11. Alexander Mackay, *The Western World; or, Travels in the United States in 1846–47: Exhibiting Them in Their Latest Development, Social, Political, and Industrial; Including a Chapter on California. With a New Map of the United States, Showing Their Recent Territorial Acquisitions, and a Map of California,* vol. 1 (1849; reprint, New York: Negro Universities Press, 1968), 151.

12. Carolina Seabury, *The Diary of Carolina Seabury, 1854–1863,* ed. Suzanne L. Bunkers (Madison: University of Wisconsin Press, 1991), 28–29.

13. Letter from Lucy Carpenter to William Blanding, November 23, 1848, folder 7,

William Blanding Papers, SCL; Frederick Law Olmsted, *A Journey in the Seaboard Slave States in the Years 1853–1854, with Remarks on Their Economy*, 2 vols. (1856; reprint, New York: Putnam's Sons, 1904), 1:20.

14. Wolfgang Schivelbusch, *The Railway Journey* (New York: Urizen, 1979), 25, 66; Louis Heuser Diary Transcription, p. 28, WHS; Seabury, *Diary*, 28; entry for February 9, 1844, in Henry Benjamin Whipple, *Bishop Whipple's Southern Diary, 1843–1844*, ed. Lester B. Shippee (1937; reprint, New York: Da Capo, 1968), 74; entry for December 28, 1841, in Charles Lyell, *Travels in North America in the Years 1841–2, with Geological Observations on the United States, Canada, and Nova Scotia* (1845; reprint, New York: Arno, 1977), 123.

15. Letter from Anna Calhoun Clemson to Maria Simkins, March 11, 1838, in Sublette, "Letters of Anna Calhoun Clemson," 154; John Herritage Bryan to Ebenezer Pettigrew, September 20, 1846, in Sarah McCulloh Lemmon, ed., *The Pettigrew Papers*, vol. 2, *1819–1843* (Raleigh: North Carolina Department of Cultural Resources, Division of Archives and History, 1988), 443; Mary Elizabeth Pearson Boyce, *And God Still Spares Us: The Diary of Mary Elizabeth Pearson Boyce, September 28, 1854–November 2, 1855* (Hickory Grove, S.C.: Broad River Basin Historical Society, 1993), 47; entry for June 30, 1851, Jeremiah C. Harris Diary, vol. 1, VHS.

16. Schivelbusch, *Railway Journey*, 66.

17. "Some Notes of a Southern Excursion," *Southern Literary Messenger* 17 (April 1851): 248.

18. The contract further stipulated that "neither the said Van Ness or the news boys employed by him shall solicit passengers for any particular hotel." Finally, Van Ness released the company from any responsibility if his employees were injured or killed, even "by negligence of the employees of the Road." Contract between the Greenville and Columbia Railroad and W. W. Van Ness, August 25, 1856, folder marked "1850–1879," Greenville and Columbia Railroad, Ms84-067, NSHC.

19. Smith, *Sensing the Past*, 8–13.

20. Samuel D. McGill, *Narrative of Reminiscences in Williamsburg County* (Columbia, S.C.: Bryan Printing, 1897), 80.

21. Samuel Gaillard Stoney, ed., "The Memoirs of Frederick Adolphus Porcher," *South Carolina Historical and Genealogical Magazine* 46 (October 1945): 204; George P. Rawick, ed., *The American Slave: A Composite Autobiography*, supplement, series 1, vol. 10, *Mississippi Narratives, Part 5* (Westport, Conn.: Greenwood, 1977), 1980; N. J. Bell, *Southern Railroad Man: Conductor N. J. Bell's Recollections of the Civil War Era*, ed. James A. Ward (DeKalb: Northern Illinois University Press, 1993), 6.

22. Samuel Melanchthon Derrick, *Centennial History of South Carolina Railroad* (Columbia, S.C.: State Company, 1930), 309; W. E. Baxter, *America and the Americans* (London: Routledge, 1855), 44.

23. R. H. Whitaker, *Whitaker's Reminiscences, Incidents and Anecdotes: Recollections of Other Days and Years; or, What I Saw and Heard and Thought of People Whom I Knew, and What They Did and Said* (Raleigh, N.C.: Edwards and Broughton, 1905), 80; Ulrich Bonnell Phillips, *A History of Transportation in the Eastern Cotton Belt to 1860* (1908; reprint, New York: Octagon Books, 1968), 165.

24. Olmsted, *A Journey in the Seaboard Slave States*, 2:1, 20.

25. *Speeches of Maj. B. F. Perry & Hon. J. D. Allen on the Blue Ridge Railroad* (n.p.: n.d.),

21; Steven G. Collins, "System, Organization, and Agricultural Reform in the Antebellum South, 1840–1860," *Agricultural History* 75 (Winter 2001): 1; Charles Fraser, *Reminiscences of Charleston* (1854; reprint, Charleston, S.C.: Garnier, 1969), 16; letter from Jennie Speer to Elizabeth Speer, February 8, 1853, in Allen Paul Speer and Janet Barton Speer, eds., *Sisters of Providence: The Search for God in the Frontier South (1843–1858)* (Johnson City, Tenn.: Overmountain Press, 2000), 153. For southerners' adaptation to the sounds of progress, see Mark M. Smith, *Listening to Nineteenth-Century America* (Chapel Hill: University of North Carolina Press, 2001), 44–45.

26. *Charleston Courier*, February 4, 1840, 3; Mrs. John A. Logan, *Reminiscences of a Soldier's Wife: An Autobiography* (1913; reprint, Carbondale: Southern Illinois University Press, 1997), 72; letter from Lucy Carpenter to William Blanding, November 23, 1848, folder 7, William Blanding Papers, SCL; Frederic Trautmann, ed., "South Carolina through a German's Eyes: The Travels of Clara von Gerstner, 1839," *South Carolina Historical Magazine* 85 (July 1984): 232.

27. Letter from William Elliott to Ann Elliott, June 11, 1853, in Beverly Scafidel, "The Letters of William Elliott" (Ph.D. diss., University of South Carolina, 1978), 721; Seabury, *Diary*, 31–32.

28. John E. Crowley, *The Invention of Comfort: Sensibilities and Design in Early Modern Britain and Early America* (Baltimore: Johns Hopkins University Press, 2001), 292; Frederick Douglass, *Life and Times of Frederick Douglass, Written by Himself* (1893; reprint, New York: Library of America, 1994), 644.

29. Charles A. Hentz, *A Southern Practice: The Diary and Autobiography of Charles A. Hentz, M.D.*, ed. Steven M. Stowe (Charlottesville: University Press of Virginia, 2000), 64; Olmsted, *Journey in the Seaboard States*, 1:59.

30. Seabury, *Diary*, 30; entry for December 27, 1845, in Albrect Karl Koch, *Journey through a Part of the United States of North America in the Years 1844 to 1846*, ed. Ernst A. Stadler (Carbondale: Southern Illinois University Press, 1972), 125.

31. Lydia Spencer Lane, *I Married a Soldier* (1893; reprint, Albuquerque: University of New Mexico Press, 1987), 81; letter from Ann Maury to Ann Tunstall Hite, November 9, 1843, section 4, Hite Family Papers, VHS; letter from Mary Hering Middleton to Eliza Middleton Fisher, December 21, 1839, in Eliza Middleton Fisher, *Best Companions: Letters of Eliza Middleton Fisher and Her Mother, Mary Hering Middleton, from Charleston, Philadelphia, and Newport, 1839–1846*, ed. Eliza Cope Harrison (Columbia: University of South Carolina Press, 2001), 84.

32. Virginia Clay-Clopton, *A Belle of the Fifties: Memoirs of Mrs. Clay of Alabama, Covering Social and Political Life in Washington and the South, 1853–66* (1905; reprint, Tuscaloosa: University of Alabama Press, 1999), 20; John Shaw, *A Ramble through the United States, Canada, and the West Indies* (London: Hope and Company, 1856), 198, 206; Arthur Cunynghame, *A Glimpse of the Great Western Republic* (London: Richard Bentley, 1851), 274–75.

33. Entry for March 16, 1838, in Mary E. Moragne, *The Neglected Thread: A Journal from the Calhoun Community, 1836–1842*, ed. Delle Mullen Craven (Columbia: University of South Carolina Press, 1951), 85–87.

34. Smith, *Mastered by the Clock*, 79. Time consciousness was not purely a characteristic of the post–Civil War era, as implied by John F. Kasson, *Civilizing the Machine: Technology and Republican Values in America, 1776–1900*, rev. ed. (New York: Hill and Wang, 1999),

188–89. The experience of improved travel time expanded across space as railroads moved West; see Thomas G. Andrews, " 'Made by Toile'? Tourism, Labor, and the Construction of the Colorado Landscape," *Journal of American History* 92 (December 2005): 847.

35. Letter from William Blanding to "My Dear R" [Rachel Blanding, his wife], November 10, 1834, folder 4, William Blanding Papers, SCL.

36. Kellar, "Diary of James D. Davidson," 373.

37. Boyce, *God Still Spares Us*, 53.

38. Entry for May 28, 1835, SCRR Minutebook; entry for March 16, 1838, in Moragne, *Neglected Thread*, 85; Boyce, *God Still Spares Us*, 39; entry for April 17, 1856, James D. Frederick Diary, box 1, James Daniel Frederick Papers, HL; voucher entitled "To the following payts for [illeg.] of travellers caused by delays in the arrival, departures & running of trains," folder 8, box 2, RFPRR-VHS.

39. Letter from William Elliott to Ann Elliott, October 5, 1839, in Scafidel, "Letters of William Elliott," 413; "Sketches of Summer Visits," by "Cecilia," *Southern Literary Messenger* 23 (November 1856): 363. See also Smith, *Mastered by the Clock*, 89.

40. Letter from Ann Maury to Ann Tunstall Hite, November 9, 1843, section 4, Hite Family Papers, VHS; McGill, *Narrative*, 81–82.

41. William J. Grayson, *Witness to Sorrow: The Antebellum Autobiography of William J. Grayson*, ed. Richard J. Calhoun (Columbia: University of South Carolina Press, 1990), 121; letter from James Shannon to William Dearing, January 23, 1840, folder 4, box 1, Rev. Dr. James Shannon Collection, HL (the Charleston and Hamburg was an early name for the SCRR); letter from Emily Sinkler to Mary Wharton, August 15, 1850, in Emily Wharton Sinkler, *Between North and South: The Letters of Emily Wharton Sinkler, 1842–1865*, ed. Anne Sinkler Whaley LeClercq (Columbia: University of South Carolina Press, 2001), 136; "Mount Vernon—A Pilgrimage," by E. Kennedy, *Southern Literary Messenger* 18 (January 1852): 53; Frederick Law Olmsted, *A Journey in the Back Country in the Winter of 1853–4*, vol. 2 (1860; reprint, New York: Putnam's Sons, 1907), 179.

42. Serious interest from the medical profession on railroads and health did not emerge until the postbellum era; see Barbara Young Welke, *Recasting American Liberty: Gender, Race, Law, and the Railroad Revolution, 1865–1920* (New York: Cambridge University Press, 2001). Most of the historiography on this topic has focused on Britain. See, for example, Ralph Harrington, "The Railway Journey and the Neuroses of Modernity," in *Pathologies of Travel*, ed. Richard Wrigley and George Revill (Atlanta: Rodopi, 2000), 229–59.

43. Letter from Wilson Lumpkin to "Miss M. W. Lumpkin," July 14, 1856, Wilson Lumpkin Collection, HL; letter from C. C. Jones to "My dear Sons," November 17, 1851, folder 9, box 2, Charles Colcock Jones Collection, HL; letter from Henry Robertson Dickson to Zebulon Vance, October 26, 1852, in Zebulon Baird Vance, *The Papers of Zebulon Baird Vance*, ed. Frontis W. Johnson, vol. 1, *1843–1862* (Raleigh, N.C.: State Department of Archives and History, 1963), 18; letter from P. M. Kollock to George Kollock, July 1, 1839, in Edith Duncan Johnston, ed., "The Kollock Letters, 1799–1850 (Part V)," *Georgia Historical Quarterly* 31 (September 1947): 213–14.

44. Eugene Alvarez, *Travel on Southern Antebellum Railroads, 1828–1860* (University: University of Alabama Press, 1974), 126; John H. White Jr., *The American Railroad Passenger Car* (Baltimore: Johns Hopkins University Press, 1978), 203. See also Welke, *Recasting American Liberty*, 253–60.

45. Entry for May 15, 1850, RDRR Minutebook; entry for October 19, 1853, Southside Minutebook-1. For additional examples, see *Report of the Chief Engineers, Presidents and Superintendents of the South-Western R. R. Co., of Georgia, from No. 1 to 22, Inclusive, with the Charter and Amendments Thereto* (Macon, Ga.: J. W. Burke, 1869), 181; *Annual Report to the Stockholders of the Petersburg R. Road Company, 1856* (Petersburg, Va.: G. Ellyson, 1856), 6; *Reports of the Presidents and Superintendents of the Central Railroad and Banking Co. of Georgia, from No. 20 to 32 Inclusive, and the Amended Charter of the Company* (Savannah, Ga.: G. N. Nichols, 1868), 145; and *Sixth Annual Report of the President and Directors of the Savannah, Albany and Gulf Rail Road Company, to the Stockholders* (Savannah, Ga.: Edward J. Purse, 1860), 18.

46. Colin Divall and George Revill, "Cultures of Transport: Representation, Practice and Technology," *Journal of Transport History*, 3d ser., 26 (March 2005): 105; Carolyn Thomas de la Peña, " 'Bleaching the Ethiopian': Desegregating Race and Technology through Early X-Ray Experiments," *Technology and Culture* 47 (January 2006): 45–46.

47. Sarah H. Gordon, *Passage to Union: How the Railroads Transformed American Life, 1829–1929* (Chicago: Ivan R. Dee, 1996), 84; Welke, *Recasting American Liberty*, 255–56; Phillips, *History of Transportation*, 165; entry for May 28, 1835, SCRR Minutebook.

48. Entry for September 5, 1835, SCRR Minutebook; Mark M. Smith, "Finding Deficiency: On Eugenics, Economics, and Certainty," *American Journal of Economics and Sociology* 64 (July 2005): 891; entry for September 10, 1835, SCRR Minutebook. But the company continued to pursue a course of segregation within the exceptions mentioned above, noting in 1837 that "it is expected by October to have twenty or more freight cars, four passenger cars, and four servant, mail and baggage cars, all on eight wheels." *Semi-Annual Report of the Direction of the South-Carolina Canal and Rail Road Company. July 10, 1837. Accepted and Ordered to Be Printed* (Charleston, S.C.: James S. Burges, 1837), 5.

49. The rules also stated that conductors did not have the authority to accept African American passengers "unless they are placed there by our own Agents." *Report of the President and Directors of the East Tennessee and Georgia Rail Road Company, to the Stockholders in the Same, at Their Annual Convention, Held at Athens, Sept. 3, 1856* (Athens, Tenn.: Printed by Sam. P. Ivins, Office of the "Post," 1857), [30].

50. Letter from John McRae to J. J. Jones, December 6, 1858, JML-6; letter from John McRae to "Dear Doctor," December 12, 1858, JML-6. Evidently the slave, Charlotte, was about fourteen years of age. Letter from John McRae to "My dear Doctor," January 27, 1858, JML-7.

51. Franz Anton Ritter von Gerstner, *Early American Railroads*, ed. Frederick C. Gamst (Stanford: Stanford University Press, 1997), 694; James Redpath, *The Roving Editor, or, Talks with Slaves in the Southern States*, ed. John R. McKivigan (University Park: Pennsylvania State University Press, 1996), 94.

52. Olmsted, *Journey in the Seaboard States*, 1:19–20.

53. Entry for October 27, 1856, RDRR Minutebook.

54. Douglass, *Life and Times*, 643–45.

55. Entry for October 30, 1835, SCRR Minutebook; Peter Kolchin, *American Slavery, 1619–1877* (New York: Hill and Wang, 1993), 118; *Semi-Annual Report of the South-Carolina Canal and Rail-Road Company. Accepted Jan. 18, 1839* (Charleston, S.C.: A. E. Miller, 1839), 11; entries for June 13, 1832, and February 1 and December 15, 1835, TCDRR Minutebook.

56. Von Gerstner, *Early American Railroads*, 680, 689; Charles W. Turner, "Railroad Ser-

vice to Virginia Farmers, 1828–1860," *Agricultural History* 22 (October 1948): 242; entry for February 10, 1858, Southside Minutebook-2; letter from M. B. Carter to Dennis M. Foulks, November 28, 1859, Virginia and Tennessee Railroad, Ticket Agent's Letter Book, Ms81-073, NSHC.

57. Cunynghame, *Glimpse of the Great Western Republic,* 257. See also Steven Deyle, *Carry Me Back: The Domestic Slave Trade in American Life* (New York: Oxford University Press, 2005), 260. Steamboats faced a similar problem. See Thomas C. Buchanan, *Black Life on the Mississippi: Slaves, Free Blacks, and the Western Steamboat World* (Chapel Hill: University of North Carolina Press, 2004), chap. 4.

58. James was interviewed in Maryland but was enslaved in Virginia. George P. Rawick, ed., *The American Slave: A Composite Autobiography,* vol. 16, *Kansas, Kentucky, Maryland, Ohio, Virginia and Tennessee Narratives* (Westport, Conn.: Greenwood, 1972), 39 (of the Maryland narratives).

59. Petition of Abraham Rencher and Charles Manly, December 1850, General Assembly, Session Records, Senate Committee Reports, November 1850–January 1851, North Carolina Department of Archives and History, Raleigh, North Carolina, filmed as a part of *Race, Slavery, and Free Blacks: Petitions to Southern Legislatures, 1777–1867* (Bethesda, Md.: University Publications of America, 1998), reel 7, frames 351–52. For another claim against the Raleigh and Gaston, see ibid., frames 530–31.

60. *The Liberator,* November 28, 1856, quoted in Robert S. Starobin, *Industrial Slavery in the Old South* (New York: Oxford University Press, 1970), 90.

61. Entry for April 20, 1842, Minutebook of the Directors (November 11, 1841–March 23, 1844), vol. 3, BARC-WR; George Bliss, *Historical Memoir of the Western Railroad* (Springfield, Mass.: Samuel Bowles, 1863), 92; White, *American Railroad Passenger Car,* 466; *Sixth Annual Report of the Directors of the Pennsylvania Rail Road Company, to the Stockholders. February 7, 1853* (Philadelphia: Crissy & Markley, 1853), 52–53; *Eleventh Annual Report of the Directors of the Pennsylvania Railroad Company, to the Stockholders. February 1, 1858* (Philadelphia: Crissy & Markley, 1858), 38; entry for November 30, 1841, Minutebook of the Directors (November 11, 1841–March 23, 1844), vol. 3, BARC-WR. See also the entries for May 25, 1842, and March 22, 1843, in the same volume. For emigrants on the Housatonic Railroad, see N. P. Burall to Addison Gilmore, November 30, 1846, folder labeled "Letters mainly to Gilmore 1846 re: Conn & Passompsic RR," case 5, BARC.

62. White, *American Railroad Passenger Car,* 208; letter from Lucy Carpenter to William Blanding, November 23, 1848, folder 7, William Blanding Papers, SCL; Dolly Lunt Burge, *The Diary of Dolly Lunt Burge, 1848–1879,* ed. Christine Jacobson Carter (Athens: University of Georgia Press, 1997), 100; entry for February 27, 1857, in Elizabeth Lindsay Lomax, *Leaves from an Old Washington Diary, 1854–1863,* ed. Lindsay Lomax Wood ([New York]: Books, 1943), 66–67.

63. Letter from Ann Maury to Ann Tunstall Hite, November 9, 1843, section 4, Hite Family Papers, VHS; entry for November 7, 1859, in Thomas Hubbard Hobbs, *The Journals of Thomas Hubbard Hobbs,* ed. Faye Acton Axford (University: University of Alabama Press, 1976), 222.

64. Letter from Anna Calhoun Clemson to Floride Clemson, June 27, 1858, in Sublette, "Letters of Anna Calhoun Clemson," 704.

65. Entry for July 29, 1852, in James Henry Hammond, *Secret and Sacred: The Diaries of*

James Henry Hammond, a Southern Slaveholder, ed. Carol Bleser (New York: Oxford University Press, 1988), 256; Fraser, *Reminiscences,* 104–5; letter from William Woods Holden to David S. Reid, July 19, 1850, in William Woods Holden, *The Papers of William Woods Holden,* ed. Horace W. Raper and Thornton W. Mitchell, vol. 1, *1841–1868* (Raleigh: Division of Archives and History, North Carolina Department of Cultural Resources, 2000), 40.

66. "Madame de Stael," by Jane T. Lomax, *Southern Literary Messenger* 8 (March 1842): 231; "The Premature Use of Books in the Education of Children," by W. F. B., *Southern Literary Messenger* 13 (May 1847): 315; letter from Henry Bird to Mary Fox, January 21, 1833, section 1, folder 8, Bird Family Papers, VHS.

67. *The Bouquet* 2, no. 41 (October 7, 1843): 161. The magazine was published in Charleston, South Carolina.

68. *The Bouquet* 2, no. 9 (February 25, 1843): 35. The overall tone of this article is sarcastic, because the writer intended to denigrate fads such as mesmerism and the Feejee mermaid, but the role of railroads as a marker of progress is not changed by the writer's intent.

69. George P. Rawick, ed., *The American Slave: A Composite Autobiography,* vol. 16, *Kansas, Kentucky, Maryland, Ohio, Virginia, and Tennessee Narratives* (Westport, Conn.: Greenwood, 1972), 1–2 (of the Virginia narratives).

70. "Short Chapters," by "Patrick Pedant," *Southern Literary Messenger* 6 (May 1840): 373.

71. "Recollections of Sully," by L. I. L., *Southern Literary Messenger* 17 (April 1851): 224–25.

72. "A Night in a Haunted House," *Southern Literary Messenger* 21 (June 1855): 374.

73. "Remarks on Various Late Poets: Elizabeth B. Barrett," *Southern Literary Messenger* 11 (April 1845): 243 (the phrase "resonant steam-eagles" is found in the poem "Lady Geraldine's Courtship"); review of "Southern Passages and Pictures," *Southern Literary Messenger* 17 (May 1851): 291; review of *You Have Heard of Them, Southern Literary Messenger* 21 (January 1855): 64. The book under review was C. G. Rosenberg, *You Have Heard of Them* (New York: Redfield, 1854).

Chapter Seven • Communities

1. Wiebe E. Bijker, *Of Bicycles, Bakelites, and Bulbs: Toward a Theory of Sociotechnical Change* (Cambridge, Mass.: MIT Press, 1995), chap. 2.

2. Entries for January 28 and February 13, 1836, Blair Bolling Diary, Bolling Family Papers, VHS, *RASP,* series M, part 5, reel 9, frames 783–84.

3. Entry for June 28, 1842, in James Henry Hammond, *Secret and Sacred: The Diaries of James Henry Hammond, a Southern Slaveholder,* ed. Carol Bleser (New York: Oxford University Press, 1988), 97; George P. Rawick, ed., *The American Slave: A Composite Autobiography,* supplement, series 1, vol. 6, *Mississippi Narratives, Part 1* (Westport, Conn.: Greenwood, 1977), 297; George P. Rawick, ed., *The American Slave: A Composite Autobiography,* supplement, series 1, vol. 8, *Mississippi Narratives, Part 3* (Westport, Conn.: Greenwood, 1977), 1128; Addie Lou Brooks, "The Building of the Trunk Line Railroads in West Tennessee, 1852–1861," *Tennessee Historical Quarterly* 1 (1942): 112–13.

4. Entry for May 14, 1858, East Tennessee and Virginia Railroad ETVRR Minutebook-1; *Charleston Courier,* June 29, 1840, 2.

5. Franz Anton Ritter von Gerstner, *Early American Railroads,* ed. Frederick C. Gamst (Stanford: Stanford University Press, 1997), 706; Allen W. Trelease, *The North Carolina Rail-*

road, 1849–1871, and the Modernization of North Carolina (Chapel Hill: University of North Carolina Press, 1991), 37–38; entry for October 14, 1858, Basil Armstrong Thomasson, *North Carolina Yeoman: The Diary of Basil Armstrong Thomasson, 1853–1862*, ed. Paul D. Escott (Athens: University of Georgia Press, 1996), 217.

6. Sabbatarians in the South have not received sustained attention. Generally, see Richard R. John, "Taking Sabbatarianism Seriously: The Postal System, the Sabbath, and the Transformation of American Political Culture," *Journal of the Early Republic* 10 (Winter 1990): 517–67; Forest L. Marion, "The Gentlemen Sabbatarians: The Sabbath Movement in the Upper South, 1826–1836" (Ph.D. diss., University of Tennessee, 1998); James R. Rohrer, "Sunday Mails and the Church-State Theme in Jacksonian America," *Journal of the Early Republic* 7 (Spring 1987): 53–74; and Alexis McCrossen, *Holy Day, Holiday: The American Sunday* (Ithaca, N.Y.: Cornell University Press, 2000). Traditionally, the South has been portrayed as opposed to organized moral reform movements. See Bertram Wyatt-Brown, "Prelude to Abolitionism: Sabbatarian Politics and the Rise of the Second Party System," *Journal of American History* 58 (September 1971): 336–37.

7. John W. Quist, *Restless Visionaries: The Social Roots of Antebellum Reform in Alabama and Michigan* (Baton Rouge: Louisiana State University Press, 1998), 72; "Discourse on the Times," by Rev. A. D. Pollock, *Southern Literary Messenger* 3 (June 1837): 349; J. D. Hufham, *Memoir of Rev. John L. Prichard, Late Pastor of the First Baptist Church, Wilmington, N.C.* (Raleigh, N.C.: Hufham and Hughes, 1867), 104–5. Indeed, from a comparative glance at other cultures, it does not appear that railroads as a technology are *inherently* antireligious; Ian Kerr has noted that railroads increased the number of South Asians who were able to make religious pilgrimages in the early nineteenth and late twentieth centuries. Ian J. Kerr, "Reworking a Popular Religious Practice: The Effects of Railways on Pilgrimage in 19th and 20th Century South Asia," in *Railways in Modern India*, ed. Ian J. Kerr (New Delhi: Oxford University Press, 2001), 313–14.

8. Entry for June 13, 1835, SCRR Minutebook.

9. Entry for April 6, 1836, SCRR Minutebook; entries for February 6 and 11, 1836, SCRR Minutebook; entry for March 29, 1837, SCRR Minutebook.

10. *Proceedings of the Stockholders of the Louisville, Cincinnati & Charleston Rail-Road Company, and the South-Western R. Road Bank, at Their Annual Meeting, in the Hall of the Bank, on the 22d, 23d, 24th, and 25th of Nov. 1842* (Charleston, S.C.: A. E. Miller, 1842), 23, 24. More resolutions were introduced, and ultimately tabled. *LCCRR Report, November 1842*, 30.

11. *Semi-Annual Report of the South-Carolina Canal and Rail-Road Company, Accepted 20th of July, 1843* (Charleston, S.C.: Miller and Browne, 1843), 13–14.

12. Forest Marion has argued that Sabbatarians in the upper South shifted their argument from "moral-religious" concerns to "social-economic" ones by the mid-1830s. Marion, "Gentlemen Sabbatarians," 216.

13. *Proceedings of the Stockholders of the South-Carolina Railroad Company, and of the South-Western Railroad Bank, at at* [sic] *Their Annual Meeting in the Hall of the Bank, on the on the* [sic] *8th and 9th of February, 1853* (Charleston, S.C.: A. E. Miller, 1853), 10–11, quotation on 11.

14. *SCRR Report, February 1853*, 13, 17.

15. *Proceedings of the Stockholders of the South-Carolina Rail Road Company, and of the South-Western Railroad Bank at Their Annual Meeting in the Hall of the Bank, on the 14th and*

15th of February 1854 (Charleston, S.C.: A. E. Miller, 1854), 9, 12. They are identified as Protestant Episcopal on 7.

16. *Annual Reports of the President and Directors and the General Superintendent of the South Carolina Railroad Company, for the Year Ending 31st December, 1856* (Charleston, S.C.: Walker, Evans & Co., 1857), 8; *Annual Reports of the President and Directors and the General Superintendent of the South Carolina Railroad Company, for the Year Ending December 31st, 1857* (Charleston, S.C.: Walker, Evans & Co., 1858), 8.

17. Von Gerstner, *Early American Railroads,* 691; entries for May 14 and 18, and September 2, 1858, ETVRR Minutebook-1. For other southern examples, see *Proceedings of the Charlotte and South Carolina Rail Road Company, at Their Second Annual Meeting, at Winnsboro, 10th October, 1849. Also the Annual Reports of the President, Chief Engineer and Treasurer, &c. &c* (Columbia, S.C.: A. S. Johnston, 1849), 26; and *Proceedings of the Stockholders of the South-Carolina Rail Road Company and of the South-Western Rail Road Bank, at Their Annual Meeting, in the Hall of the Bank, on the 13th and 14th of February, 1855* (Charleston, S.C.: A. E. Miller, 1855), 5.

18. *Third Annual Report of the Directors of the Pennsylvania Rail-Road Company to the Stockholders, October 31, 1849* (Philadelphia: Crissy & Markley, 1850), 9, 19; entry for August 18, 1844, in Albrect Karl Koch, *Journey through a Part of the United States of North America in the Years 1844 to 1846,* ed. Ernst A. Stadler (Carbondale: Southern Illinois University Press, 1972), 32.

19. Entries for April 7 and May 5, 1835, Minutebook of the Director and Stockholders (1835–1837), vol. 2, BARC-BW; entry for April 23, 1841, Minutes of the Directors (January 8, 1836–October 16, 1841), vol. 2, BARC-WR; entry for April 10, 1857, Minutebook of the Directors (1849–1896), Norwich and Worcester Railroad, case 4, vol. 4 of Penn Central Collection, BHC.

20. *Charleston Courier,* March 19, 1831, 2. For a history of the Post Office Department, see Richard R. John, *Spreading the News: The American Postal System from Franklin to Morse* (Cambridge, Mass.: Harvard University Press, 1995).

21. *Annual Report of the Board of Directors of the South Carolina Canal & Rail Road Company to the Stockholders, with Accompanying Documents, Submitted at Their Meeting, May 7, 1832* (Charleston, S.C.: William S. Blain, 1832), 4–5. Even later in the antebellum era, companies were anxious to secure the business of the Post Office when their roads were opening. As the RDRR neared completion, the board of directors authorized a committee to meet with the postmaster general and arrive at a contract for carrying the mail. Entry for June 13, 1850, RDRR Minutebook.

22. *Annual Report of the Direction of the South Carolina Canal and Rail Road Company, to the Stockholders, May 6th, 1834, with Accompanying Documents* (Charleston, S.C.: W. S. Blain, 1834), 4–5; entry for May 22, 1835, SCRR Minutebook. Northern railroads also experienced difficulty. See memorandum from George Bliss to the Board of Directors, June 8, 1841, folder labeled "Western Railroad, Clerks File, 1841," case 1, BARC; and entries for February 19, 1842, and March 21, 1844, Minutebook of the Directors (November 11, 1841–March 23, 1844), vol. 3, BARC-WR.

23. Entries for November 26 and December 3, 1835, and January 18, 1836, SCRR Minutebook. The debate continued to rage for some time. See the entries for August 13, October 18, and November 21, 1836; March 29, 1837; December 11, 20, 21, and 31, 1838; and January 3, 1839, all in SCRR Minutebook. Citizens also intervened in the North; see George Bliss, *Historical Memoir of the Western Railroad* (Springfield, Mass.: Samuel Bowles, 1863), 67; and

entry for December 20, 1842, Minutebook of the Directors (November 11, 1841–March 23, 1844), vol. 3, BARC-WR.

24. *Semi-Annual Report of the Direction of the South-Carolina Canal and Rail-Road Company, to July, 1838* (Charleston, S.C.: Burges & James, 1838), 5–6, quotation on 6.

25. *Semi-Annual Report of the South-Carolina Canal and Rail-Road Company. Accepted Jan. 18, 1839* (Charleston, S.C.: A. E. Miller, 1839), 6. Other companies also complained about the Post Office's demands to control the schedule. The Petersburg Railroad found itself being required to run trains at night at additional expense in 1841. *Annual Report of the Petersburg Rail-Road Company, March 1, 1841* (Petersburg, Va.: Petersburg Intelligencer Job Office, 1841), 6.

26. *SCRR Report, January 1839*, 5–7, 9.

27. *Semi-Annual Report to the Stockholders of the South-Carolina Canal and Rail-Road Company. Made on the Third Monday in January, (the 17th) in Conformity with a By-Law of the Company* (Charleston, S.C.: A. E. Miller, 1842), 12–13.

28. Letter from C. C. Jones to C. C. Jones Jr., November 8, 1854, folder 4, box 3, Charles Colcock Jones Collection, HL; entry for the South Carolina Railroad, South Carolina, vol. 6, p. 142, R. G. Dun & Co. Collection, BHC.

29. Letter from Charles William Ashby to Joseph Holt, August 29, 1849, VHS.

30. Mary Elizabeth Pearson Boyce, *And God Still Spares Us: The Diary of Mary Elizabeth Pearson Boyce, September 28, 1854–November 2, 1855* (Hickory Grove, S.C.: Broad River Basin Historical Society, 1993), 35–36; letter from Anna Calhoun Clemson to Patrick Calhoun, December 3, 1842, in Julia Wright Sublette, "The Letters of Anna Calhoun Clemson, 1833–1873" (Ph.D. diss., Florida State University, 1993), 319–20, 645; George P. Rawick, ed., *The American Slave: A Composite Autobiography*, vol. 2, *South Carolina Narratives, Part 1* (Westport, Conn.: Greenwood, 1972), 104.

31. *SCRR Report, January 1839*, 10; *Proceedings of the Annual Convention of Stockholders in the East Tenn. and Georgia Railroad Company, Held at Athens, Tennessee. Wednesday the 1st Day of September, 1858* (Knoxville: Chas. A. Rice, Book and Job Printer, Gay Street, 1858), 64.

32. *Semi-Annual Report of the Direction of the South-Carolina Canal and Rail Road Company. July 10, 1837. Accepted and Ordered to Be Printed* (Charleston, S.C.: James S. Burges, 1837), 6; Lawrence T. McDonnell, "Money Knows No Master: Market Relations and the American Slave Community," in *Developing Dixie: Modernization in a Traditional Society*, ed. Winfred B. Moore Jr., Joseph F. Tripp, and Leon G. Tyler (Westport, Conn.: Greenwood, 1988), 32; Bill Cecil-Fronsman, *Common Whites: Class and Culture in Antebellum North Carolina* (Lexington: University Press of Kentucky, 1992), 125.

33. *Proceedings of the Stockholders of the South-Carolina Rail-Road Company, and the South-Western Rail-Road Bank, at Their Annual Meeting, in the Hall of the Bank, on the 10th, 11th, and 12th February, 1846* (Charleston, S.C.: Miller and Browne, 1846), 20–21.

34. Chad Coffman and Mary Eschelbach Gregson, "Railroad Development and Land Value," *Journal of Real Estate Finance and Economics* 16 (March 1998): 191; letter from Thomas Napier to J. P. DeVeaux, October 13, 1847, Napier Letterbook; letters from Thomas Napier to James Hibben on January 6 and 26, 1848, Napier Letterbook; Tom Downey, *Planting a Capitalist South: Masters, Merchants, and Manufacturers in the Southern Interior, 1790–1860* (Baton Rouge: Louisiana State University Press, 2006), 96–97; Charles W. Turner, "The Louisa Railroad, 1836–1850," *North Carolina Historical Review* 24 (January 1947): 55; Andrew Leary O'Brien, *The Journal of Andrew Leary O'Brien* (Athens: University of

Georgia Press, 1946), 60. For Virginia, see Kenneth W. Noe, *Southwest Virginia's Railroad: Modernization and the Sectional Crisis* (Urbana: University of Illinois Press, 1994), 50; and Charles W. Turner, "Railroad Service to Virginia Farmers, 1828–1860," *Agricultural History* 22 (October 1948): 246.

35. Entries for November 19, 1835, and April 20, 1840, SCRR Minutebook.

36. Entry for December 6, 1836, in Herbert A. Kellar, ed., "A Journey through the South in 1836: Diary of James D. Davidson," *Journal of Southern History* 1 (August 1935): 373; James Baughman, "A Southern Spa: Antebellum Lake Pontchartrain," *Louisiana History* 3 (Winter 1962): 9, 15; *Thirtieth Semi-Annual Report of the South-Carolina Canal and Rail-Road Company, to January 1st, 1843, Accepted at the Adjourned Meeting, 26th February, 1843* (Charleston, S.C.: Miller and Browne, 1843), 25; *Fifteenth Annual Meeting of the Petersburg Railroad Company, March 3, 1845* (Richmond, Va.: Shepherd and Colin, 1845), 5; *Reports of the Presidents and Superintendents of the Central Railroad and Banking Co. of Georgia, from No. 20 to 32 Inclusive, and the Amended Charter of the Company* (Savannah, Ga.: G. N. Nichols, 1868), 167.

37. Entry for November 13, 1854, Southside Railroad, Minutebook of the Stockholders (March 21, 1854–December 9, 1868), Ms81-063, NSHC; entry for 1837, Elliott Lemuel Story Diary, vol. 1, VHS; entry for December 28, 1841, in Charles Lyell, *Travels in North America in the Years 1841–2, with Geological Observations on the United States, Canada, and Nova Scotia* (1845; reprint, New York: Arno, 1977), 129.

38. *Annual Report of the Board of Directors of the South Carolina Canal and Rail-Road Company to the Stockholders, with Accompanying Documents Submitted at Their Meeting, May 6, 1833* (Charleston, S.C.: A. E. Miller, 1833), 6; Thomas Hubbard Hobbs, *The Journals of Thomas Hubbard Hobbs*, ed. Faye Acton Axford (University: University of Alabama Press, 1976), 180.

39. Franklin L. Riley, "Extinct Towns and Villages of Mississippi," *Publications of the Mississippi Historical Society* 5 (1920): 337, 374, 381; Henry A. M. Smith, "Some Forgotten Towns in Lower South Carolina," *South Carolina Historical Magazine* 14 (July 1913): 146; Frederick Beck Gates, "Building the 'Empire State of the South': Political Economy in Georgia, 1800–1860" (Ph.D. diss., University of Georgia, 2001), 120–21. See also Howard G. Adkins, "The Historical Geography of Extinct Towns in Mississippi" (Ph.D. diss., University of Tennessee, 1972).

40. Von Gerstner, *Early American Railroads*, 688–89; William Cronon, *Nature's Metropolis: Chicago and the Great West* (New York: Norton, 1991), 372; Jeffrey P. Roberts, "Railroads and the Downtown: Philadelphia, 1830–1900," in *The Divided Metropolis: Social and Spatial Dimensions of Philadelphia, 1800–1975*, ed. William W. Cutler III and Howard Gillette Jr. (Westport, Conn.: Greenwood, 1980), 33. The lack of union stations was problematic elsewhere. Transferring from the Philadelphia and Washington to the Baltimore and Ohio in Baltimore involved a trip to a separate station. Robert J. Brugger, *Maryland: A Middle Temperament, 1634–1980* (Baltimore: Johns Hopkins University Press, 1988), 253.

41. James Redpath, *The Roving Editor, or, Talks with Slaves in the Southern States*, ed. John R. McKivigan (University Park: Pennsylvania State University Press, 1996), 118; Gates, "Building," 218; Lacy K. Ford Jr., *Origins of Southern Radicalism: The South Carolina Upcountry, 1800–1860* (New York: Oxford University Press, 1988), 262.

42. James Mallory diary quoted in Walter M. Jackson, *The Story of Selma* (Birmingham, Ala.: n.p., 1954), 113; entry for September 20, 1838, SCRR Minutebook.

43. Frederick Law Olmsted, *A Journey in the Seaboard Slave States in the Years 1853–1854*,

with Remarks on Their Economy, vol. 1 (1856; reprint, New York: Putnam's Sons, 1904), 404. To be sure, not everyone appreciated the benefits, as Olmsted notes on 405.

44. Entry for February 2, 1854, SURR Minutebook; entry for April 7, 1854, VTRR Minutebook. In the North as well, groups lobbied the railroad to have turnouts serve them. A committee of the Western Railroad reported in 1840 that a group of Shakers near Pittsfield had made application for a turnout. Finding that the Shakers "offered to be at the whole expense including the erection of the building, as the local business will be considerable, & they have been very friendly to the road, the com[mitte]e think the interests of the Corporation will be promoted by such a turnout at that place, & that recommend its allowance." "Report—in part—on depots west of the Connt River," dated June 19, 1840, in folder labeled "Western Railroad, Clerks File, 1840," case 1, BARC.

45. Depots in America are finally beginning to receive the attention of social and cultural historians. See Laura Elaine Milsk, "Meet Me at the Station: The Culture and Aesthetics of Chicago's Railroad Terminals, 1871–1930" (Ph.D. diss., Loyola University, 2003).

46. Letter from John McRae to James Gadsden, October 19, 1847, JML-2.

47. Solon Robinson, *Solon Robinson: Pioneer and Agriculturalist,* ed. Herbert Anthony Kellar, vol. 2, *1846–1851* (Indianapolis: Indiana Historical Bureau, 1936), 391–92; Louis Heuser Diary Transcription, p. 28, WHS; entry for August 5, 1854, VTRR Minutebook.

48. Henry Benjamin Whipple, *Bishop Whipple's Southern Diary, 1843–1844,* ed. Lester B. Shippee (1937; reprint, New York: Da Capo, 1968), 94; entry for October 18, 1855, Jeremiah C. Harris Diary, vol. 3, VHS; entry for March 8, 1838, in Mary E. Moragne, *The Neglected Thread: A Journal from the Calhoun Community, 1836–1842,* ed. Delle Mullen Craven (Columbia: University of South Carolina Press, 1951), 79; Hobbs, *Journals,* 218. For the impact of the railroad on luggage design, see Deborah Sampson Shinn, introduction to *Bon Voyage: Designs for Travel,* ed. J. G. Links (New York: Cooper-Hewitt Museum, Smithsonian Institution's National Museum of Design, 1986), 14.

49. Entry for December 6, 1855, James D. Frederick Diary, box 1, James Daniel Frederick Collection, HL; Thomas W. Chadwick, ed., "The Diary of Samuel Edward Burges, 1860–1862," *South Carolina Historical and Genealogical Magazine* 48 (April 1947): 70.

50. This anticipates the developments John Stilgoe describes in the postbellum era. John R. Stilgoe, *Metropolitan Corridor: Railroads and the American Scene* (New Haven, Conn.: Yale University Press, 1983), 193.

51. Peter Howell, *The Life and Travels of Peter Howell, Written by Himself, in Which Will Be Seen Some Marvelous Instances of the Gracious Providence of God* (Newbern, N.C.: H. W. Mayhew, 1849), 159–60; Lilla Mills Hawes, ed., "The Memoirs of Charles H. Olmstead, Part II," *Georgia Historical Quarterly* 43 (March 1959): 63; John W. Bear, *The Life and Travels of John W. Bear, "The Buckeye Blacksmith." Written by Himself* (Baltimore: Binswanger and Company, 1873), 110.

52. Entries for August 13 and December 31, 1835, and March 3, 1836, SCRR Minutebook; entry for May 18, 1858, ETVRR Minutebook-1; Robert S. Starobin, *Industrial Slavery in the Old South* (New York: Oxford University Press, 1970), 79.

53. Entry for November 6, 1837, SCRR Minutebook. The railroad reached a compromise with him the following year. Entry for September 20, 1838, SCRR Minutebook.

54. Entry for March 29, 1837, SCRR Minutebook.

55. Entry for May 21, 1838, SCRR Minutebook. The railroad's presence in the city also

found opposition in the North. John Jervis recalled opposition in Albany, New York, when he worked on the Mohawk and Hudson Railroad: "The citizens of Albany were not generally favorable to this railway at the time of its commencement. To this I must say there were some eminent exceptions. This unfavorable sentiment in Albany was opposed to any location in the streets of the city, and this led the location of the line to the south side of the city. The Albany and Schenectady Turnpike Company, the stock of which was principally held in Albany, was the nucleus of this opposition." John Bloomfield Jervis, *The Reminiscences of John B. Jervis, Engineer of the Old Croton,* ed. Neal FitzSimons (Syracuse, N.Y.: Syracuse University Press, 1971), 105.

56. Entries for April 20 and May 8, 1838, SCRR Minutebook.

57. Entry for September 14, 1854, RDRR Minutebook.

58. James W. Ely Jr., *Railroads and American Law* (Lawrence: University Press of Kansas, 2001), 118; Downey, *Planting a Capitalist South,* 181–82; entries for May 20 and June 15, 1836, SCRR Minutebook; entry for August 7, 1856, ETVRR Minutebook-1; entry for January 11, 1859, ETVRR Minutebook-2.

59. *Tenth Annual Report of the President and Other Officers of the East Ten. and Va. Railroad Co. to the Stockholders, at Their Annual Meeting at Greeneville, November 24, 1859* (Jonesborough, Tenn.: W. A. Sparks & Co., 1859), 11–12. Other Tennessee railroads also railed against the legislature, see *ETGRR Report, September 1858,* 26.

60. Entry for February 19, 1857, MCRR Minutebook; entry for January 5, 1854, VTRR Minutebook.

61. Entries for July 10, 1856, and November 10, 1857, Southside Minutebook-2.

62. Entry for July 25, 1855, VTRR Minutebook. In 1860 the CRRG killed a white man on March 4, and one black man apiece on July 3 and August 18. *CRRG Reports, 20 to 32 Inclusive,* 175–76.

63. Jenny Bourne Wahl, "The Bondsman's Burden: An Economic Analysis of the Jurisprudence of Slaves and Common Carriers," *Journal of Economic History* 53 (September 1993): 518, 508.

64. Ruth Schwartz Cowan, "The Consumption Junction: A Proposal for Research Strategies in the Sociology of Technology," in *The Social Construction of Technological Systems: New Directions in the Sociology and History of Technology,* ed. Wiebe E. Bijker, Thomas P. Hughes, and Trevor J. Pinch (Cambridge, Mass.: MIT Press, 1987), 279–80; letter from John McRae to T.[?] J. Summer, February 20, 1853, JML-6.

65. O'Brien, *Journal,* 35; entry for December 26, 1855, James D. Frederick Diary, box 1, James Daniel Frederick Collection, HL; George P. Rawick, ed., *The American Slave: A Composite Autobiography,* supplement, series 1, vol. 5, *Indiana and Ohio Narratives* (Westport, Conn.: Greenwood, 1977), 336–67 (although enslaved in Georgia, Fambro was interviewed in Ohio), 445 (although enslaved in Kentucky, Smith was interviewed in Ohio); letter from Robert Habersham to Barnard Elliot Habersham, December 5, 1840, folder 11, box 1, Barnard Elliot Habersham Papers, HL.

66. Entries for May 18 and September 2, 1858, ETVRR Minutebook-1; entry for March 8, 1859, ETVRR Minutebook-2.

67. Entry for April 10, 1837, Hiwassee Minutebook.

68. Entry for October 1, 1835, SCRR Minutebook; entry for August 25, 1838, SCRR Minutebook.

69. *Proceedings of the Stockholders of the South-Carolina Rail-Road Company and the South-Western Rail-Road Bank, at Their Annual Meeting, in the Hall of the Bank, on the 11th, 12th and 13th February 1845* (Charleston, S.C.: Burges and James, 1845), 16; entry for June 3, 1848, "The Schirmer Diary," *South Carolina Historical Magazine* 80 (April 1979): 192; *Charleston Courier,* April 18, 1850, 2; entry for May 20, 1853, RDRR Minutebook; entry for April 9, 1856, Southside Minutebook-2; entry for October 9, 1860, ETVRR Minutebook-2.

70. John H. White Jr., "The Railroad Pass: Perk or Plunder?" *Railroad History* 182 (Spring 2000): 59–71; R. H. Whitaker, *Whitaker's Reminiscences, Incidents and Anecdotes: Recollections of Other Days and Years; or, What I Saw and Heard and Thought of People Whom I Knew, and What They Did and Said* (Raleigh, N.C.: Edwards and Broughton, 1905), 59; entry for November 1, 1851, Jeremiah C. Harris Diary, vol. 1, VHS; entry for May 20, 1836, SCRR Minutebook; entry for April 13, 1835, TCDRR Minutebook; George P. Rawick, ed., *The American Slave: A Composite Autobiography,* supplement, series 1, vol. 5, *Indiana and Ohio Narratives* (Westport, Conn.: Greenwood, 1977), 333–34.

71. Entries for March 10 and April 2, 1851, December 21, 1855, and October 15, 1857, RDRR Minutebook. The CSCRR denied clergy free passes in 1852; see *Proceedings of the Stockholders of the Charlotte & South Carolina Railroad Company, at Their Fifth Annual Meeting, at Charlotte, N. C., on the 17th and 18th of November, 1852. Also the Annual Reports of the President, Chief Engineer and Treasurer* (Columbia, S.C.: Johnston & Cavis, 1852), 51.

72. For a summary of opinion, see John E. Clark Jr., *Railroads in the Civil War: The Impact of Management on Victory and Defeat* (Baton Rouge: Louisiana State University Press, 2001), 18–19. For an interpretation stressing the inadequacy of the southern network, see George Rogers Taylor and Irene D. Neu, *The American Railroad Network, 1861–1890* (1956; reprint, New York: Arno, 1981), chap. 5. For twentieth-century links among railroads and other types of transportation, see Richard Saunders Jr., *Main Lines: Rebirth of the North American Railroads, 1970–2002* (DeKalb: Northern Illinois University Press, 2003), 38–39, 206–9, 223.

73. Entry for October 1, 1835, SCRR Minutebook; entries for August 11, 1852, and June 11, 1856, RDRR Minutebook.

74. Letter from John A. Lancaster, Jos. M. Sheppard, and C. Robinson to "E. Porter & Co," February 8, 1836, folder 1, box 1, RFPRR-VHS.

75. Entry for August 9, 1856, ETVRR Minutebook-1; broadside, ca. 1855, Charlotte and South Carolina Railroad Collection, SCL.

76. Wayne Cline, *Alabama Railroads* (Tuscaloosa: University of Alabama Press, 1997), 41; Charles Ripley Johnson, "Railroad Legislation and Building in Mississippi, 1830–1840," *Journal of Mississippi History* 4 (October 1942): 196; entry for December 15, 1835, TCDRR Minutebook.

77. *Proceedings of the Greenville and Columbia Rail Road Company, at the Extra Meeting, 1st December, 1848, and Annual Meeting, 7th and 8th May, 1849* (Columbia, S.C.: John G. Bowman, 1849), 28; *CRRG Reports, 20 to 32 Inclusive,* 127.

78. *Charleston Courier,* January 3, 1840, 1; January 14, 1840, 1.

79. James Gadsden to James Edward Calhoun, October 30, 1844, James Gadsden Papers, SCL; letter from John McRae to D. K. Minor, December 1, 1846, JML-1.

80. *Report of the Chief Engineers, Presidents and Superintendents of the South-Western R. R. Co., of Georgia, from No. 1 to 22, Inclusive, with the Charter and Amendments Thereto* (Macon, Ga.: J. W. Burke, 1869), 245; Charles W. Turner, "Virginia Ante-Bellum Railroad Disputes and Problems," *North Carolina Historical Review* 27 (July 1950): 320.

81. Entries for October 22, 1835, and June 7, 1839, SCRR Minutebook.

82. The extent of such cooperation calls into question Welke's claim that midcentury railroads were purely local enterprises. Barbara Young Welke, *Recasting American Liberty: Gender, Race, Law, and the Railroad Revolution, 1865–1920* (New York: Cambridge University Press, 2001), 250–51. The companies were New Jersey Railroad; Philadelphia, Wilmington and Baltimore Railroad; Baltimore and Ohio Railroad; Orange and Alexandria Railroad; Richmond and Danville Railroad; Norfolk and Petersburg Railroad; Southside Railroad; East Tennessee and Virginia Railroad; East Tennessee and Georgia Railroad; Western and Atlantic Railroad; Memphis and Charleston Railroad; Montgomery and West Point Railroad; New Orleans, Jackson and Great Northern Railroad; Nashville and Chattanooga Railroad; and the Louisville and Nashville Railroad. "Monthly Report" for December 1859, Virginia and Tennessee Railroad, Ticket Agent's Letter Book, Ms81-073, NSHC.

83. *Seventh Annual Report of the President and Directors to the Stockholders of the Virginia and Tennessee Railroad Company, October 25th and 26th, 1854* (Lynchburg: "Virginian" Job Printing Establishment, 1854), 13.

84. Entry for May 8, 1837, SCRR Minutebook; entry for November 13, 1855, RDRR Minutebook; *Proceedings of the Stockholders of the Charlotte and South Carolina Railroad Company, at Their Eleventh Annual Meeting, Held at Columbia, S. C., Ninth day of February, 1859; Also, the Annual Reports of the President, Treasurer, and Engineer and Superintendent* (Columbia, S.C.: R. W. Gibbes, 1859), 25. See also *Proceedings of the Stockholders of the Charlotte and South Carolina Railroad Company, at Their Seventh Annual Meeting at Winnsboro', S. C., on the Sixth and Seventh of February, 1855, Also the Annual Reports of the President, Treasurer and General Superintendent* (Columbia, S.C.: R. W. Gibbes, 1855), 4.

85. *Proceedings of the Conventions of Stockholders of the Charlotte and South Carolina Rail Road, Held at Chesterville, S. C., on Thursday and Friday, January 13 and 14, 1848. Also, Report of the Experimental Surveys, from Columbia to Winnsborough* (Columbia, S.C.: A. G. Summer, 1848), 17; *Proceedings of the Stockholders of the Charlotte & South Carolina Railroad Company at Their Ninth Annual Meeting at Columbia, S.C., on the Fourth of February 1857; and Also, the Annual Reports of the President, Treasurer, and General Superintendent* (Columbia, S.C.: R. W. Gibbes, 1857), 5; Colleen A. Dunlavy, *Politics and Industrialization: Early Railroads in the United States and Prussia* (Princeton: Princeton University Press, 1994), 173; *Semi-Annual Report of the President to the Stockholders of the South-Carolina Railroad Company, with the Statements of the Auditor, Showing an Accurate View of Its Fiscal Affairs for the Past Six Months, Ending 30th June 1853* (Charleston, S.C.: A. E. Miller, 1853), 5.

86. *Minutes of Proceedings of the Stockholders of the Greenville and Columbia R. R. Co., at Their Annual Meeting, Held at Abbeville C. H., Wednesday and Thursday, the Eleventh and Twelfth of July, 1855* (Columbia, S.C.: R. W. Gibbs, 1855), 8; *SWRR Reports, 1 to 22 Inclusive*, 244.

Epilogue

1. Entry for November 30, 1859, RDRR Minutebook; entry for December 28, 1860, in R. Conover Bartram, ed., "The Diary of John Hamilton Cornish, 1846–1860," *South Carolina Historical Magazine* 64 (April 1963): 157; Reuben Davis, *Recollections of Mississippi and Mississippians*, rev. ed. ([Hattiesburg]: University and College Press of Mississippi, 1972), 402.

2. William J. Grayson, *Witness to Sorrow: The Antebellum Autobiography of William J.*

Grayson, ed. Richard J. Calhoun (Columbia: University of South Carolina Press, 1990), 238; George P. Rawick, ed., *The American Slave: A Composite Autobiography*, vol. 12, *Georgia Narratives, Part 1* (Westport, Conn.: Greenwood, 1972), 274–75; George P. Rawick, ed., *The American Slave: A Composite Autobiography*, vol. 5, *Texas Narratives, Part 3* (Westport, Conn.: Greenwood, 1972), 122.

3. Grayson, *Witness to Sorrow*, 195; George P. Rawick, ed., *The American Slave: A Composite Autobiography*, supplement, series 1, vol. 4, *Georgia Narratives, Part 2* (Westport, Conn.: Greenwood, 1977), 614; Sarah Morgan Dawson, *The Civil War Diary of Sarah Morgan*, ed. Charles East (Athens: University of Georgia Press, 1991), 242.

4. Brian Schoen, "Alternatives to Dependence: The Lower South's Antebellum Pursuit of Sectional Development through Global Interdependence," in *Global Perspectives on Industrial Transformation in the American South*, ed. Susanna Delfino and Michele Gillespie (Columbia: University of Missouri Press, 2005), 75; Robert G. Angevine, *The Railroad and the State: War, Politics, and Technology in Nineteenth-Century America* (Stanford: Stanford University Press, 2004), 63; John E. Clark Jr., *Railroads in the Civil War: The Impact of Management on Victory and Defeat* (Baton Rouge: Louisiana State University Press, 2001). See also Wiebe E. Bijker, "Understanding Technological Culture through a Constructivist View of Science, Technology, and Society," in *Visions of STS: Counterpoints in Science, Technology, and Society Studies*, ed. Stephen H. Cutcliffe and Carl Mitcham (Albany: State University of New York Press, 2001), 26.

5. Robert C. Black III, *The Railroads of the Confederacy* (1952; reprint, Chapel Hill: University of North Carolina Press, 1998), 189–91; Edward L. Ayers, *The Promise of the New South: Life after Reconstruction* (New York: Oxford University Press, 1992), 10–11.

6. Gustavus W. Dyer, John Trotwood Moore, Colleen Mores Elliott, and Louise Armstrong Moxley, eds., *The Tennessee Civil War Veterans Questionnaires*, vol. 2, *Confederate Soldiers (Caldwell–Fuston)* (Easley, S.C.: Southern Historical Press, 1985), 753; William Harden, *Recollections of a Long and Satisfactory Life* (1934; reprint, New York: Negro Universities Press, 1968), 77; Lilla Mills Hawes, ed., "The Memoirs of Charles H. Olmstead, Part II," *Georgia Historical Quarterly* 43 (March 1959): 66; David Gregg McIntosh, *Reminiscences of Early Life in South Carolina*, ed. Horace Fraser Rudisill (Florence, S.C.: St. David's Society, 1985), 2; R. H. Whitaker, *Whitaker's Reminiscences, Incidents and Anecdotes: Recollections of Other Days and Years; or, What I Saw and Heard and Thought of People Whom I Knew, and What They Did and Said* (Raleigh, N.C.: Edwards and Broughton, 1905), 59; N. J. Bell, *Southern Railroad Man: Conductor N. J. Bell's Recollections of the Civil War Era*, ed. James A. Ward (DeKalb: Northern Illinois University Press, 1993), 5.

7. George P. Rawick, ed., *The American Slave: A Composite Autobiography*, vol. 3, *South Carolina Narratives, Part 4* (Westport, Conn.: Greenwood, 1972), 38–39; George P. Rawick, ed., *The American Slave: A Composite Autobiography*, vol. 2, *South Carolina Narratives, Part 1* (Westport, Conn.: Greenwood, 1972), 127; George P. Rawick, ed., *The American Slave: A Composite Autobiography*, supplement, series 2, vol. 3, *Texas Narratives, Part 2* (Westport, Conn.: Greenwood, 1979), 741.

8. Mrs. John A. Logan, *Reminiscences of a Soldier's Wife: An Autobiography* (1913; reprint, Carbondale: Southern Illinois University Press, 1997), 71–72; W. C. Curtis, *Reminiscences* (Southport, N.C.: Herald Job Office, 1905), 1.

9. Whitaker, *Reminiscences,* 81, 291.

10. See Klaus Benesch, "Technology and American Culture: An Introduction," *Amerikastudien/American Studies* 41, no. 3 (1996): 335–36.

11. Leo Marx, *The Machine in the Garden: Technology and the Pastoral Ideal in America,* rev. ed. (New York: Oxford University Press, 2000), 205.

Primary Literature

One of the best sources for railroad history are the annual reports generated by the corporations. While the language in these reports is usually upbeat and attempts to present a positive assessment, companies also revealed a fair amount about their failures. These reports can be found in repositories across the South; I principally used reports located at the South Caroliniana Library in Columbia, the Virginia Historical Society in Richmond, and the Virginia Polytechnic and State Institute in Blacksburg. While these reports have often been mined by historians for the financial data contained therein, they are often rich in social and cultural material as well.

The manuscript minutes for boards of directors meetings are also extant for several companies. These minutebooks provide information about the daily operation of companies and debates on topics such as slavery, time management, worker compensation, landownership, cooperation with other corporations, and the other topics discussed throughout this book. The most important collection that I consulted was held at Virginia Polytechnic and State Institute in Blacksburg, which contained the materials relating to predecessor companies of Norfolk Southern. After I completed my dissertation, the collection was returned to the Norfolk Southern Corporation in Norfolk, Virginia. I am grateful to Kyle Davis of Norfolk Southern for helping me adjust my citations; researchers should be able to locate the materials I used in the new archive. I was able to do important comparative work on northern railroads thanks to the extensive business history collections at the Baker Library of Harvard Business School. One collection there, the Stone and Harris collection, also contained material on the company's southern operations.

Two other major manuscript collections are worthy of note. The Richmond, Fredericksburg and Potomac Railroad Company Records (1833–1909) at the Virginia Historical Society are remarkably detailed. Small scraps of paper and receipts allow for the reconstruction of slave labor and community relations for this particular corporation. The letterbooks of John McRae, held at the Wisconsin Historical Society in Madison, offer a thorough run of letters from a civil engineer who spent a significant portion of his career in the South and retired there when he stopped working as an engineer. Finally, travelers' descriptions of travel and railroads that I quoted in the text were culled from a range of manuscripts and printed diaries and letters.

Secondary Literature

The scholarly literature on southern railroads dates back to Ulrich Bonnell Phillips's *A History of Transportation in the Eastern Cotton Belt to 1860* (1908; reprint, New York: Octagon Books, 1968), which, despite Phillips's racial blinders, remains a useful work on railroads in South Carolina and Georgia. Other scholarly works of note from the pre–World War II period include Cecil Kenneth Brown, *A State Movement in Railroad Development: The Story of North Carolina's First Effort to Establish an East and West Trunk Line Railroad* (Chapel Hill: University of North Carolina Press, 1928); Thomas D. Clark, *A Pioneer Southern Railroad from New Orleans to Cairo* (Chapel Hill: University of North Carolina Press, 1936); Samuel Melanchthon Derrick, *Centennial History of South Carolina Railroad* (Columbia, S.C.: State Company, 1930); Balthasar Henry Meyer, ed., *History of Transportation in the United States before 1860* (Washington, D.C.: Carnegie Institute of Washington, 1917); and N. P. Renfro, *The Beginning of Railroads in Alabama* (Auburn, Ala.: n.p., 1910).

General works on railroad history, and its economic history, include Robert Fogel's *Railroads and American Economic Growth: Essays in Economic History* (Baltimore: Johns Hopkins University Press, 1964); Albert Fishlow, *American Railroads and the Transformation of the Ante-Bellum Economy* (Cambridge, Mass.: Harvard University Press, 1965) and his more recent essay, "Internal Transportation in the Nineteenth and Early Twentieth Centuries," printed in *The Cambridge Economic History of the United States*, vol. 2, edited by Stanley L. Engerman and Robert E. Gallman (New York: Cambridge University Press, 2000); John F. Stover, *Iron Road to the West: American Railroads in the 1850s* (New York: Columbia University Press, 1978); and George Rogers Taylor, *The Transportation Revolution, 1815–1860* (New York: Rinehart, 1951). Two other important works are James A. Ward's *Railroads and the Character of America, 1820–1887* (Knoxville: University of Tennessee Press, 1986) and Eugene Alvarez's *Travel on Southern Antebellum Railroads, 1828–1860* (University: University of Alabama Press, 1974), which survey the travel experience, including attention to time, accidents, car construction, and other topics. Examinations of individual southern states or corporations include Wayne Cline, *Alabama Railroads* (Tuscaloosa: University of Alabama Press, 1997); Kenneth Noe, *Southwest Virginia's Railroad: Modernization and the Sectional Crisis* (Urbana: University of Illinois Press, 1994); Merl Reed, *New Orleans and the Railroads: The Struggle for Commercial Empire, 1830–1860* (Baton Rouge: Louisiana State University Press, 1966); Allen W. Trelease, *The North Carolina Railroad, 1849–1871, and the Modernization of North Carolina* (Chapel Hill: University of North Carolina Press, 1991); and Gregg Turner, *A Short History of Florida Railroads* (Charleston, S.C.: Arcadia, 2003). Many graduate students have also adopted individual corporations as subjects of study in their master's theses and doctoral dissertations. I have relied heavily on their expertise, as will be clear from individual citations in the notes.

In recent years, historians are beginning to survey a wider range of topics than simply economic performance. Although many of these books focus on the postbellum era, their insights can be fruitfully applied to the antebellum years. Mark Aldrich's *Death Rode the Rails: American Railroad Accidents and Safety, 1828–1965* (Baltimore: Johns Hopkins University Press, 2006) is a fascinating analysis of accidents and includes some material on the pre–Civil War years. John E. Clark Jr.'s *Railroads in the Civil War: The Impact of Management on Victory and Defeat* (Baton Rouge: Louisiana State University Press, 2001) is a comparative study of northern and southern management techniques during the war. The essays in Maury Klein's *Unfinished Business: The Railroad in American Life* (Hanover, N.H.: University Press of New England, 1994) include a call for historians to consider the impact of railroads on society. Two recent works have examined railroads and gender: Barbara Young Welke's *Recasting American Liberty: Gender, Race, Law, and the Railroad Revolution, 1865–1920* (New York: Cambridge University Press, 2001) and Amy G. Richter's *Home on the Rails: Women, the Railroad, and the Rise of Public Domesticity* (Chapel Hill: University of North Carolina Press, 2005).

A number of historians have also begun investigating more closely the relationship of railroad development and politics. Robert G. Angevine's *The Railroad and the State: War, Politics, and Technology in Nineteenth-Century America* (Stanford: Stanford University Press, 2004) discusses the relationship of the military with railroads through the nineteenth century, including the role that army engineers played in early railroad projects. Colleen A. Dunlavy's *Politics and Industrialization: Early Railroads in the United States and Prussia* (Princeton: Princeton University Press, 1994) is an outstanding example of comparative history and explores how different political systems affect railroad development. James W. Ely Jr.'s *Railroads and American Law* (Lawrence: University Press of Kansas, 2001) is a wide-ranging study that examines land grants, regulation, liability, accidents, and a host of other topics. John Lauritz Larson's *Internal Improvement: National Public Works and the Promise of Popular Government in the Early United States* (Chapel Hill: University of North Carolina Press, 2001) is a compelling study of the relationship between republicanism and public works. The challenge of regulating new technologies is analyzed in Steven Usselman's *Regulating Railroad Innovation: Business, Technology, and Politics in America, 1840–1920* (New York: Cambridge University Press, 2002).

Although they do not necessarily discuss the South in detail, important works on technology and American culture include Richard D. Brown's *Modernization: The Transformation of American Life, 1600–1865* (New York: Hill and Wang, 1976); John R. Stilgoe, *Metropolitan Corridor: Railroads and the American Scene* (New Haven, Conn.: Yale University Press, 1983); John F. Kasson, *Civilizing the Machine: Technology and Republican Values in America, 1776–1900* (rev. ed., New York: Hill and Wang, 1999); and Leo Marx's classic *The Machine in the Garden: Technology and the Pastoral Ideal in America* (rev. ed., New York: Oxford University Press, 2000).

The nature of the southern economy has long been the topic of historiographic debate. Many scholars have argued that the South's reliance on slavery prevented it from modernizing. I have found more compelling the work of scholars who attempt to draw out the complexity of the southern economy, both its capitalistic acquisitiveness and reliance on slave labor. Any investigation of the southern economy must necessarily grapple with the works of Eugene Genovese and Elizabeth Fox-Genovese. In writing this book I found myself turning most often to Genovese's *The Political Economy of Slavery* (2d ed., Middletown, Conn.: Wesleyan University Press, 1989) and *The Slaveholders' Dilemma: Freedom and Progress in Southern Conservative Thought, 1820–1860* (Columbia: University of South Carolina Press, 1992), as well as the jointly authored *Fruits of Merchant Capital* (New York: Oxford University Press, 1983). In *Slaveholders' Dilemma*, Genovese presents the argument that slaveholder's sought their own "alternate route to modernity."

Other historians have resisted an "either-or" distinction for the antebellum economy; refusing to paint the "capitalist North" and "slaveholding South" as polar opposites. Joyce E. Chaplin's *An Anxious Pursuit: Agricultural Innovation and Modernity in the Lower South, 1730–1815* (Chapel Hill: University of North Carolina Press, 1993) lays the groundwork in the eighteenth century for how antebellum planters would continue to press for advanced and creative solutions to their problems. Walter Johnson has demonstrated the links of commodification, markets, and slavery in *Soul by Soul: Life Inside the Antebellum Slave Market* (Cambridge, Mass.: Harvard University Press, 1999). Tom Downey has illustrated the web of agricultural, commercial, and industrial development in antebellum South Carolina in *Planting a Capitalist South: Masters, Merchants, and Manufacturers in the Southern Interior, 1790–1860* (Baton Rouge: Louisiana State University Press, 2006).

Other works critical to my understanding of the southern economy include Peter A. Coclanis's *The Shadow of a Dream: Economic Life and Death in the South Carolina Low Country, 1670–1920* (New York: Oxford University Press, 1989); Fred Bateman and Thomas Weiss, *A Deplorable Scarcity: The Failure of Industrialization in the Slave Economy* (Chapel Hill: University of North Carolina Press, 1981); Sean Patrick Adams's *Old Dominion, Industrial Commonwealth: Coal, Politics, and Economy in Antebellum America* (Baltimore: Johns Hopkins University Press, 2004); John Majewski's *A House Dividing: Economic Development in Pennsylvania and Virginia before the Civil War* (New York: Cambridge University Press, 2000); Gavin Wright's *Slavery and American Economic Development* (Baton Rouge: Louisiana State University Press, 2006); and Peter Kolchin's *A Sphinx on the American Land: The Nineteenth-Century South in Comparative Perspective* (Baton Rouge: Louisiana State University Press, 2003). *Global Perspectives on Industrial Transformation in the American South* (Columbia: University of Missouri Press, 2005), edited by Susanna Delfino and Michele Gillespie, is the first volume in the New Currents in the History of Southern Economy and Society series, which promises to publish newer scholarship on the southern economy.

One particular portion of the antebellum southern economy—slavery—has an enormous literature all of its own. The works listed above on the history of individual southern corporations generally include information on slavery on those particular railroads. I have also relied on works that focus on nonagricultural slavery. Robert S. Starobin's *Industrial Slavery in the Old South* (New York: Oxford University Press, 1970) is a wide-ranging study of industrial slavery in a variety of contexts, including railroads. Charles B. Dew's *Bond of Iron: Master and Slave at Buffalo Forge* (New York: Norton, 1994) is a remarkably detailed study of slave labor, which provides fascinating information about how individual slaves were remunerated for their labor. Theodore Kornweibel Jr.'s article "Railroads and Slavery" (*Railroad History* 189 [2003]: 34–59) gives an overview of the use of slave labor on antebellum railroads. Jonathan D. Martin's *Divided Mastery: Slave Hiring in the American South* (Cambridge, Mass.: Harvard University Press, 2004) is a long-overdue study of this important topic.

Some theoretical works have been important to my understanding of the history of technology. Although I have not necessarily incorporated their terminology into my own writing, their ideas have proved useful. In particular, I relied on Wiebe E. Bijker's *Of Bicycles, Bakelites, and Bulbs: Toward a Theory of Sociotechnical Change* (Cambridge, Mass.: MIT Press, 1995); Wiebe E. Bijker and John Law, eds., *Shaping Technology/Building Society: Studies in Sociotechnical Change* (Cambridge, Mass.: MIT Press, 1992); Wiebe Bijker, Thomas P. Hughes, and Trevor J. Pinch, eds., *The Social Construction of Technological Systems: New Directions in the Sociology and History and Technology* (Cambridge, Mass.: MIT Press, 1987); and

Merritt Roe Smith and Leo Marx, eds., *Does Technology Drive History? The Dilemma of Technological Determinism* (Cambridge, Mass.: MIT Press, 1994). Historical works on early civil engineering include Daniel Hovey Calhoun, *The American Civil Engineer: Origins and Conflict* (Cambridge, Mass.: Technology Press, Massachusetts Institute of Technology, 1960); Raymond H. Merritt, *Engineering in American Society, 1850–1875* (Lexington: University Press of Kentucky, 1969); Elting E. Morison, *From Know-How to Nowhere: The Development of American Technology* (New York: Basic, 1974); and Terry S. Reynolds, ed., *The Engineer in America: A Historical Anthology from Technology and Culture* (Chicago: University of Chicago Press, 1991).

I have benefited a great deal from the work of historians and social scientists who study time. Barbara Adam's *Timewatch: The Social Analysis of Time* (Cambridge, Mass.: Polity, 1995) provides a theoretical framework for the history of time. Mark M. Smith's *Mastered by the Clock: Time, Slavery, and Freedom in the American South* (Chapel Hill: University of North Carolina Press, 1997) has been an enormously important work for my own study, not only for its particular information about time management in the South, but also for its articulation of the South's creation of an "alternate route" different from that in the North. Two other useful studies are Ian R. Bartky's *Selling the True Time: Nineteenth-Century Timekeeping in America* (Stanford: Stanford University Press, 2000) and Alexis McCrossen's *Holy Day, Holiday: The American Sunday* (Ithaca, N.Y.: Cornell University Press, 2000).

I have depended on three comprehensive books by John White Jr. for much of my technical knowledge about early railroads: *American Locomotives: An Engineering History, 1830–1880* (2d ed., Baltimore: Johns Hopkins University Press, 1997); *The American Railroad Freight Car: From the Wood-Car Era to the Coming of Steel* (Baltimore: Johns Hopkins University Press, 1993); and *The American Railroad Passenger Car* (Baltimore: Johns Hopkins University Press, 1978).

Finally, my understanding of railroads has been informed in part by reading literature on railroads in other countries and time periods. Historians of Europe have long been discussing the social impact of railroads in more depth than historians of America. In addition, reading about topics such as labor relations and engineering challenges helped me understand that the South was not necessarily unique in suffering from these problems. Works of particular interest include Colin Divall and George Revill, "Cultures of Transport: Representation, Practice and Technology," *Journal of Transport History,* 3d ser., 26 (March 2005): 99–111; Günter Dinhobl, ed., *Eisenbahn/Kultur = Railway/Culture* (Innsbruck: Studien Verlag, 2004); Matthew J. Payne, *Stalin's Railroad: Turksib and the Building of Socialism* (Pittsburgh: University of Pittsburgh Press, 2001); Ralph Harrington, "The Railway Journey and the Neuroses of Modernity," in *Pathologies of Travel,* edited by Richard Wrigley and George Revill (Atlanta: Rodopi, 2000); and Wolfgang Schivelbusch, *The Railway Journey* (New York: Urizen, 1979).

accidents, 184; causes, 127–30; complaints about, 179–81; fatal, 50, 68, 85; investigations, 129–31; passenger response, 131–33; prevention, 88, 90; severity, 127, 131; and slaves, 64, 235n42

Aiken, South Carolina: hotels in, 173; and South Carolina Railroad, 36, 88, 100, 109, 125, 165, 169, 179, 185; travel in, 124, 147; workshops in, 68

Alabama, 18, 33, 114, 176; canals in, 12; early railroads in, 11, 15, 23, 187; railroad statistics, 4, 5; travel in, 132. *See also* Tuscaloosa

Alabama and Tennessee Railroad, 189

alcohol, 63, 72–73, 97–98, 223n47

Allen, Horatio, 33, 53, 90, 128; and laborers, 56; as promoter, 14, 19; training, 32, 34; willingness to experiment, 35–36, 37, 95–96

American Railroad Journal, 12, 14–16, 22–23, 35, 189

Arkansas, 5

Atlanta, Georgia, 174

Augusta, Georgia, 182; and early railroads, 31; and South Carolina Railroad, 24, 174; travel in, 136, 143, 147, 149

Baltimore, Maryland, 154, 157, 188

Baltimore and Ohio Railroad, 34, 52; and labor, 55; and steam power, 31; travel on, 143

Baltimore and Philadelphia Railroad, 34

Bangs, Anson, 43–44, 50–51

Bell, N. J., 35, 69, 82, 141, 195

bells, 94, 132, 141, 142, 148. *See also* sound; whistles

Berry, Fannie, 158–59

Bird, Henry, 32, 37, 97–98, 158

Birnie, William, 41, 57, 72

Blanding, William, 146–47

Bliss, George, 38

Blue Ridge Railroad (of South Carolina) (BRRR), 2, 28–29, 70–71, 73, 216n62, 221n35; and landowners, 18; and slave laborers, 42; and state funding, 19, 20. *See also* Stumphouse Mountain

Blue Ridge Railroad (of Virginia), 81, 159

boards of directors, 48, 94, 186. *See also specific companies*

Bolling, Blair, 93–94, 131–32, 162–63

boosters, 6, 11–12; challenges faced, 18–25; early boosters, 12–16; as part of a national movement, 15; principal arguments of, 13–16, 27–29. *See also* Perry, Benjamin

Boston, Massachusetts, 156

Boston and Lowell Railroad, 92, 98

Boston and Providence Railroad, 33, 34

Boston and Worcester Railroad: and alcohol, 98; and civil engineers, 215n55; and landowners, 26; and nature, 38; and time, 90, 168

Boyce, Ker, 48

Boyce, Mary, 99, 140, 148, 172

Branchville, South Carolina, 41, 47, 88, 89, 147, 174

bridges, 184, 187

Broome, Anne, 172

Brunswick and Florida Railroad, 62

Buck, Nelson, 163

bureaucracy, 46–48, 64, 84, 86–87; use of preprinted forms, 34, 46, 63

Burges, Samuel, 1–2, 5, 10, 133, 178–79

California, 4, 5

Camden, South Carolina, 152; boosterism in, 27; and South Carolina Railroad, 39, 86, 128, 177, 190; travel in, 156 canals, 12–13, 17, 20, 31, 217n5. *See also* water transportation

Carpenter, Lucy, 139, 143, 156

Central Railroad of Georgia (CRRG), 34, 174, 194; and accidents, 131; and contractors, 41; and disease, 126; and freight, 117, 233n24, 234n25; and labor, 55, 71–72; and landowners, 17; and rolling stock, 101, 103; and slavery, 57, 58; and state funding, 20; and time, 96; and water transportation, 188; and white labor, 68

Charleston, South Carolina, 17, 28, 35, 47, 48, 70, 142, 163; boosterism in, 11, 13, 16, 27; and Civil War, 192, 193; competition with other cities, 23, 24, 31; damage from railroads, 180; and disease, 125–26; and Sabbatarianism, 165–67; and South Carolina Railroad, 67, 68, 84, 88, 93, 97, 100, 109, 117, 125, 152, 156, 173, 174, 185, 188, 190; travel in, 136, 141, 147, 149, 150, 182; and U.S. Post Office, 169–70

Charleston and Hamburg Railroad. *See* South Carolina Railroad

Charleston and Savannah Railroad, 2, 23; and civil engineers, 33, 49; and rolling stock, 101; and slavery, 62, 77; travel on, 178–79

Charlotte and South Carolina Railroad (CSCRR), 27; cooperation with other railroads, 190; and freight, 113, 119; and landowners, 21; and slavery, 58, 61, 63

Cheraw and Darlington Railroad, 2, 136

Chesbrough, E. S., 33, 34–35

Cincinnati, Ohio, 13, 16

City Point Railroad, 58, 108, 168

City Railroad (Philadelphia), 26, 175

civil engineers: and contractors, 39–44, 47, 49–53; and experimentation, 35–37; federal provision of, 32; mobility of, 8, 32–33; and politics, 47–48; and the public, 48–49; skills of, 32, 37; training of, 31–35, 53; workload of, 46–47, 53, 85–86

Civil War, 2, 3, 53, 192–94, 237n10

Clemson, Anna Calhoun, 99, 136, 139–40, 157, 172

clocks and watches: and boards of directors, 94; and conductors, 88–89, 99; and engineers, 89; and passengers, 91, 147; at stations, 87–88. *See also* time

Columbia, South Carolina, 17, 38, 41, 115, 119; boosterism in, 27, 36; celebration in, 163; and Sabbatarians, 166; and South Carolina Railroad, 90, 117, 167

Columbus, Georgia, 24, 188

comfort, 143–44

conductors, 58, 185; and accidents, 128, 130, 133; and aural cues, 141; authority of, 98–99, 127, 153–54, 155; and counterfeit money, 99; and freight, 107, 109–10; interaction with passengers, 145, 152, 156–57; judging passengers' race, 152–54; and time management, 88–89, 98–99

Connecticut, 103, 133; labor in, 56; opposition to railroads in, 25; railroad statistics, 4, 5

construction, 17, 84; celebration of completion, 162–64; challenges, 37–39, 45, 47, 50–53, 212n16; contracts for, 39–40, 43, 51, 52; equipment for, 47, 51–52, 70; techniques, 11, 35–37, 50

contractors, 62, 64; and civil engineers, 39–44, 47; fiscal responsibilities of, 43–46; planters as contractors, 40, 41–42

contracts: for railroad construction, 39–40, 43, 45–46, 51, 52; for slaves, 62–63, 77, 79

Cooke, Edward St. George, 46–47, 53

cotton, 18, 45, 77, 103, 139, 141, 198, 207n21; and agricultural cycle, 88, 121, 123; in booster rhetoric, 12, 14; and depot design, 177; as freight, 93, 106, 112–13, 115, 117, 126, 176, 189; and labor, 109; price of, 22, 56, 172

Craven, A. W., 39, 40, 70

Crozet, Claudius, 81

Cunynghame, Arthur, 111–12, 145, 155

Davidson, James, 123, 136, 138, 147, 173–74

Davis, Young Winston, 218n9

Delaware, 4, 5, 6

Delaware and Hudson Railroad, 33

depots, 139, 140, 175, 176, 188, 192; activity at, 178–79, 192; design of, 177; slave sale at, 111–12

Derricotte, Ike, 192

disease, 38–39, 125–26

Douglass, Frederick, 143–44, 154

East Tennessee and Georgia Railroad, 172, 184, 189, 194; and accidents, 131, 235n42; and alcohol, 97; and contractors, 70; and landowners, 22; and rolling stock, 104; and slavery, 152; and telegraph, 90

East Tennessee and Virginia Railroad (ETVRR), 184, 185, 189; and accidents, 74; and alcohol, 97; celebration of completion, 163; cooperation with stagecoaches, 187; and labor, 72, 76; and landowners, 21; and livestock, 180–81; and Sabbatarians, 168; and telegraph, 90; and time, 92; and vandals, 179

Elliott, William, 143, 148

Emerson, Ralph Waldo, 3

engineers, 58; and accidents, 127, 130; and aural cues, 141; and time, 88–89

equipment and rolling stock: freight, 100, 103–4, 107; origin of, 100, 101, 103–4; passenger, 100–103, 104, 230n52. *See also* locomotives

experimentation, 1, 11, 31, 34–37

Fambro, Hanna, 58, 59, 184, 185–86

federal government, 32, 171

firemen: and accidents, 74, 128, 129; slaves as, 58, 68, 77

Fleming, L. D., 34

Florida, 22–23, 73; canals in, 12; railroad statistics, 4, 5, 6

food. *See* taste

Fraser, Charles, 142, 158

Frederick, James, 148, 178, 183–84

Fredericksburg, Virginia, 23–24, 27, 72, 96

free blacks: as laborers, 58, 128, 129; as passengers, 152, 154

freight, 90, 136, 176, 231n1; bidirectional traffic, 112–21; night work, 97; and private turnouts, 94, 112, 176–77; rolling stock, 100, 103–4; rules, 107; and Sabbatarians, 165, 166, 167; storage, 107–8, 109–10, 177; types, 28, 29, 106, 113–15; work requirements, 106–9

funding, 12; difficulty of securing, 18–21; from individuals, 18–19, 20, 21; from state governments, 6, 19–21

Gadsden, James, 48, 102, 112, 188–89, 228n23

Galena and Chicago Union Railroad, 25

Georgia, 24, 66; early railroads in, 11; freight in, 114; laborers in, 69, 82; railroad statistics, 3, 4, 5, 6; state funding of railroads in, 21; travel in, 139. *See also* Atlanta; Augusta; Columbus; Macon; Savannah

Georgia Railroad, 33; incorporation of, 31; night operation, 96; rolling stock, 102; and slavery, 61; and time, 94; and white laborers, 189

Gerstner, Franz Anton Ritter von: in Georgia, 103; in Louisiana, 93; in South Carolina, 71; in Virginia, 68, 96, 108, 152, 168, 175, 227n22

Girard and Mobile Railroad. *See* Mobile and Girard Railroad

Glass, John, 69, 73, 82, 95

Glass, Maria, 138

Glenn, Robert, 110

Grayson, William, 149

Greenville, South Carolina, 28, 115

Greenville and Columbia Railroad, 2, 35, 140; cooperation with other railroads, 190, 191; and freight, 114–15, 119; and landowners, 17; and water transportation, 188; and weather, 125

Gregg, William, 20, 27

Gwynn, Walter, 52

Hamburg, South Carolina, 11, 195; and South Carolina Railroad, 84, 87, 90, 117, 165, 169, 170; and time, 88; weather in, 124–25

Hamburgh Railroad. *See* South Carolina Railroad

Hammond, James Henry, 158; opposition to state spending, 20, 24–25, 27, 207n21; organizer of celebration, 163

Harden, William, 101, 194

Harris, Jeremiah, 132, 140, 178

Haupt, Herman, 34, 56

Heuser, Louis, 101, 139, 178

Hiwassee Railroad. *See* East Tennessee and Georgia Railroad

Hobbs, Thomas, 103, 174, 178

Horry, Elias, 17, 169

horses, 91, 94, 136; and civil engineers, 70; and construction, 71; pulling trains, 26, 31, 180; reaction to trains, 141

hotels and taverns, 143, 144, 157, 173–74

Howze, Isham, 142

Illinois, 4, 5, 6, 16

Illinois Central Railroad, 57

Indiana, 4, 5, 16

Iowa, 4, 5

James, Mary Moriah Anne Susanna, 155

Jarnagan, Andrew Jackson, 163

Jervis, John, 32, 34

Jones, Charles Colcock, 114, 150, 171

Keenan, Sally Layton, 138

Kentucky, 1, 4, 5, 110

Knoxville, Tennessee, 13, 50

Kollock, George, 132, 150

labor, 69–70; amusement, 82; apprenticeships, 67–68, 221n35; difficulty of acquiring, 55–57; housing for, 71–72; racial integration, 68–69; size of work forces, 50, 52–53, 55–56; skilled, 66–68; slave vs. free, 6, 66–67; strikes, 81–82; unskilled, 66–67; wages, 56, 76, 77, 79–82; work time, 52, 64, 94–97, 108–9, 166, 168. *See also* slave laborers; white laborers

land: acquired by railroads, 17–18, 21–22, 26, 48–49; economic impact of railroads on, 28, 173; viewed by passengers, 139

landowners: complaints during construction, 26, 49; turning over land, 17–18, 21–22, 26, 48–49

Latrobe, Benjamin Henry, 34

livestock, 180–81. *See also* accidents

locomotives, 99–100, 103; sensory impact of, 1, 138; steam vs. horse power, 26, 31. *See also* equipment and rolling stock

Louisa Railroad, 24, 41, 55, 150, 173

Louisiana, 4, 5, 12, 178

Louisville, Cincinnati and Charleston Railroad (LCCRR): and civil engineers, 33, 46, 47; construction of, 38, 39–40, 41–42; coordination with other companies, 24; and freight, 115, 121; and labor, 76; and landowners, 49; night trains on, 96; and Sabbatarians, 165; and slavery, 61. *See also* South Carolina Railroad

Lyell, Charles, 101, 139, 174

Mackay, Alexander, 125, 139

Macon, Georgia, 156, 175

Macon and Western Railroad, 66

mail. *See* U.S. Post Office Department

Maine, 4, 5

management, 84, 85–87

Maryland, 4, 5, 15. *See also* Baltimore

Massachusetts, 41, 103; civil engineers, 33; early railroads in, 11; opposition to railroads in, 25, 26; railroad statistics, 3, 4, 5, 6. *See also* Boston

Maury, Ann, 144–45, 148–49, 157

May, Ann, 138

McGill, Samuel, 141, 149

McRae, John, 7, 71, 189; and boosters, 22, 27; and bureaucracy, 86–87; and competition, 23, 24; and contractors, 39–46; and depot design, 177; and disease, 39; and experiments, 31; and landowners, 21–22, 48–50; and nature, 38; and networking, 34–35; and pedestrians, 182;

as planter, 152, 157; and politics, 48; and rolling stock, 100–102, 103; and slavery, 56, 64–66, 77; and speed, 91–92; and time, 89, 95, 96–97; and training, 32

Memphis and Charleston Railroad, 189; celebration of completion, 163; and livestock, 181; and Sabbatarians, 168; and slavery, 75; and turnouts, 94

Michigan, 4, 5

Minor, D. K., 27, 30, 35, 189; and southern railroads, 15–16, 22–23

Mississippi, 73, 187; canals in, 12; Civil War, 192; early railroads in, 11; railroad statistics, 3, 4, 5; towns abandoned, 175; travel in, 142, 144

Mississippi and Pearl River Railroad, 58

Mississippi and Tennessee Railroad, 175

Mississippi Central Railroad, 28, 59, 61

Missouri, 4, 5

Mobile and Girard Railroad, 191

Mobile and Ohio Railroad, 33, 82, 163, 192

modernization, 158–59, 182, 197–98; and sensory perception, 1, 138, 140–46; and slave society, 2, 6–8, 10

Mohawk and Hudson Railroad, 248n55

Montgomery and West Point Railroad, 13, 58, 66

Moore, Jerry, 192–93

Moragne, Mary, 101, 145–46, 148, 178

Morgan, Sarah, 193

Napier, Thomas, 23, 27, 173

Nashua and Lowell Railroad, 92, 98, 234n30

Nashville, Tennessee, 11, 110

nature, 106; and problems with construction, 37–38; and problems with operations, 123–26, 234n30; and time, 94–97, 123–26

New Hampshire, 4, 5, 25

New Jersey, 4, 5, 6

New Orleans and Ohio Railroad, 33

New York, 70, 100, 136; canals in, 12; and competition, 14–15, 23; railroad statistics, 3, 4, 5, 6

New York and Erie Railroad, 14–15

New York, New Haven and Hartford Railroad, 26–27

night trains, 95–97, 127, 156, 170, 182

Norfolk and Petersburg Railroad, 90

North Carolina, 28, 68, 164, 191; boosterism in, 15, 18–19, 21, 27; canals in, 12; and competition, 23; economy in, 14, 172; law, 75; railroad statistics, 4, 5; slavery in, 155; travel in, 101,

110, 125, 132, 145. *See also* Raleigh; Weldon; Wilmington
North Carolina Central Railroad (NCCRR), 190
North Carolina Railroad (NCRR): and alcohol, 73; cooperation with other railroads, 190; and landowners, 17, 204; and slavery, 58; and telegraph, 90
Northeastern Railroad, 2, 23, 81–82, 175
Northwestern Virginia Railroad, 48
Norwich and Worcester Railroad, 168
Nott, Samuel, 38, 44

Ocmulgee and Flint Railroad, 58
Ohio, 4, 5, 6, 16, 25. *See also* Cincinnati
Olmsted, Frederick Law: on aural cues, 142; on freight, 112, 176; on hotels, 150; on labor, 108–9; on passengers, 144, 153; on rolling stock, 103; and slavery, 111, 139, 153
Orange and Alexandria Railroad, 43, 70, 72, 171
Orangeburg, South Carolina, 39, 163

passengers, 7, 36, 106, 136; children as, 144, 146, 152, 154–55; and class distinctions, 150–51, 156; comments on other passengers, 110, 139–40, 144, 153; entertainment on train, 140; European, 101, 111–12, 131, 139, 150–51, 155, 178 (*see also* Cunynghame, Arthur; Heuser, Louis; Lyell, Charles; Mackay, Alexander); fares, 154–55, 168, 184–86; and gender distinctions, 101, 151, 153, 156–57, 177, 178; growth in number, 136–37, 149–50; perception of time, 91, 146–49; and racial distinctions, 151–56; reaction to accidents, 127, 131–33; reaction to first trip, 1, 137–38, 194–96; reasons for travel, 150, 184–85, 193; rolling stock, 100–103, 104, 230n52. *See also* slaves: as passengers
pedestrians, 9, 136; during railroad travel, 110, 133, 144–45; using railroad tracks, 182–84
Pendleton, South Carolina, 18
Pennsylvania, 32, 33, 158; and competition, 14, 25–26; railroad statistics, 3, 4, 5, 6. *See also* Philadelphia
Pennsylvania Railroad: and competition, 14–15; and labor, 56, 82; management, 34; passengers on, 156; and Sabbatarians, 168; and speed, 92; and time, 90
Perry, Benjamin, 29, 114–15, 119
Petersburg, Virginia, 97, 176; depots in, 175, 186; travel in, 150, 188, 195

Petersburg Railroad, 68, 81, 152–53, 174, 188
Philadelphia, Pennsylvania, 26, 100, 103, 148, 150
Philadelphia and Reading Railroad, 44, 56
Philadelphia and Trenton Railroad, 26
Philadelphia and Washington Railroad, 25
planters: as contractors, 40, 41–42; investing in railroads, 18; investing in slaves, 2; and modernity, 5, 6–7; as shippers, 12, 13, 112, 176, 190
Pontchartrain Railroad, 93, 174

Rail-Road Advocate, 12
railroads: compared to other types of transportation, 12–13, 24, 135–36, 149, 189, 236n1; comparing northern and southern, 2–6, 8–9, 14–15, 38, 44, 53–54, 56–57, 82, 84, 90, 98–99, 175, 196–97, 223n55, 248n44, 248n55; competition among, 22–25, 26–27; cooperation among, 189–91; economic benefits of, 13–15, 112–15, 172–73; expansion, 5–6, 11, 27; and law, 25–26, 74, 75, 180, 182; in literature, 158–61; in memory, 1, 194–96; as metaphor, 157–58, 160; moral benefits of, 172; opposition to, 8, 9, 20, 22, 25–26, 164–68, 248n55; and preserving the Union, 12, 15–16; and slavery, 8, 12; and stagecoaches, 142, 147, 187; as symbol of progress, 159; system, 14–16, 24, 29, 185–87, 189–91, 193–94; unintended uses, 9, 179, 182–84; and water transportation, 12–13, 23, 170, 187–89. *See also individual companies;* slavery; slave trade
Raleigh, North Carolina, 170, 176
Raleigh and Gaston Railroad, 18–19, 40, 155, 179, 190
Reconstruction, 194
Redpath, James, 72, 153, 176
R. G. Dun, 19, 51, 52, 125, 171
Rhode Island, 3, 4, 5
Richmond, Virginia, 41, 47, 63, 190; depot in, 175; disease in, 39; prohibition of locomotives in, 91; Sabbatarians in, 164; travel in, 149, 156
Richmond and Danville Railroad (RDRR): and accidents, 129–31, 180; Civil War, 192; conductors on, 99; and contractors, 41, 45, 47, 81; coordination, 190; and freight, 107, 108, 113; funding, 19–20; and labor, 55, 57, 71, 72, 74; and nature, 37; passengers on, 185, 186; and public roads, 187; and rolling stock, 151; and slavery, 66, 153–54; and telegraph, 90; and time, 133; and U.S. Post Office Department, 245n21

Richmond and Petersburg Railroad, 93, 154, 160, 227n22

Richmond, Fredericksburg and Potomac Railroad (RFPRR): and accidents, 96; celebration of completion, 162; and competition, 23–24; cooperation with stagecoaches, 187; cooperation with water transportation, 189; and engineers, 67; and labor, 72; and landowners, 48; night trains on, 102–13; and slavery, 59, 63, 66, 75, 76–81, 154; and speed, 91; travel on, 148–49; wages, 76, 77, 80

Robinson, Moncure, 44

Robinson, Solon, 98, 178

rolling stock. *See* equipment and rolling stock

Rosboro, Al, 138, 195

Sabbatarians, 9, 88, 164–69, 190

Savannah, Georgia, 11, 82, 119, 174, 179

Savannah and Albany Railroad (SARR), 57, 92–93

Schenectady and Troy Railroad, 44

Seabury, Carolina, 132–33, 139, 143, 144

Self-Instructor, 27–28, 33–34

senses. *See* comfort; sound; taste; vision

Simms, William Gilmore, 160

slave laborers, 6, 52, 110, 205n11; advantages of, 41–42, 57, 61–62, 83; agricultural, 14, 176; bonds and contracts, 62–63, 75, 77, 79; cost, 58–62, 65, 76–81, 220n31; death of, 75–76; disadvantages of, 56, 59–60; extent of slave labor, 8, 55–56, 57, 58; gender of, 57–58; health care, 75; hired, 59–60, 61–66, 77–83; insured by owners, 64; owned by corporations, 59–62, 83; payments received, 66, 79–81; restrictions on, 63, 65–66; and time, 63, 64; and weather, 37; work performed, 58–59, 108–9, 112. *See also* labor; white laborers

slavery, 2–3, 6–7, 8–9, 29, 197–98. *See also* slave laborers; slaves; slave trade

slaves: descriptions of railroads, 138, 141, 158–59, 195; escaping via railroads, 111, 143–44, 154, 155–56; as passengers, 8, 110–12, 151–56, 157, 193, 235n42, 237n10; perceived effects of railroads on, 21, 139, 166–67, 175–76

slave trade, 110–12

sleep, 102–3, 107, 144–45, 150, 156, 178, 182

Smith, Anna, 184

Smith, Berry, 141

sound, 21, 111, 145–46, 159, 163, 167; aural cues, 89, 94, 112, 138, 141–42, 166–67, 182, 192–93;

music, 82, 109, 110, 163; new sounds of the railroad, 141; representing progress, 28, 142. *See also* bells; whistles

South Carolina, 48, 142, 195; boosterism in, 27; canals in, 12; and competition, 23; early railroads in, 11; economic impact of railroads in, 113, 114–15, 172–74, 176; freight, 108–9; landowners in, 17–18; railroad statistics, 4, 5; slavery in, 110–11; state funding to railroads in, 20–21; towns created in, 174; towns destroyed in, 175; travel in, 1–2, 25, 133, 138, 139, 185–86, 188; weather in, 124–25. *See also* Aiken; Branchville; Camden; Charleston; Columbia; Greenville; Hamburg; Orangeburg; Pendelton; Spartanburg; Stumphouse Mountain; Sumter; Walhalla

South Carolina Canal and Railroad Company. *See* South Carolina Railroad

South Carolina Railroad (SCRR), 3, 7, 187; and accidents, 74, 75–76, 91, 127, 128, 131; and alcohol, 73; and bureaucracy, 46, 197; celebration of completion, 10, 163; and civil engineers, 32, 33; and competition, 23, 24; conductors on, 39, 42–44; construction of, 35–37; contractors on, 70; cooperation with other transportation companies, 188–90; and counterfeit money, 99; depots of, 177; and disease, 125–26; early operations, 11, 13, 53, 195; economic impact of, 15, 172–73; experiments on, 31, 35–37; and freight, 107–10, 112, 117, 119, 121, 123, 176; and labor, 55, 56, 67–68, 71, 76; and landowners, 17, 179–80; and livestock, 180; and nature, 37–38, 124–25; night trains on, 96–97; and politics, 48; and rolling stock, 102–3; and Sabbatarians, 165–68; and slavery, 58–61, 66, 151–52, 154; and sound, 141–42; and state funding, 19; and stockholders, 18; and telegraph, 89–90; and time, 13, 87, 88–90, 91, 93, 94, 95, 105, 146–47, 148; and towns, 174; travel on, 2, 91, 98, 108–9, 123, 137–38, 139, 143, 146–47, 149, 156, 174–75, 185; and U.S. Post Office Department, 169–71; and vandalism, 179; wages, 77. *See also* Allen, Horatio; Gadsden, James; Louisville, Cincinnati and Charleston Railroad; McRae, John

Southern Literary Messenger, 140, 148, 149–50, 158, 159–61

Southside Railroad: cooperation with other transportation companies, 190; depots of, 174; fares,

185; and freight, 114; and labor, 71, 95; and livestock, 181; and rolling stock, 103, 104, 151; and
slavery, 154–55

Southwestern Railroad: and accidents, 131; competition with other transportation companies,
189; cooperation with other transportation companies, 191; and freight, 114, 119; and labor, 55,
56, 71; and rolling stock, 104

Spartanburg, South Carolina, 147

Spartanburg and Union Railroad (SURR): and
bureaucracy, 86; conductors on, 98; and contractors, 40, 42; depots of, 176; and freight, 108;
and landowners, 17–18; and slavery, 58–59; and
time, 89

speed: advantage of railroads, 90; passenger
comment on, 1, 138, 145–46, 147, 159, 194;
and safety, 90–93; as symbol, 158; and time,
13, 90–91; town regulations, 91–92

stagecoaches, 186; cooperation with railroads,
142, 187; railroad's superiority over, 135–36,
149

Stark, George, 92

Stark, Thomas, 42–43

state governments, 6, 17, 19, 25. *See also individual states*

steamboats. *See* water transportation

Stone and Harris, 39, 41. *See also* Birnie, William

Stroyer, Jacob, 110–11

Stumphouse Mountain (South Carolina), 19, 50–
53, 55, 70. *See also* Blue Ridge Railroad (of
South Carolina)

Sumter, South Carolina, 48

Talcott, Andrew, 37, 45–46

taste, 142–43, 163

taverns. *See* hotels and taverns

telegraph, 89–90

Tennessee, 14, 70; canals in, 12; economic impact of railroads in, 115; opposition to railroads
in, 208n34; railroads statistics, 4, 5; slavery in,
155–56; travel in, 136. *See also* Knoxville;
Nashville

Tennessee and Alabama Railroad, 33

Texas, 4, 5, 193

Thomaston and Barnesville Railroad, 17

Thomson, John Edgar, 33, 61, 210n7

Thomsson, Basil, 163–64

Thornton, Tim, 193

time, 197; clock vs. natural time, 9, 12, 87–99,

108–9, 121, 147; coordination among companies, 9, 189–90; coordination within a company, 9, 85, 87–89; and economic benefits, 13;
and efficiency, 2, 13, 28, 84, 90, 109, 114, 136,
147; multiplicity, 9, 87–88, 94–95, 108, 121,
134, 169; punctuality, 12, 87, 88–89, 90, 93,
94, 112, 143, 147–48, 169–70, 236n1; regularity, 9, 87, 90, 93, 171, 181; rules, 88–89, 90,
94, 105, 148; and safety, 88–89, 91, 97, 104,
132–33; scheduling, 9, 89, 93–94, 96, 170–71,
187, 189–90; value debated, 93, 164. *See also*
clocks and watches; night trains; Sabbatarians;
U.S. Post Office Department

towns, 24–25, 174–75

trade, 20; with the southern interior, 13–14, 23;
with the West, 14, 16, 24

Tupper, Thomas, 76, 125–26

Tuscaloosa, Alabama, 28

Tuscumbia, Courtland and Decatur Railroad, 22,
93, 154, 185, 187, 188

U.S. Post Office Department, 9, 96, 245n21; and
Sunday delivery, 165–66, 167; and time, 169–71,
181, 190

vandalism, 179

Vermont, 4, 5, 44

Virginia: canals in, 12, 158; civil engineers in, 32,
33; depots in, 179; economic impact of railroads in, 14, 174; labor in, 56; and nature, 124;
railroad statistics, 4, 5, 6; slavery in, 155; state
funding for railroads in, 19, 23; travel in, 110,
132, 139, 168, 178. *See also* Fredericksburg;
Petersburg; Richmond

Virginia and Tennessee Railroad (VTRR), 189–90;
and accidents, 74–75, 128–29, 182; depots of,
178; fares, 155; and freight, 110, 113–14, 119;
housing for labor, 72; and livestock, 181; and
slavery, 55–56, 64, 81, 155; and time, 89;
turnouts on, 176–77

Virginia Central Railroad, 52, 68

vision, 138–41, 145–46, 171, 172

Walhalla, South Carolina, 18

walking. *See* pedestrians

Washington, D.C., 136, 156, 195–96

water transportation: cooperation with railroads,
187–89; railroad's superiority over, 12–13, 121,
126, 136, 148. *See also* canals

weather. *See* nature

Webster, Daniel, 179

Weldon, North Carolina, 176

Western and Atlantic Railroad, 21, 176, 189

Western North Carolina Railroad, 61–62, 71

Western Railroad (of Massachusetts): and alcohol, 98; and bureaucracy, 85; fares, 156; and labor, 82; and landowners, 26; and nature, 38; night trains on, 156; and Sabbatarians, 268; turnouts on, 248n44

West Feliciana Railroad, 11

Wharton, Joseph, 110, 131

Whipple, Henry, 127, 139

whistles, 89, 138, 141–42, 157, 166, 182, 192; as symbol of progress, 142. *See also* bells; sound

Whitaker, R. H., 141, 142, 185, 194–96

white laborers, 6, 55, 60, 66–68; death of, 68, 74–75; ethnic rivalries among, 69; imported, 57, 67, 68, 189; injuries to, 74, 223n55; length of hire, 76; wages, 67, 72–73, 77, 80–82. *See also* labor; slave laborers

Wilmington, North Carolina, 23, 164, 170, 190; travel in, 144, 248, 196

Wilmington and Manchester Railroad (WMRR), 2, 23, 34, 72, 190

Wilmington and Raleigh Railroad / Wilmington and Weldon Railroad, 163, 190, 196

Winchester and Potomac Railroad, 227n22

Wisconsin, 4, 5